TWENTY BEAUTIFUL YEARS OF BOTTOM PHYSICS

TWENTY BEAUTIFUL YEARS OF BOTTOM PHYSICS

Proceedings of the b20 Symposium

Chicago, Illinois   June-July 1997

EDITORS
Ray A. Burnstein
Daniel M. Kaplan
Howard A. Rubin
Illinois Institute of Technology

American Institute of Physics

AIP CONFERENCE
PROCEEDINGS 424

Woodbury, New York

Editors:

Ray A. Burnstein, Daniel M. Kaplan
and Howard A. Rubin
Illinois Institute of Technology
3101 S. Dearborn Sreet
Chicago, IL 60616

EMAIL: Burnstein@fnal.gov
  Kaplan@fnal.gov
  Rubin @fnal.gov

Authorization to photocopy items for internal or personal use, beyond the free copying permitted under the 1978 U.S. Copyright Law (see statement below), is granted by the American Institute of Physics for users registered with the Copyright Clearance Center (CCC) Transactional Reporting Service, provided that the base fee of $15.00 per copy is paid directly to CCC, 222 Rosewood Drive, Danvers, MA 01923. For those organizations that have been granted a photocopy license by CCC, a separate system of payment has been arranged. The fee code for users of the Transactional Reporting Service is: 1-56396-745-6/ 98 /$15.00.

© 1998 American Institute of Physics

Individual readers of this volume and nonprofit libraries, acting for them, are permitted to make fair use of the material in it, such as copying an article for use in teaching or research. Permission is granted to quote from this volume in scientific work with the customary acknowledgment of the source. To reprint a figure, table, or other excerpt requires the consent of one of the original authors and notification to AIP. Republication or systematic or multiple reproduction of any material in this volume is permitted only under license from AIP. Address inquiries to Office of Rights and Permissions, 500 Sunnyside Boulevard, Woodbury, NY 11797-2999; phone: 516-576-2268; fax: 516-576-2499; e-mail: rights@aip.org.

L.C. Catalog Card No. 98-70174
ISBN 1-56396-745-6
ISSN 0094-243X
DOE CONF- 970623

Printed in the United States of America

CONTENTS

Preface ix
International and Local Organizing Committees x
Symposium Program xi

DISCOVERY OF THE b QUARK

Heavy Quark Spectroscopy Before the Discovery of Υ 3
    K. Gottfried
CP Violation in a Six-Quark Model 15
    M. Kobayashi
Milestones en Route to the Beauty Quark 26
    L. M. Lederman
The Discovery of the b Quark at Fermilab in 1977:
The Experiment Coordinator's Story 29
    J. K. Yoh
Observing $\Upsilon \to \mu\mu$ 43
    J. P. Rutherfoord
b-Quark Studies at DESY 50
    D. Wegener

EXPERIMENTS AT e^+e^- COLLIDERS

CESR and CLEO — The Early Years 65
    A. Silverman
The Discovery of the B Mesons 75
    S. Stone
Hidden and Open Beauty in CUSB 85
    J. Lee-Franzini
Upsilon Physics at MD-1 in Novosibirsk 97
    S. E. Baru, A. E. Blinov, V. E. Blinov, A. E. Bondar,
    A. D. Bukin, V. R. Groshev, Yu. I. Eidelman, V. A. Kiselev,
    S. G. Klimenko, G. M. Kolachev, S. I. Mishnev, A. P. Onuchin,
    V. S. Panin, V. V. Petrov, I. Ya. Protopopov, A. G. Shamov,
    V. A. Sidorov, Yu. I. Skovpen, A. N. Skrinsky, V. A. Tayursky,
    V. I. Telnov, Yu. A. Tikhonov, G. M. Tumaikin, A. E. Undrus,
    A. I. Vorobiov, and V. N. Zhilich
First Measurements of the b Lifetime 106
    J. A. Jaros
B Physics with CESR and CLEO 113
    R. A. Poling
Beauty and the Beast: Hadronic B Decay and QCD 125
    K. Honscheid
b Physics at SLD 147
    D. Su

Status of KEK-B and BELLE 161
   D. R. Marlow

THE b QUARK AND QCD

Realizing the Potential of Quarkonium.................................... 173
   C. Quigg
Hadronic Production of Heavy Quarks..................................... 189
   M. Cacciari
The Beauty of Quarkonia 198
   A. El-Khadra
Exclusive Hadronic B Decays... 207
   B. Stech
The b Quark and Symmetries of the Strong Interaction 217
   M. B. Wise

CKM PARAMETERS, B MIXING, AND CP VIOLATION

B Mixing on the Lattice: f_B, f_{B_s}, and Related Quantities................... 227
   C. Bernard, T. Blum, T. DeGrand, C. DeTar, S. Gottlieb,
   U. M. Heller, J. Hetrick, C. McNeile, K. Rummukainen, R. Sugar,
   D. Toussaint, and M. Wingate
Beautiful CP Violation.. 235
   I. Dunietz
EPR in B Physics and Elsewhere.. 246
   H. J. Lipkin

HADROPRODUCTION EXPERIMENTS

B Physics at CDF 251
   A. B. Wicklund
B Physics at D0... 269
   R. Lipton
HERA-B, An Experiment to Study CP Violation at the HERA
Proton Ring Using an Internal Target 279
   W. Schmidt-Parzefall
BTeV: A Heavy Quark Experiment at the Tevatron 287
   P. McBride
Studies of CP Violation in B-Meson Decays at LHC 298
   T. Nakada
B Physics in Central Geometry at LHC 308
   Y. Lemoigne

THEORY AND PROSPECTS

New Physics and the Unitarity Triangle 321
     D. London
Probing New Physics in the *B* System 328
     J. L. Hewett
New Physics and Enhanced Gluonic Penguin 339
     G. W. S. Hou
Testing *CPT* with *B* Mesons 345
     V. A. Kostelecký
The Mystery of Flavor 354
     R. D. Peccei

CONFERENCE SUMMARY

20 Years of Beauty Physics and 50 Years of Search for Discoveries 367
     A. I. Sanda

Symposium Participants 381
Author Index 383

Preface

Multihundred-million-dollar "beauty factories" are under construction in Stanford and Tsukuba. In Hamburg, DESY is building a beauty facility (HERA-B) that has 427 collaborators. Fermilab and CERN plan major collider detectors for beauty physics (BTeV and LHC-B), which is already the bread-and-butter task of the accelerator at Cornell. All of these activities are motivated by the rich potentialities of the fifth quark, the beauty, bottom, or b quark.

These pages present a snapshot of the current state of b physics. Notable are several first-hand accounts detailing how we got to this excited state. Whether b physics will simply de-excite (with the emission of post-doc radiation?) or proceed to higher excitations – even a population inversion?? – might also be gleaned herein.

The Symposium *Twenty Beautiful Years of Bottom Physics* (of which this volume is a record) was held June 29–July 2, 1997, at Illinois Institute of Technology in Chicago. It marked the Twentieth Anniversary of the 1977 discovery of the b quark, as well as the 75th birthday of IIT's Nobel Laureate Leon Lederman, who led the group that made the discovery.

Today b physics is a growth industry, and there is hope for a fundamental breakthrough. This, then, was an opportune time to survey the experimental and theoretical terrain. The $b20$ Symposium brought together physicists who played key roles at the time of the discovery and younger researchers who have lately come to b prominence to recall the history, review the status, and consider the prospects.

Although the schedule was full and the weather not always cooperative, we had an enjoyable cruise down the Chicago River, seeing many fine examples of the city's famed architecture. Also, we partook of a memorable symposium dinner, highlighted by the presentation of awards to Kobayashi (and, *in absentia*, Maskawa) on behalf of the b-quark discoverers (the Fermilab E288 Collaboration), and by Rocky Kolb's hilarious "roast" of Leon.

The Local Committee would like to thank all who participated in the $b20$ Symposium, especially the speakers and moderators who conveyed both recollections and current interests so well. Particular thanks go to the people behind the scenes whose able assistance was indispensable: Fermilab's Cynthia Sazama, Angela Gonzalez, and Adrienne Kolb, and Alak Chakravorty. We are grateful to John Peoples for Fermilab's support. Partial funding was provided by the U.S. Department of Energy, the National Science Foundation, and the LeCroy Corporation.

The Editors
Ray Burnstein
Daniel Kaplan
Howard Rubin

INTERNATIONAL ADVISORY COMMITTEE

Edmund L. Berger	Argonne National Laboratory
Leslie L. Camilleri	CERN
David G. Cassel	Cornell University
Luigi Di Lella	CERN
Estia Eichten	Fermilab
Mary K. Gaillard	Lawrence Berkeley Laboratory
Haim Harari	Weizmann Institute
David G. Hitlin	California Institute of Technology
Nathan Isgur	Jefferson Laboratory
Matias Moreno	IFUNAM
Jonathan L. Rosner	University of Chicago
Walter Schmidt-Parzefall	DESY
Berthold Stech	Universität Heidelberg

LOCAL ORGANIZING COMMITTEE

Ray A. Burnstein	Illinois Institute of Technology
Daniel M. Kaplan	Illinois Institute of Technology
Howard A. Rubin	Illinois Institute of Technology

Symposium:
Twenty Beautiful Years of Bottom Physics (*b20*)
June 29 - July 2, 1997
Illinois Institute of Technology
Chicago, IL, USA

Sunday (6/29)	12:00 - 6:15
Registration and light luncheon	12:00 - 1:00
Discovery of the b Quark - I Chair: J. Peoples (FNAL)	1:00 - 3:30
Welcome to IIT - T. Morrison (IIT)	1:00 - 1:15
Heavy Quark Spectroscopy Before the Discovery of the Upsilon - K. Gottfried (Cornell)	1:15 - 2:00
CP Violation in a Six-Quark Model - M. Kobayashi (KEK)	2:00 - 2:45
Milestones in the Twenty Years of Bottom Physics - L. M. Lederman (IIT)	2:45 - 3:30
- Coffee Break -	3:30 - 3:45
Discovery of the b Quark - II Chair: P. Johnson (IIT)	4:00 - 6:15
The Discovery of the b Quark - J. K. Yoh (FNAL)	4:00 - 4:30
Through a Glass Darkly: Observing Upsilon $\to \mu\mu$ - J. P. Rutherfoord (Arizona)	4:30 - 5:00
b-Quark Studies at DESY - D. Wegener (Dortmund)	5:00 - 5:30
Upsilon panel discussion - Leaders: H. Frisch (Chicago), A. Kolb (FNAL)	5:30 - 6:15
- Reception/Buffet Dinner -	6:15 - 8:30
Featuring the Multiwire Proportional Chamber Ensemble (aka the SuperString Quartet) in a program of mostly Mozart	
Monday (6/30)	9:00 - 3:15
Experiments at e^+e^- Colliders - I Chair: P. Franzini (Rome)	9:00 - 10:35
CESR and CLEO: The Early Years - A. Silverman (Cornell)	9:00 - 9:25
Discovery of B Mesons - S. Stone (Syracuse)	9:25 - 9:55
Hidden and Open Beauty in CUSB - J. Lee-Franzini (Frascati)	9:55 - 10:35
- Coffee Break -	10:35 - 10:55
Experiments at e^+e^- Colliders - II Chair: J. Slaughter (Yale)	10:55 - 11:35
Upsilon Physics at MD-1 in Novosibirsk - A. P. Onuchin (Budker Inst.)	10:55 - 11:15
First Measurements of the B Lifetime - J. A. Jaros (SLAC)	11:15 - 11:35
The b Quark and QCD - I Chair: E. Berger (ANL)	11:35 - 12:45
Realizing the Potential of Quarkonium - C. Quigg (FNAL)	11:35 - 12:15
Hadronic Production of Heavy Quarks - M. Cacciari (DESY)	12:15 - 12:45
- Lunch -	12:45 - 1:45

Experiments at e^+e^- Colliders - III Chair: C. Brown (FNAL)	1:45 - 3:15
B Physics at CLEO - R. A. Poling (Minnesota)	1:45 - 2:30
Beauty and the Beast: What Hadronic B Decays Tell Us - K. Honscheid (Ohio State)	2:30 - 3:15
- Break -	3:15 - 3:30
Buses leave for boat trip	**3:30**
River Tour of Chicago Landmarks	**4:00 - 5:30**

Tuesday (7/1)   **9:00 - 6:30**

The b Quark and QCD - II Chair: E. Eichten (FNAL)	**9:00 - 10:45**
Lattice QCD for Heavy Quarks - A. El-Khadra (Illinois)	9:00 - 9:30
Heavy-Meson Decays - B. Stech (Heidelberg)	9:30 - 10:00
The b Quark and Symmetries of QCD - M. Wise (Caltech)	10:00 - 10:45
- Coffee Break -	**10:45 - 11:05**
Experiments at e^+e^- Colliders - IV Chair: D. Christian (FNAL)	**11:05 - 12:35**
B Physics at LEP - V. Sharma (UCSD)	11:05 - 11:50
B Physics at SLD - D. Su (SLAC)	11:50 - 12:35
- Lunch -	12:35 - 1:45
CKM Paramaters, B Mixing, CP Violation, and Rare Decays Chair: H. Lipkin (ANL)	**1:45 - 3:45**
The Brown Muck of Beauty -- The Beauty of the Brown Muck - I. I. Bigi (Notre Dame)	1:45 - 2:30
B-Meson Mixing on the Lattice: f_B, f_{B_s}, and Related Quantities - C. Bernard (Wash. U)	2:30 - 3:00
Beautiful CP Violation - I. Dunietz (FNAL)	3:00 - 3:45
- Coffee Break -	**3:45 - 4:15**
B Factories Chair: M. Adams (UIC)	**4:15 - 5:15**
BABAR: The Elephant's Tale - W. R. Innes (SLAC)	4:15 - 4:45
BELLE - D. R. Marlow (Princeton)	4:45 - 5:15
Current Hadroproduction Experiments Chair: R. Raja (FNAL)	**5:15 - 6:30**
B Physics at CDF - A. B. Wicklund (ANL)	5:15 - 6:00
B Physics at D0 - R. Lipton (FNAL)	6:00 - 6:30
- Reception and Banquet -	**6:30 - 10:00**
After-dinner talk:	
New York to New Age in 75 Years - E. W. Kolb (FNAL & Chicago)	

Wednesday (7/2) — 9:00 - 4:00

Future Hadroproduction Experiments Chair: J. Rosen (NU)	**9:00 - 10:45**
B Physics at HERA-B - W. Schmidt-Parzefall (DESY)	9:00 - 9:30
BTeV - P. McBride (FNAL)	9:30 - 10:00
B Physics at LHC - T. Nakada (PSI)	10:00 - 10:30
B Physics in Central Geometry at LHC - Y. Lemoigne (Saclay)	10:30 - 10:45
- **Coffee Break** -	**10:45 - 11:05**
Theory and Prospects - I Chair: T. Erber (IIT)	**11:05 - 12:30**
New Physics and the Unitarity Triangle - D. London (Montreal)	11:05 - 11:45
Probing New Physics in the B Sector - J. L. Hewett (SLAC)	11:45 - 12:30
- **Lunch** -	**12:30 - 1:45**
Theory and Prospects - II Chair: S. Pakvasa (UH)	**1:45 - 4:00**
New Physics and Enhanced Gluonic Penguins - W. S. Hou (NTU)	1:45 - 2:05
Testing CPT with B Mesons - V. A. Kostelecky (Indiana)	2:05 - 2:40
The Mystery of Flavor - R. D. Peccei (UCLA)	2:40 - 3:15
Summary talk - A. I. Sanda (Nagoya)	3:15 - 4:00
- **End of Symposium** -	**4:00**

DISCOVERY OF THE *b* QUARK

Heavy Quark Spectroscopy Before the Discovery of Υ

Kurt Gottfried

Laboratory of Nuclear Studies, Cornell University, Ithaca NY 14853

The exploitation of the quark concept grew into an industry after quarks were invented in 1964 by Gell-Mann and Zweig. Despite successesful applications that relied on non-relativistic quantum mechanics (NRQM), this aspect of the quark model was widely regarded with suspicion, and for good reason: the non-relativistic approximation was unjustified in what we now call the light hadrons. Many, such as I, considered it below their dignity to engage in such activity.

All that changed with the November Revolution of 1974. The successful interpretation of J/ψ and ψ' as bound states of a heavy quark-antiquark pair made it plausible to apply NRQM to the internal structure of a hadronic system. And with the discovery of Υ three years later, what had been plausible became credible.

That NRQM appears to be valid at this distance scale is remarkable – the Υ states are some five orders of magnitude smaller than hydrogen. In the obsession with the unknown that is at the heart of particle physics, this confirmation of basic if old physics must not be forgotten.

I gather that my role here is that of amateur historian. Professional historians warn "don't write Whig history," i.e., don't write accounts that folds in knowledge not available at the time. I will obey and *compare only original theoretical predictions with the current experimental data*. This episode was a watershed in hadronic spectroscopy, in that *quantitative* predictions based on an economical and rather well defined theory were, for the first time, strikingly successful. Heavy quark spectroscopy thus played a significant role in establishing the Standard Model by demonstrating the power of the quark concept and basic ideas of QCD. In this context, therefore, a Whig account would be doubly damnable, as bad history and as bad public relations.

Prehistory

The November Revolution, though the handiwork of experiment, exploded on a plot of well-prepared theoretical soil. Naturally, this was not the only plot tilled by theorists, but I will speak only of what quickly flourished.

Two insights were central: that colored quarks interacting via a gauge field offered the best hope for a fundamental theory of the strong interaction; and that a respectable theory of the weak interaction could not be constructed with only three quark flavors $q \equiv (u, d, s)$. The solution to the latter problem, discovered by Glashow, Iliopoulos and Maiani, was to add a fourth quark c. That c had not been seen was taken to mean that it was too heavy to have been seen – the standard strategic planning assumption of high energy physics.[1]

From these concepts, Appelquist and Politzer[2] inferred the essential features of heavy quark-antiquark $(Q\bar{Q})$ spectroscopy *before* the discovery of J[3] and ψ[4].

Because Bjorken scaling was known to set in by about 2 GeV, they proposed that perturbative QCD would already be a fair guide to $e\bar{e}$ annihilation at energies of this order, and by that token, at energies well above the threshold for producing mesons bearing c or $\bar{c}$; and above all, that in the threshold region, bound 1^{--} states of the $c\bar{c}$ system would be seen as narrow resonances in $e\bar{e}$ annihilation because the 3-gluon Zweig-forbidden decay into light hadrons would be relatively rare.

Here pure QED provides a useful analogy in $e\bar{e}$ collisions just *below* the $\mu\bar{\mu}$ threshold.[5] The $\mu\bar{\mu}$ system has an infinite number of hydrogenic bound states with masses $2m_\mu - (E_0/n^2), n = 1, 2, \ldots$ Those with the quantum numbers of the photon produce enormous sharp resonances in elastic $e\bar{e}$ scattering and in $e\bar{e} \to 3\gamma$.

In essence, Appelquist and Politzer applied this lesson to the conjectured charmed quark. After appropriate modifications they used the QED formulae from positronium for the leptonic widths and $\Gamma(\psi_n \to 3\gamma)$ for the hadronic widths. They knew that the QED analogy could not predict the properties of $c\bar{c}$ bound states because the QCD interaction would only be Coulombic at distances small compared to the expected size of this system. Nevertheless, they anticipated that $c\bar{c}$ would at least have a pair of positronium-like bound states, 1^1S and 1^3S, and possibly excited bound states as well.[6] They called their system charmonium.

Charmonium Spectroscopy

That the huge J/ψ spike *is* the lowest 1^- charmonium state was, of course, immediately advocated at Harvard,[7] then the home base of Appelquist and Politzer. At Harvard[8] and Cornell[9] the excitation spectrum of charmonium was studied in short order, spurred on by the discovery of a second astonishing spike, ψ'.[10] We got into this act at $t = J/\psi + \epsilon$ because Tom Appelquist gave a stirring seminar at Cornell about a month before Day 1 of the Revolution, and our hometown oracle, Ken Wilson, was a charm enthusiast.

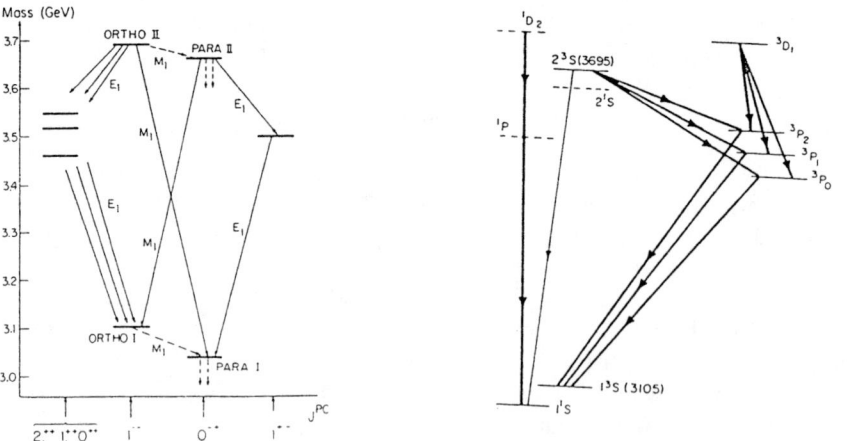

FIG. 1. The charmonium spectrum as predicted by the Harvard[8] and Cornell[9] groups, on the left and right respectively.

The bound state spectra predicted in the Harvard and Cornell papers appear to be identical at first sight (see Fig. 1), as they should because both assumed that ψ' was the first excited S-state.[11] Their most important prediction was a P-multiplet fed from ψ' and then decaying into J/ψ by E1 transitions with appreciable rates.

Not all of these predictions were identical, however, because they arose from somewhat different attitudes. The Harvard group was conservative in that it did not assume any explicit $c\bar{c}$ potential, nor that the system is nonrelativistic. Hence it restricted itself to rough estimates. We were less cautious, and constructed an explicit non-relativistic model. To that end, we chose an *ad hoc* but QCD inspired potential $V(r)$ that is Coulombic as $r \to 0$ and linearly confining as $r \to \infty$. The parameters in V and the quark mass m_c were chosen to fit the $1S - 2S$ splitting and $\Gamma(1^3S \to e\bar{e})$, and by imposing the condition that $(v/c)^2$ be "small." The latter proved to be a tight constraint; the best we could do was $(v/c)^2 \simeq (1/5)$. This led to the prediction that the charm threshold would be at $\simeq 3.9$ GeV, and that the lowest $^3D_1$ state might be below threshold and visible in $e\bar{e}$ annihilation because of $S - D$ mixing. By contrast, the Harvard paper speculated that $2S = \psi'$ might be just above threshold, and therefore did not consider $^3D_1$. A second significant difference was that the Cornell group predicted a very slow rate for the M1 transition $\psi' \to \eta_c\gamma$ because in a nonrelativistic sysem the radial wave functions of 2^3S and 1^1S are orthogonal, whereas no such statement can be made without the nonrelativistic approximation.

The photon transitions to the P-states were only discovered some six months later,[12] and what follows is therefore immune to Whiggery. Three distinct P-states were observed, but their center-of-gravity (c.o.g.) was about 100 MeV higher than predicted. Clearly, the potential was not quite right.

Turning to the E1 transitions, their rates are given by

$$\Gamma_{fi} \propto \alpha e_Q^2 (2J_f + 1)|E_{fi}|^2 k^3, \tag{1}$$

where E_{fi} is the E1 matrix element and e_Q the quark charge. In a nonrelativistic system, E_{fi} depends only on the orbital angular momenta of states, not their total angular momenta; furthermore, it is insensitive to the photon energy k unless an ostensibly allowed transition is supressed by a selection rule, as in $\psi' \to \eta_c$.

Next I compare the original predictions E_{th} for these matrix elements, made *before* the P-states were found, with those extracted from the *current* data.[13]

		Γ_γ(keV)	k	E_{exp}/E_{th}
2S → P$_J$	$J = 2$	22 ± 3	127	1.02 ± 0.08
	$J = 1$	24 ± 4	171	0.88 ± 0.06
	$J = 0$	26 ± 4	261	0.84 ± 0.06
P$_J$ → 1S	$J = 2$	270 ± 33	430	0.81 ± 0.15
	$J = 1$	240 ± 41	389	0.89 ± 0.08
	$J = 0$	92 ± 41	303	0.80 ± 0.18

There are two noteworthy features. First, the measured matrix elements E_{exp} have at most a weak variation with J within a multiplet, which supports the non-relativistic assumption; and second, the magnitude of the predicted elements are remarkably close to the measured values. Disagreement had to be expected because relativistic corrections are substantial, and the potential was not quite right as it did not predict the P-state c.o.g. properly.

The M1 transition rates are also given by Eq. 1 when E_{fi} is replaced by the appropriate matrix element M_{fi}. For $2^3S \to 1^1S$, M_{fi} is a spin-flip factor multiplied by $\int \psi_{1S}^*(r) j_0(kr) \psi_{2S}(r) \, d^3r$. In any nonrelativistic system the multipole expansion ($kr \ll 1$) is valid, but when j_0 is replaced by 1 this integral vanishes. Hence for $\psi' \to \eta_c \gamma$, $M_{fi} \propto k^2$ and $\Gamma \propto k^7$, which leads to a severe supression of the rate below that expected from a dimensional estimate, such as in the Harvard paper. We predicted a rate of ~ 1 keV using the $J/\psi - \psi'$ mass difference for k. Now we know the mass of η_c, so this should be increased by $(639/589)^7$, which gives an "original" prediction of ~ 2 keV. The measured rate is (0.78 ± 0.19) keV. As this matrix element is very sensitive to the wavefunctions, this disagreement is quite acceptable. The crucial point, of course, is that such an "anomalously" low rate is to be expected from the orthogonality of the *non-relativistic* wavefunction.

This successful prediction of the radiative transition rates in charmonium is, surprisingly enough, something of a surprise. Such a comparison with the original calculations seems not to have been made for over two decades, at which time the experimental rates were not so well known. And by the time the data improved various embellishments of the original simple model were in vogue.

Charmed Meson Production

The realization that ψ' must be close to the charm threshold W_c led us to study how coupling to open charmed states would impact the $c\bar{c}$ states. This investigation had led to an intriguing conclusion *before* the P-states were even discovered: the $^3D_1$ state should be visible in $e\bar{e}$ annihilation just above threhold as a quite prominent narrow resonance which decays *only* into the lowest open-charm channel $D\bar{D}$.

The problem to be addressed concerns two subspaces of the Hilbert space shown in Fig. 2: $\mathcal{H}_c$ spanned by the naive charmonium states $c\bar{c}$, and $\mathcal{H}_d$ spanned by the charmed meson states. In the naive model, there is no coupling between these subspaces, whereas Zweig-allowed couplings will cause the originally discrete states in $\mathcal{H}_c$ to decay if they are above W_c, and also produce shifts of states below W_c.

The coupling between the two subspaces has a novel and important feature: states with the same parity and total angular momentum but differing orbital angular momentum (e.g., $^3D_1$ and ^3S), which are uncoupled in the sector $\mathcal{H}_c$, will become coupled via open and closed decay channels. This mechanism can generate a non-zero wavefunction at the origin for ostensible $^3D_1$ states and make them accessible to $e\bar{e}$ annihilation. As indicated by Fig. 3, what is the 1^3D_1 state in the absence of coupling between the two sectors becomes the following continuum state of energy W when the coupling is incorporated:

$$\begin{aligned} |1^3D_1; W\rangle \; = \; & \alpha_1(W)|1^3D_1; c\bar{c}\rangle \; + \; \alpha_2(W)|2^3S_1; c\bar{c}\rangle \; + \; \alpha_3(W)|3^3S_1; c\bar{c}\rangle + \ldots \\ & + \; \beta_1(W)|D\bar{D}; W\rangle \; + \; \beta_2(W)|D\bar{D}^*; W\rangle + \ldots \end{aligned} \quad (2)$$

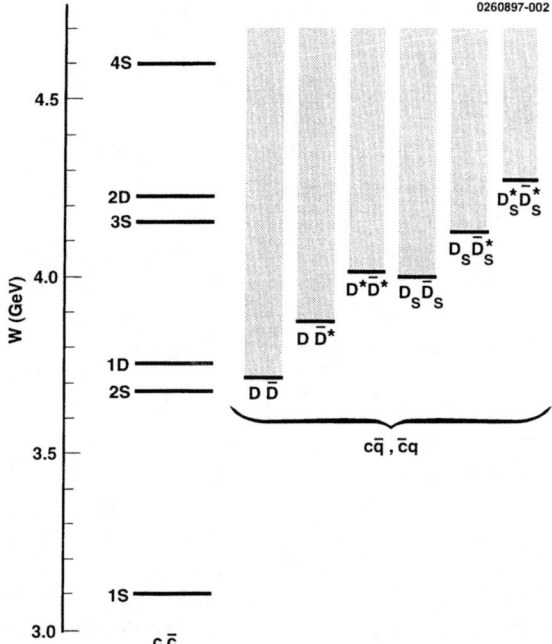

FIG. 2. The naive charmonium model lives in the $c\bar{c}$ subspace $\mathcal{H}_c$, while the low-lying open charm states form the subspace $\mathcal{H}_d$. The complete Hilbert space is far more complicated, of course, but the coupled channel model ignores that fact.

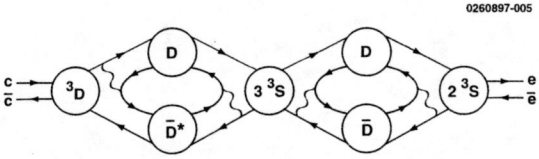

FIG. 3. Coupling to charmed meson states can mix a $^3D_1(c\bar{c})$ state, which has vanishing wave function at the origin, with ^3S states that do not vanish there, and so allow the former to be visible in $e\bar{e}$ annihilation. The coupled channel model sums all graphs of the type shown.

Of course, $S-D$ coupling could be produced by an explicit $c\bar{c}$ tensor force, but the fine structure splitting of the ^3P states shows that such a force is at most secondary in comparison to the effective tensor force T_{eff} induced by the decay channels. This spin-dependent T_{eff} can be generated out of interactions that have no explicit spin dependence because the pseudoscalar and vector charmed mesons have different masses. Hence this mechanism is most significant for $c\bar{c}$ states lying between the different thresholds in Fig. 2, as is the case for 1^3D_1.

An interaction that connects the two subspaces in Fig. 2 is needed. To obtain predictions not contingent on additional parameters we made the desperate assumption that the the static interaction $V(r)$ can be generalized to describe pair creation and annihilation. That is, the interaction in both sectors was assumed to be

$$H_I = \frac{1}{2} \cdot \frac{3}{4} \sum_{a=1}^{8} \int :\rho_a(\mathbf{r})V(|\mathbf{r}-\mathbf{r}'|)\rho_a(\mathbf{r}'): d^3r\, d^3r', \qquad (3)$$

where ρ_a is the octet of color densities for *all* quarks, heavy and light. When these densities are written out in terms of creation and destruction operators, H_I separates into many terms. The one in the $c\bar{c}$ sector just reproduces the simple charmonium model if all other terms are ignored. But there are also terms that couple the $c\bar{c}$ states to the charmed meson sector, and thanks to our drastic assumption, this coupling entails no additional parameters.

The very first coupled channel results are shown in Fig. 4,[14] and were obtained *before* the discovery of the P states. Those in Fig. 5 appeared somewhat thereafter in a paper[15] that closed with *"it is therefore important to search for the $^3D_1$ resonance shown ... as it would provide a copious source of slow $D\bar{D}$ pairs."* The subsequent discovery of charm[16] was at higher energy and therefore involved more difficult kinematics and backgrounds than at this resonance.

When the $^3D_1$ resonance, now called $\psi(3700)$, was finally found,[17] its mass and hadronic width were in excellent quantitative agreement with the predictions, as Fig. 6 demonstrates. In view of the "desparate" assumption embodied in Eq. 3, this level of agreement is, to some degree, a happy accident, although the insensitivity to the somewhat different thresholds assumed in Figs. 4 & 5 shows that the prediction's success was no mere happenstance. In any event, the fact that this state does indeed decay only into *slow* $D\bar{D}$ pairs, without any background from D^*, has made it the favorite site for studies of weak decays of charm.

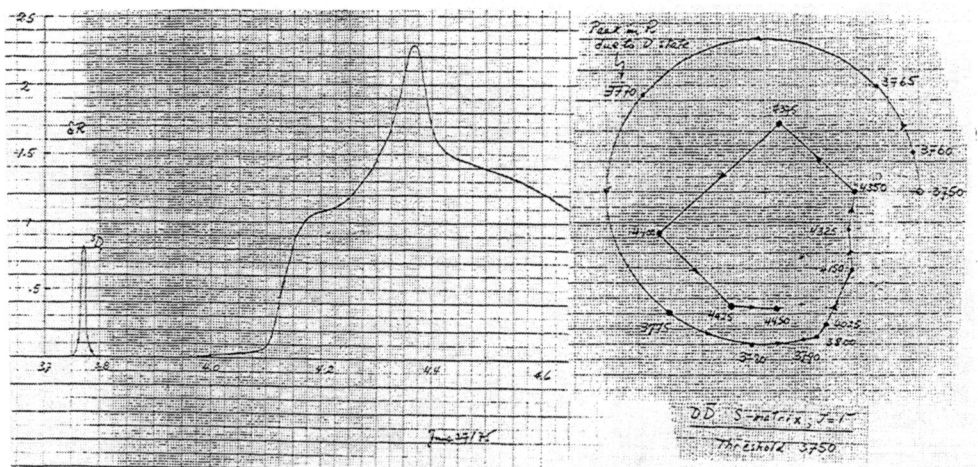

FIG. 4. Preliminary results from the coupled channel model for the $^3D_1$ resonance just above charm threshold, here taken as 3750 MeV. δR is the contribution of charmed mesons to $R = \sigma(e\bar{e} \to \text{hadrons})/\sigma(e\bar{e} \to \mu\bar{\mu})$.

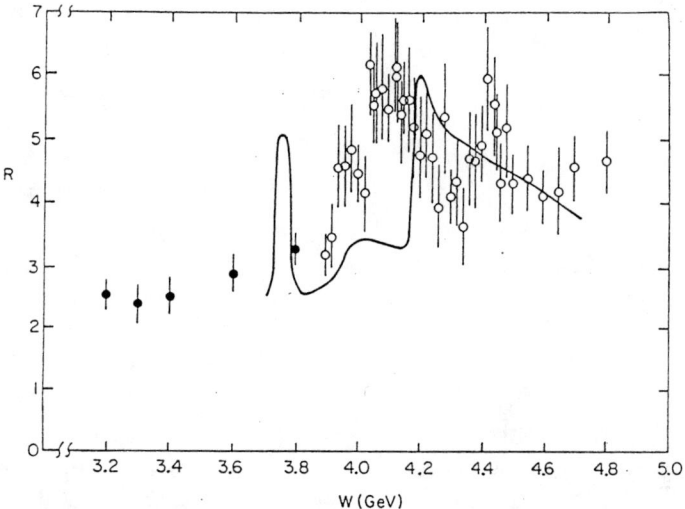

FIG. 5. R as predicted in Ref. 15 showing the $^3D_1$ resonance at 3775 MeV, assuming that charm threshold is at 3700 MeV.

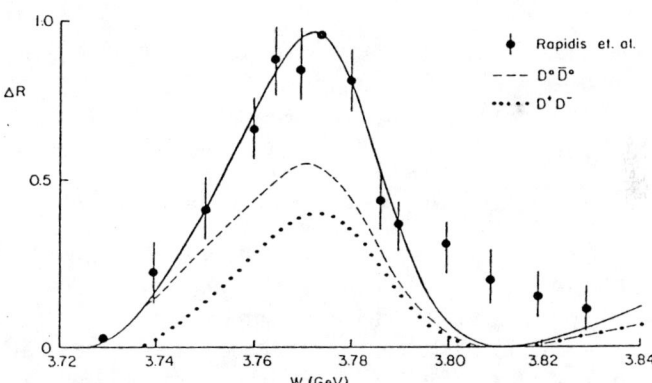

FIG. 6. $\psi(3770)$: comparison of the data[17] with the coupled channel model.[15][18] Only the height of the theoretical curve was adjusted.[19]

The cross section for charmed meson production in the resonance region reflects the radial nodes of the bound $c\bar{c}$ states, and could provide compelling qualitative evidence for the validity of quantum mechanics at this distance scale. The qualification "could" in the preceding sentence alludes to the paucity of relevant data; as of now, the evidence is suggestive but not compelling. This then is a topic that deserves greater experimental attention.

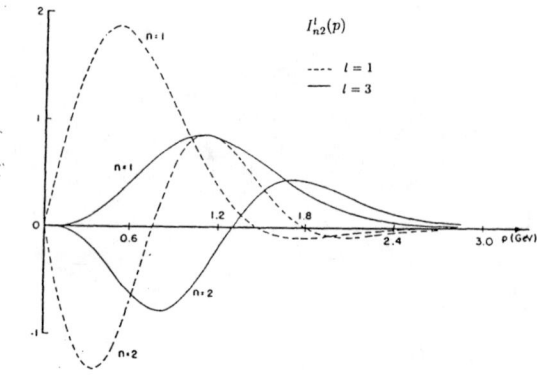

FIG. 7. P- and F-wave amplitudes for the transition $n^3 D_1 \to c\bar{q} + \bar{c}q$.

How $c\bar{c}$ states enter into $e\bar{e} \to C\bar{C}'$, where C and C' are charmed mesons, can be inferred from Fig. 3 by reading the graph from right to left and ending it at one of the open channels. Every such graph will have an overlap integral $I_{nL}^l(p)$ involving three wave functions: one each for C and C' in a state of relative angular momentum l, and one for a $c\bar{c}$ bound state ψ_{nL}, with the latter's nodes leaving a characteristic fingerprint (see Fig. 7) on the momentum dependence of the amplitude for $\psi_{nL}(c\bar{c}) \to C\bar{C}'$.[20]

While comprehensive tests of the model are not now possible, data at $W = 4028$ MeV[21] do offer evidence for nodes in $\psi_{nL}(c\bar{c})$. A traditional approach to $e\bar{e} \to C\bar{C}'$ would involve a form factor $f(p)$ that decreases monotonically, and leads to a cross section $\sigma \propto S\, p^3 |f(p)|$, where S is a spin factor:

$$S = 7:4:1 \text{ for } D^*\bar{D}^* : (D\bar{D}^* + \bar{D}D^*) : D\bar{D}. \tag{4}$$

In the following table the measured ratios are compared to a traditional model with $|f|^2 = \exp(-p/0.5\text{MeV})$, and to the coupled channel model. Clearly, the former is an utter failure while the latter does have the same qualitative features as the data. It is noteworthy that the coupled channel model accounts for the small $(D\bar{D}^* + \bar{D}D^*)$ to $D\bar{D}$ ratio even though the former is well above threshold whereas the latter is not. The nodes in the overlap integrals are the reason for the model's ability to account for this surprising feature, as is evident from Fig. 8.

	$D^*\bar{D}^*$	$D\bar{D}^* + \bar{D}D^*$	$D\bar{D}$
W_{thresh}	4013	3871	3729
p	140	450	630
traditional	1	17	1
coupled channel	1	1.4	0.005
Goldhaber et al [21]	1.0 ± 0.1	0.85 ± 0.09	0.10 ± 0.06

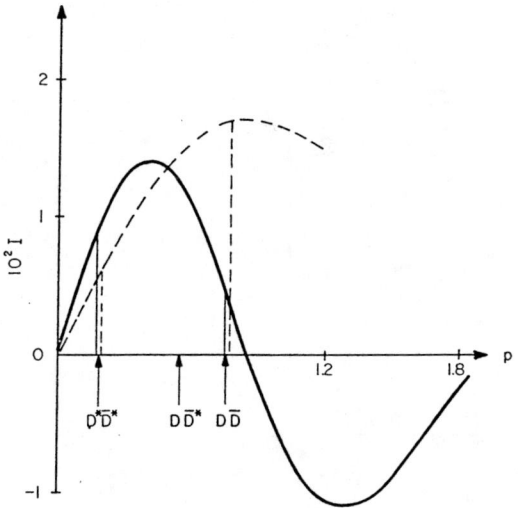

FIG. 8. The dominant state in charm production at W=4.028 GeV is 3^3S. This shows the P-wave amplitude for $3^3S \to c\bar{q} + \bar{c}q$ according to the coupled channel model (solid) and for a traditional model with a monotonic form factor (dashed); and the three channel momenta.

The Υ Spectrum Before Its Discovery

In the aftermath of the November Revolution, and the discovery of τ, talk of still heavier quarks became commonplace. I was charged with writing a theoretical section for the first CESR proposal,[22] which led me to realize that if such quarks exist there would be a richer set of bound $Q\bar{Q}$ states (Fig. 9). We then calculated

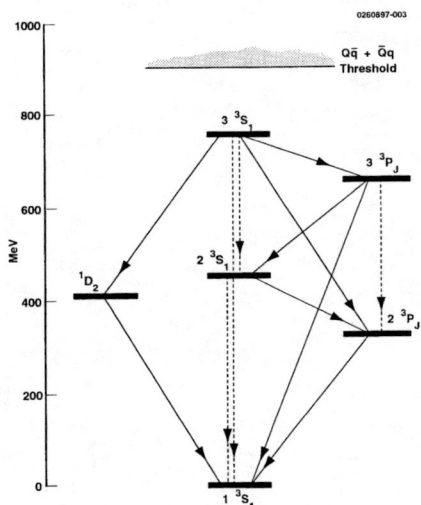

FIG.9. Conjectured $Q\bar{Q}$ spectrum in 1976 CESR proposal; E1 (solid), $\pi\pi$ (dashed lines).[22]

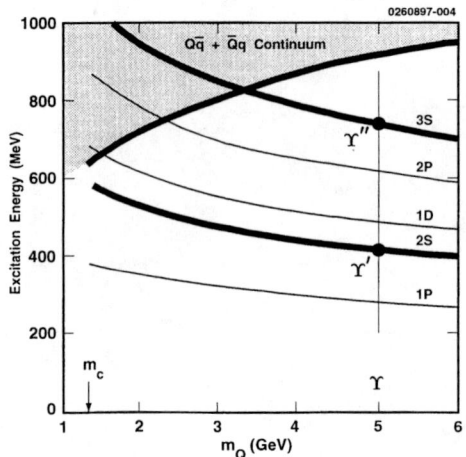 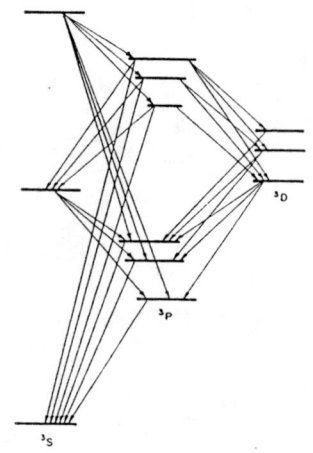

FIG. 10. $Q\bar{Q}$ excitation spectrum as a function of m_Q, and E1 transitions when $m_Q = 5$ GeV.

this spectrum as a function of m_Q using a potential fitted to charmonium.[23] Thanks to the spurious and forgotten high-y anomaly in ν scattering, which was ascribed to a new 5 GeV quark, we displayed the electromagnetic and hadronic spectra for this value of m_Q! (Figs. 10).

When Υ was subsequently discovered at about this mass,[24] the number of bound states agreed with our prediction,[25] but the energies of the excited S-states were substantially off. The heavier $b\bar{b}$ system is sensitive to a larger range of distances and therefore imposes more demands on the form of the potential than does $c\bar{c}$.[26]

E1 matrix elements for $3S \to 2P$ and $2S \to 1P$, as calculated *before* the discovery of Υ, are compared to *today's* data in the following table. Once again, the agreement is excellent – better than reasonable!

		Γ_γ(keV)	k	$E_{\text{exp}}/E_{\text{th}}$
	$J = 2$	3.00 ± 0.45	87	1.07 ± 0.08
$\mathbf{3S \to 2P_J}$	$J = 1$	2.97 ± 0.43	100	1.12 ± 0.08
	$J = 0$	1.42 ± 0.25	123	0.98 ± 0.09
	$J = 2$	2.90 ± 0.61	110	1.14 ± 0.12
$\mathbf{2S \to 1P_J}$	$J = 1$	2.95 ± 0.61	131	1.14 ± 0.12
	$J = 0$	1.89 ± 0.53	162	1.15 ± 0.16

Because the width of the P states is not known, their radiative rates are not available. The data does, however, yield the following ratio of matrix elements:

$$\left| \frac{E(2P \to 2S)}{E(2P \to 1S)} \right| = \left(\frac{k^3_{2P \to 1S}}{k^3_{2P \to 2S}} \cdot \frac{\Gamma(2P \to 2S)}{\Gamma_{\text{tot}}(2P)} \cdot \frac{\Gamma_{\text{tot}}(2P)}{\Gamma(2P \to 1S)} \right)^{\frac{1}{2}}. \quad (5)$$

The experimental values are

$$|E_{2\to 2}/E_{2\to 1}|_{\exp}: \quad 8.7 \pm 0.9 \ (J=2), \quad 9.6 \pm 1.2 \ (J=1), \quad 15.1 \pm 1.6 \ (J=0),$$

in rought agreement with the theoretical ratio of 11.

Conclusion

Our first paper, written under the spell of the November Revolution, ended with

> "If future experiments reveal a spectrum bearing a resemblence to these predictions, it could well be that charmonium is the "hydrogen atom" of strong interaction physics. For it may then be possible to subject the gauge theories of strong interactions to fairly stringent tests in a reasonably simple setting. Much of hadronic physics could then be related to charmonium spectroscopy as molecular spectra are related to that of hydrogen."

We now know that this was too grandiose, but thanks to Υ, not by much. The $b\bar{b}$ system *is* sufficiently nonrelativistic to permit stringent tests of QCD. Indeed, the most recent compilation of α_s determinations[27] shows that the most accurate value is from lattice QCD applied to the $b\bar{b}$ spectrum. While charmonium was not yet the hydrogen atom of strong interaction physics, Υ is mighty close.

I thank Kenneth Lane for advice and critical reading of the manuscript.

References

[1] M.K. Gaillard, B.W. Lee and J.L. Rosner, *Rev. Mod. Phys.* **47**, 277 (1975). This review was written before the November Revolution, but has a 12 page "note added in proof" completed January 5, 1975.

[2] T. Appelquist and H.D. Politzer, *Phys. Rev. Lett.* **34**, 43 (1975). [Received 19 Nov. 1974]

[3] J.J. Aubert et al., *ibid.* **33**, 1404 (1974). [12 Nov. 1974]

[4] J.-E. Augustin et al., *ibid.* 1406. [13 Nov. 1974]

[5] For more details, see K. Gottfried and V.F. Weisskopf, *Concepts of Particle Physics,* Vol. II, Oxford University Press (1986); pp. 313-317.

[6] Appelquist and Politzer wrote "we expect spikes," and it is unclear whether they realized how huge these spikes would be. Gaillard et al. (p. 296), while "suggesting" that "narrow peaks" in $e\bar{e}$ would be one phenomenon "indicative of charmed particles," also did not state that incredibly dramatic peaks were expected.

[7] A. De Rújula and S.L. Glashow, *ibid.* **34**, 46 (1975). [27 Nov. 1975]

[8] T. Appelquist, A. De Rújula, H.D. Politzer and S.L. Glashow, *ibid.* 365. [11 Dec. 1974]

[9] E. Eichten, K. Gottfried, T. Kinoshita, J. Kogut, K.D. Lane and T.-M. Yan, *ibid.* 369 [17 Dec. 1974]

[10] G.S. Abrams et. al., *ibid.* **33**, 1453 (1974). [25 Nov. 1974]

[11] Other early papers on the charmonium model are C.G. Callan, R.L. Kingsley, S.B. Treiman, F. Wilczek and A. Zee, *Phys. Rev. Lett.* **34**, 52 (1974); B.J. Harrington, S.Y. Park and A. Yildiz, *ibid.* 168; J. Borenstein and R. Shankar, *ibid.* 619; J.S. Kang and H.J. Schnitzer, *Phys. Rev. D* **12**, 841, 2791 (1975).

[12] W. Braunschweig et al, *Phys. Lett.* **57B**, 407 (1975). [22 July 1975]

[13] Particle Data Group, *Phys. Rev.D* **54**, No. 1 (1996).

[14] E. Eichten, in *Weak and Electromagnetic Interactions at High Energies,* Proc. Inst. Cargèse 1975, M. Lévy, J.-L. Basdevant, D. Speiser and R. Gastman (eds.), Plenum (1976). K. Gottfried, in *Proc. Inst. Particle Phys.*, R. Henzi and B. Margolis (eds.), McGill University (1975).

[15] E. Eichten, K. Gottfried, T. Kinoshita, K.D. Lane and T.-M. Yan, *Phys. Rev. Lett.* **36**, 500 (1976). [25 Nov. 1975]

[16] G. Goldhaber et al, *ibid.* **37**, 255 (1976). [14 June 1976]

[17] P.A. Rapidis et al, *ibid.* **39**, 526 (1977).

[18] K. Lane and E. Eichten, *ibid.* **37**, 477 (1976).

[19] K. Gottfried, *Proc. Intl. Symp. Lepton Photon Interactions*, DESY, Hamburg (1977).

[20] Expressions for $I_{nL}^l(p)$ are given in E. Eichten, K. Gottfried, T. Kinoshita, K.D. Lane and T.-M. Yan, *Phys. Rev. D* **17**, 3090 (1978).

[21] G. Goldhaber et al, *Phys. Lett.* **69B**, 503 (1977).

[22] Proposal to the National Science Foundation, October 1976, pp. 36-39.

[23] E. Eichten and K. Gottfried, *Phys. Lett.* **66B**, 286 (1977). [16 Nov. 1976]

[24] S. Herb et al, *Phys. Rev. Lett.* **39**, 252 (1977). [1 July 1977]

[25] J.K. Bienlein et al, *Phys. Lett.* **78B**, 360 (1978); G. Finocchario et al, *Phys. Rev. Lett.* **45** 222 (1980); J. Lee-Franzini, these Proceedings.

[26] See C. Quigg, these Proceedings.

[27] See [13], Fig. 9.1, the point "$Q\bar{Q}$ lattice;" and A. El-Khadra, these Proceedings.

CP Violation in a Six-Quark Model

Makoto KOBAYASHI

Theory Group, KEK
Tsukuba-shi, Ibaraki-ken, 305, Japan

Abstract. Author's personal recollections of the early 70's when the six-quark model was proposed to explain CP violation are presented. A brief summary of the mechanism of CP violation in the model and an analysis of the neutral K-meson system are also given.

PERSONAL RECOLLECTIONS

CP violation was found in 1964 by Cronin, Fitch, *et al.* in the decay $K_L \to \pi^+ + \pi^-$ [1]. At that time, however, I was an undergraduate student, and I was not aware of the importance of the discovery until I entered the graduate course. When I entered the graduate course of Nagoya University in 1967, CP violation was an established experimental fact, although it did not yet have any satisfactory theoretical explanation. I remember that I had an impression that in my surroundings only a few specialists in the weak interactions were interested in the CP problem.

In order to explain the general feeling of the time, I would like to quote a few paragraphs from the proceedings of a couple of high energy conferences in the late 60's.

In the review talk of the Berkeley conference in 1966, T.D.Lee remarked, "Two years ago there was almost no theory predicting CP nonconservation; now, there are far too many different kind of CP violating theories", and "It is too easy to add a small C or CP violating term to the present strong or weak interactions. The difficulty is to develop some principles which might enable one to make selections among this multitude of such possibilities".

In 1968, at the Vienna conference, S.B.Treiman described the situation as, "The precise value of the magnitude of ϵ is not as yet a matter of crucial theoretical importance, since there is no model in which one would anyhow know how to compute this quantity accurately – indeed, in most models a CP-violating parameter is introduced phenomenologically, to be adjusted to requirements". Most of the attention at that time was paid to the classification of various models.

In 1971, the Weinberg-Salam-Glashow model of the electroweak interactions began to attract widespread attention, because in this year the renormalizabilty of the non-Abelian gauge theory was proved by 'tHooft.

I soon realized that the Weinberg-Salam-Glashow theory nicely describes the ordinary weak and electromagnetic interactions. As is well known, when we apply the theory to the quark sector, we need the fourth quark to avoid the strangeness-changing neutral current. This was also nice for me, because I was discussing a four-quark model at the time.

Although it was before the discovery of the J/ψ particle, a Japanese group led by Niu had already found a new kind of event in emulsion chambers exposed to cosmic rays which suggested the existence of new particles [2]. Soon after Ogawa pointed out that they might be the particles which include the fourth quark. There was a background to his suggestion. In Japan, Sakata and his collaborators had been paying attention to the correspondence between leptons and baryons. After the discovery of two kinds of neutrinos in 1962 [3], the existence of the fourth element of the constituents of hadrons was discussed by them from time to time, first as an extension of the Sakata model and later as a more fundamental object [4].

Following Ogawa's suggestion, a few Japanese groups, including myself, started the investigation of the cosmic ray events based on the four-quark models, so that, more or less, we were familiar with the structure of the weak interactions in the four-quark scheme [5].

The Weinberg-Salam-Glashow theory was fine as a theory for the ordinary weak and electromagnetic interactions. However, this theory requires that everything must be gauge invariant. Otherwise the renormalizabilty of the theory is destroyed. So the theory put a strong constraint on other interactions too. Then, my question was whether the CP-violating interactions could be accommodated into the theory in a gauge-invariant and renormalizable manner.

I started this study with Maskawa when I moved to Kyoto University after obtaining my PhD from Nagoya University in 1972. Maskawa is also a graduate of Nagoya University. When I entered the graduate course of Nagoya, he was a postdoc there, but in 1970 he moved to Kyoto.

Soon after starting the study we found that four quarks, four leptons and a single Higgs doublet are not enough to accommodate CP violation in the theory. We have to have unknown particles in addition to the fourth quark. By introducing various new fields, we made several models which violate CP invariance. Among them, a simple and interesting one was the six-quark model. We noticed that this is an interesting possibility, but we did not think that the new particles would be discovered so quickly.

THE SIX-QUARK MODEL

Here I would like to recall how CP is violated in the six-quark model. Six quarks form three doublets of the weak SU(2) gauge group. If we denote them as

$$\begin{pmatrix} u \\ d' \end{pmatrix} \begin{pmatrix} c \\ s' \end{pmatrix} \begin{pmatrix} t \\ b' \end{pmatrix},$$

then d', s' and b' are obtained via a unitary transformation of the mass eigenstates, d, s and b,

$$\begin{pmatrix} d' \\ s' \\ b' \end{pmatrix} = V \begin{pmatrix} d \\ s \\ b \end{pmatrix},$$

and

$$V = \begin{pmatrix} V_{ud} & V_{us} & V_{ub} \\ V_{cd} & V_{cs} & V_{cb} \\ V_{td} & V_{ts} & V_{tb} \end{pmatrix}.$$

The matrix elements of V, multiplied by the universal gauge coupling g, appear as the coupling constants of the charged-current weak interactions. Their suffixes indicate the initial and final states of the transition. Those matrix elements are complex numbers in general, and their phases are therefore potential sources of CP violation. However, not all the phases are physically observable, because the phase convention of the quark fields can absorb some of them. Consequently, CP is violated only when at least one phase remains after absorbing as many as possible by the redefinition of the quark fields.

So the question is, how many phases can be absorbed into the phase convention? Let us count this number in the following way. We make the redefinition of the quark fields so that the transitions indicated by the following diagram are described by real and positive coupling constants:

$$\begin{array}{ccc} u & c & t \\ | \diagdown & | \diagdown & | \\ d & s & b \end{array}$$

Then, V looks like

$$V = \begin{pmatrix} R & R & C \\ C & R & R \\ C & C & R \end{pmatrix},$$

where the matrix elements denoted as R are real and positive and the rest are in general complex. This is the best that we can do to absorb the phase factors, because only the relative phases among the six quark fields can be used for the absorption of the phase factors of the elements of V.

17

Now we consider the parametrization of V in this phase convention. Using the unitarity of the matrix V, we can show that V is parametrized by the elements of the upper right triangle underlined in the above expression. So we find that the independent parameters are one phase and three real parameters.

Now we parametrize the triangle in the following way,

$$V = \begin{pmatrix} * & \lambda & A\lambda^3(\rho - i\eta) \\ * & * & A\lambda^2 \\ * & * & * \end{pmatrix}.$$

Then we can express all the matrix elements in terms of these four parameters in an exactly unitary manner. The result of such a procedure is the following:

$$V = \begin{pmatrix} D_1 & \lambda & A\lambda^3(\rho - i\eta) \\ -\lambda\dfrac{D_2 + A^2\lambda^4(\rho + i\eta)}{D_1} & D_2 & A\lambda^2 \\ A\lambda^3\dfrac{D_2 - (\rho + i\eta)D_0^2}{D_1 D_3} & -A\lambda^2\dfrac{D_2 + \lambda^2(\rho + i\eta)}{D_3} & D_3 \end{pmatrix},$$

where

$$D_1 = \sqrt{1 - \lambda^2 - A^2\lambda^6(\rho^2 + \eta^2)},$$

$$D_2 = \frac{-A^2\lambda^6\rho + \sqrt{D_1^2 D_3^2 - A^4\lambda^{12}\eta^2}}{1 - A^2\lambda^6(\rho^2 + \eta^2)},$$

$$D_3 = \sqrt{1 - A^2\lambda^4 - A^2\lambda^6(\rho^2 + \eta^2)},$$

$$D_0 = \sqrt{1 - \lambda^2 - A^2\lambda^4 - A^2\lambda^6(\rho^2 + \eta^2)}.$$

Expanding the D's with respect to λ, we can obtain an expression of the mixing matrix which is approximately unitary up to any order in λ. The lowest-order form coincides with the familiar Wolfenstein parametrization [6].

In this parametrization, Jarlskog's paramater [7], which is a measure of CP violation in the standard model, is given by

$$J = A^2\lambda^6\eta D_2.$$

Of course the choice of the phase convention and the parametrization are not unique. Actually the one presented here is different from what we have given in our original paper. Our original parametrization has been taken over by the Wolsfenstein parametrization. However, the Wolfenstein parametrization is not exactly unitary, although it is sufficient for most practical purposes. On the other hand, what I presented above is an exactly unitary expression with the same parameters as the Wolfenstein parametrization.

I would also like to note that the above phase convention is slightly different from what is adopted as a standard by the particle data group. In the phase convention of the particle data group, the central element, V_{cs}, is still a complex number, while it is real in the present convention, where the number of complex elements is reduced maximally.

This phase convention has another advantage. The same method can be generalized to any number of generations quite easily. From the construction we know the general rule. The diagonal elements ($V_{i,i}$) and their right-hand neighbors ($V_{i,i+1}$) are real and positive. Therefore, the general form is the following:

$$V = \begin{pmatrix} R & \underline{R} & \underline{C} & \underline{C} & \cdot \\ C & R & \underline{R} & \underline{C} & \cdot \\ C & C & R & \underline{R} & \cdot \\ C & C & C & R & \cdot \\ \cdot & \cdot & \cdot & \cdot & \cdot \end{pmatrix},$$

where the underlined elements can be chosen as free parameters. From this expression we can easily see why CP is not violated in the case of two generations and why three is a critical number. It is also easy to count the number of free parameters for the multi-generation case with this expression.

CP VIOLATION IN THE NEUTRAL K-MESON SYSTEM

Our original paper on the six-quark model was submitted in 1972 and published in 1973 in *Progress of Theoretical Physics* [8].

In 1974, the J/ψ particle was discovered, and excitement spread around the world. However, as far as CP violation is concerned, this discovery did not change the situation much, because four quarks are not sufficient for CP violation.

It was the discovery of the τ lepton that changed the situation. The discovery of the τ strongly suggested the existence of corresponding quarks. This is not only because of simple quark-lepton correspondence, but anomaly cancellation requires the existence of some counter elements, and the most natural choice is the existence of a full third generation.

In 1976, three papers discussed CP violation in the six-quark model. Pakvasa and Sugawara argued that predictions from the six-quark model would be similar to that of the superweak model [9]. Distinction of the six-quark model from the superweak model has been a long-standing problem since then and it is not solved completely yet. Maiani also discussed some properties of CP violation in the six-quark model [10].

Probably the first extensive analysis of CP violation in the six-quark model was made by Ellis, Gaillard and Nanopoulos [11]. There has been much progress from that time, in detailed calculations and both experimental and theoretical improvement of the input parameters, but the essence of the analysis has not changed from their paper.

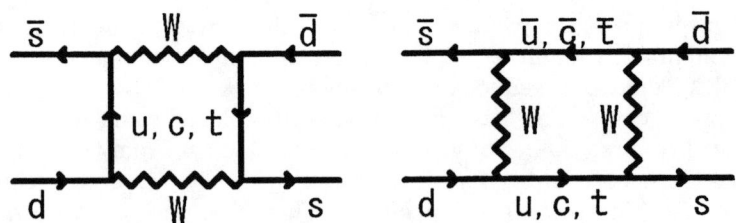

FIGURE 1. Box Diagrams

Here I recapitulate the CP-violating mechanism in the neutral K-meson system. The most important process is the $K^0 - \bar{K}^0$ transition which is generated by box diagrams(Fig.1). Through these diagrams, the dispersive part of the $K^0 - \bar{K}^0$ transition amplitude acquires an imaginary part due to the phase of the matrix elements of V, which play the role of the coupling constants of the vertices appearing in the box diagrams. The imaginary component of the dispersive part will interfere with the absorptive part of the transition amplitude, in a different manner for K^0 to $\bar{K}^0$ and $\bar{K}^0$ to K^0 transitions. As a result, the strength of the transition differs between K^0 to $\bar{K}^0$ and $\bar{K}^0$ to K^0, and therefore the weights of K^0 and $\bar{K}^0$ in the mass eigenstate become different. This is the basic mechanism of CP violation in the neutral K-meson system.

When we express the K_L state as

$$|K_L\rangle = \frac{1}{\sqrt{2}}\{(1+\epsilon)\,|K^0\rangle + (1-\epsilon)\,|\bar{K}^0\rangle\},$$

then

$$\epsilon \approx \frac{1}{2}\frac{i\mathrm{Im}M_{12}}{M_{12} - \frac{i}{2}\Gamma_{12}},$$

where M_{12} and Γ_{12} are the dispersive and absorptive parts of the transition amplitude, respectively. A detailed calculation of the box diagrams gives the following formula,

$$|\epsilon| \approx \frac{G_F^2 m_W^2 B_K F_K^2 m_K}{6\sqrt{2}\pi^2 \Delta m}\eta A^2\lambda^6[-E(x_c)\eta_c + E(x_c, x_t)\eta_{ct} + A^2\lambda^4(1-\rho)E(x_t)\eta_t],$$

where $x_{c,t} = m_{c,t}^2/m_W^2$ and η_c, η_{ct} and η_t are QCD correction factors.

Important parameters appearing in the above formula are the top-quark mass and the bag constant, B_K, as well as the mixing parameters, A, ρ and η.

The parameter B_K reflects the shape of the wave function of the K meson as a bound state of the quarks.

Now we know the top-quark mass to good accuracy, and B_K is calculated numerically from lattice QCD. The recent results of such a calculation made by the KEK group shows $B_K \approx 0.78 \pm 0.03$ [12]. The error is remarkably small, but it does not include the error due to the quenched approximation. The latter approximation is adopted in the majority of such calculations. We note that the value of B_K obtained is quite large compared to the vacuum insertion value, $1/3$.

Using this formula and the experimental value of $|\epsilon| = (2.26 \pm 0.02) \times 10^{-3}$, we can obtain a constraint on the Wolfenstein parameters. Usually this is given as an allowed region on the $\rho - \eta$ plane, together with the constraints from the $B_d - \bar{B}_d$ mixing and charmless B decays. When we express the constraint from ϵ in this fashion, the dominant sources of uncertainty are the error in B_K due to the quenched approximation and the parameter A. The accuracy of the top-quark mass is good enough, and its error has a rather minor effect.

The result is consistent with the constraints from B decays, and we can conclude that there are no serious problems as far as ϵ is concerned.

Next we consider the comparison with the superweak model. In principle, the six-quark model is different from the superweak model. In the superweak model CP is violated only in interactions which change the strangeness by two units. Therefore, in the six-quark model, if the CP-violating effects appear only through the box diagrams, then the result is equivalent to that of the superweak model. Distinction from the superweak model comes from the so-called penguin diagrams(Fig.2). They contribute to $K_L \to \pi + \pi$ decays differently for the $I = 0$ and $I = 2$ final states, giving a nonzero value for ϵ', where ϵ' is defined as

$$\eta_{+-} = \frac{A(K_L \to \pi^+\pi^-)}{A(K_S \to \pi^+\pi^-)} = \epsilon + \epsilon',$$

$$\eta_{00} = \frac{A(K_L \to \pi^0\pi^0)}{A(K_S \to \pi^0\pi^0)} = \epsilon - 2\epsilon'.$$

Strictly speaking, the present ϵ and the previous ϵ appearing in the expression for the K_L state are not the same. They can be identified only in a special phase convention, called the Wu-Yang convention. The phase convention we are using now is not the Wu-Yang type, so the two ϵ's are different. But, for practical purposes, we can identify them, because the difference is very small.

Now the problem is, how large is the contribution of the penguin diagrams? This requires a complicated analysis. In particular, when the top quark is very heavy as we now know it to be, the contribution of the electroweak penguin diagrams is not negligible compared to the QCD penguin diagrams. In the electroweak penguin diagrams a γ or Z is exchanged instead of the gluon.

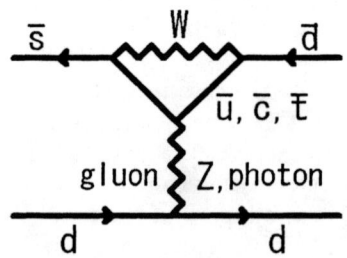

FIGURE 2. Penguin Diagrams

An important fact is that the contribution of the electroweak penguin has the opposite sign as the QCD penguin. The theoretical estimate still seems to involve a large uncertainty, and ϵ' could even be zero within the uncertainty. A conservative estimate gives [13]

$$-1.2 \times 10^{-4} < \epsilon'/\epsilon < 16.0 \times 10^{-4}.$$

(Both ϵ and ϵ' are complex numbers, but it is known that they have almost the same phases accidentally, so the ratio can be regarded as a real number in practice.)

The experimental value of ϵ' is also uncertain. The latest experimental results of ϵ'/ϵ reported by the two groups, CERN NA31 and Fermilab E731, are as follows:

$$\begin{aligned}\epsilon'/\epsilon &= (23 \pm 7) \times 10^{-4} \quad \text{(NA31)}, \\ &= (7.4 \pm 5.9) \times 10^{-4} \quad \text{(E731)}.\end{aligned}$$

B MESON

In 1977 the Υ particle was discovered. It turned out shortly that the Υ is made of b and $\bar{b}$ quarks. On the other hand, the top quark was not discovered for a long time. In the meantime, the lower bound on the top-quark mass became higher and higher. Finally, three years ago, the existence of the top quark was confirmed at Fermilab.

The implication of the heavy mass of the top quark to CP violation is twofold. One aspect was already discussed above: When the top quark is heavy, cancellations take place between the QCD penguin and the electroweak penguin, and the predicted value of ϵ' becomes very small. Basically, ϵ' is a crucial quantity to test the six-quark scheme of CP violation. But due to its smallness, it is very difficult to measure in the K-meson system, and so far we have no conclusive results.

The other important effect of the heavy top quark is large $B - \bar{B}$ mixing. The $B - \bar{B}$ mixing goes through box diagrams similar to the $K - \bar{K}$ mixing. In the case of $B - \bar{B}$, however, the top quark dominates in the intermediate states, and its heavy mass makes the $B - \bar{B}$ mixing large. Large $B_d - \bar{B}_d$ mixing was first found by ARGUS group.

The observed size of $B_d - \bar{B}_d$ mixing is

$$x_d = \frac{\Delta M}{\Gamma_B} = 0.72 \pm 0.03,$$

where ΔM is the mass difference of the two mass eigenstates, and it characterizes the speed of mixing. According to the standard theory, much larger mixing is expected in the $B_s - \bar{B}_s$ system.

The large mixing in the $B - \bar{B}$ system provides a very nice place to test the model of CP violation. A clever method of observing CP violation using $B - \bar{B}$ mixing was proposed by Bigi and Sanda [14]. Suppose both B and $\bar{B}$ states can decay directly into a final state f. Then the decay to such a final state goes through two routes, a direct decay and a decay after the mixing, and the amplitudes for those two routes will interfere. In such a case, we have a chance to see some difference between the decays from B and $\bar{B}$ states. If we see any difference between them, then it is evidence for CP violation.

$$\begin{array}{c} B \rightarrow f \\ \searrow \nearrow \\ \bar{B} \end{array} \neq \begin{array}{c} \bar{B} \rightarrow f \\ \searrow \nearrow \\ B \end{array}$$

For this type of CP violation to be observable, large mixing is essential. The best-studied example is the $B_d(\bar{B}_d) \rightarrow J/\psi + K_S$ decay.

It is interesting to note that both effects of the heavy top quark increase the importance of the B-meson system as a testing ground for CP violation.

The six-quark scheme predicts fairly large CP violation in the B-meson system. To see this, the notion of the unitarity triangle is quite useful. Since the mixing matrix, V, is unitary, different rows or columns must be orthogonal. A particularly interesting orthogonality relation is

$$V_{ud}V_{ub}^* + V_{cd}V_{cb}^* + V_{td}V_{tb}^* = 0.$$

This relation implies that the sum of three vectors in the complex plain closes and forms a triangle. This triangle is called a unitarity triangle(Fig.3).

The sides and angles of this triangle are related to various observable quantities of the B-meson system. In particular, the angle ϕ_1 will be determined by a CP asymmetry in the $B_d(\bar{B}_d) \rightarrow J/\psi + K_S$ decay.

When we express each term of the above unitarity relation with the Wolfenstein parameters, they have the same power with respect to λ. From this fact, and considering the allowed region in the $\rho - \eta$ plane, we expect that the

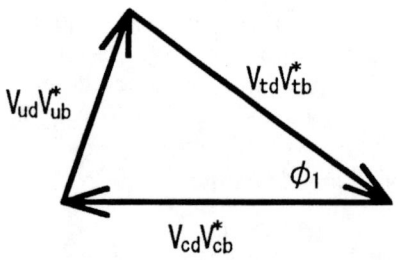

FIGURE 3. The Unitarity Triangle

unitarity triangle has a "fat shape." This implies that many CP asymmetries related to the unitarity triangle are fairly large. In order to observe these CP asymmetries, many experimental projects are in progress, and the details are discussed in other talks of this symposium. These experimental efforts may or may not prove that the six-quark scheme is correct as the source of the observed CP violation. Even if it is proved, however, it is not the end of the story. In some sense the remarks made by T. D. Lee and S. B. Treiman 30 years ago remain true. Our understanding of CP violation is still at a phenomenological level.

Even after the precise values of the mixing parameters and quark masses are determined experimentally, we will be left with questions why the parameters should have the values observed and why we have three generations. The most important thing is to clarify the principles which enable us to answer these questions. The precise experimental determination of the parameters may give us some clue to such principles. But, it is likely that these principles are hidden from the low-energy world. So, it will be a challenging task to find an experimental or theoretical breakthrough to a new stage of understanding nature.

REFERENCES

1. Christenson, J. H., Cronin, L. W., Fitch, V. L., and Turlay, R., *Phys. Rev. Lett.* **13**, 138-140 (1964).
2. Niu, K., Mikumo, E., and Maeda, Y., *Progr. Theor. Phys.* **46**, 1644-1646 (1971).
3. Danby, G. *et al.*, *Phys. Rev. Lett.* **9**, 36-44 (1962).
4. Katayama, Y., Matumoto, K., Tanaka, S., and Yamada, E., *Progr. Theor. Phys.* **28**, 675-689 (1962); Maki, Z., Nakagawa, M., and Sakata, S., *Progr. Theor. Phys.* **28**, 870-880 (1962).

5. Maki, Z., and Maskawa, T., *Progr. Theor. Phys.* **46**, 1647-1649 (1971); Hayashi, T., Kawai, E., Matsuda, M., Ogawa, S., and Shige-eda, S., *Progr. Theor. Phys.* **47**, 280-287 (1972); Kobayashi, M., Nakagawa, M., and Nitto, H., *Progr. Theor. Phys.* **47**, 982-995 (1972).
6. Wolfenstein, L., *Phys. Rev. Lett.* **51**, 1945-1947 (1983)
7. Jarlskog, C., *Phys. Rev. Lett.* **55**, 1039-1042 (1985)
8. Kobayashi, M., and Maskawa, T., *Progr. Theor. Phys.* **49**, 652-657 (1973).
9. Pakvasa, S., and Sugawara, H., *Phys. Rev.* **D14**, 305-308 (1976).
10. Maiani, L., *Phys. Lett.* **62B**, 183-186 (1976).
11. Ellis, J., Gaillard, M.K., and Nanopoulos, D. V., *Nucl. Phys.* **B109**, 213-243 (1976).
12. Hashimoto, S., private communication.
13. Buras, A. J., Jamin, M., and Lautenbacher, M. E., [hep-ph/9608365]; Buras, A. J., talk at the "Workshop on K-Physics", Orsay, May-June, 1996, [hep-ph/9609324]
14. Bigi, I. I., and Sanda, A. I., *Nucl. Phys.* **B193**, 85-108 (1981).

Milestones en Route to the Beauty Quark

Leon M. Lederman

*Fermi National Accelerator Laboratory
and Illinois Institute of Technology*

The "milestones" are illustrated by the simple diagram (Fig. 1) which makes everything perfectly clear. It begins in the 1950's at NEVIS which was a pre-factory pion machine. Pion beams were actually pioneered at NEVIS when the late John Tinlot and I discovered fringe field focussing of pions emitted from a proton target near the edge of the magnetic field of the synchrocyclotron magnet. The array of experiments that followed involved pion decay, pion scattering, muon decay (especially the subsequent electron spectrum) and the competing muon capture.

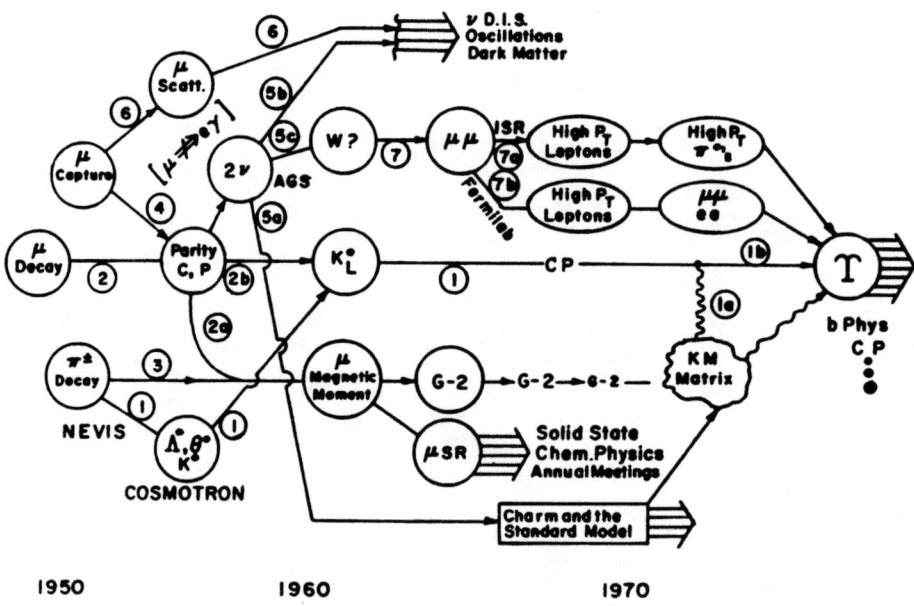

Fig. 1. Impressionistic diagram of pre-b-quark milestones.

Let me follow the pion decay path (track 1). We had built an 11" diameter Wilson Cloud Chamber to look at our new pion beam. Every photograph had 8 to 10 pions (at about 70 MeV) and in a few days we had seen more pions than the rest of the world together. Pion decay was the first subject of study. We measured the lifetime by the characteristic kinks in the track. We were able to confirm the kinematics of $\pi \to \mu + \nu$. Later, we stuck a carbon plate across the chamber and studied pion scattering. We then constructed a one meter cloud chamber designed to repeat these successes in the beams of the new 3 GeV Cosmotron, under construction at Brookhaven National Lab. I think we were the first or among the first "user groups" at BNL. Now the target were the Λ^o's and K^o's (then called θ^o) which had been seen in cosmic rays and were discovered to be pair-produced at Brookhaven.

We measured the lifetimes of both with a new level of precision and were in perfect position to test the Pais-Gell-Mann suggestion of particle mixtures predicted because in 1956, it was clear that $K^o \neq \overline{K^o}$. The argument, first made on the basis of C conservation (track 2), went through nicely with CP conservation. It predicted a long-lived piece of K^o, $\overline{K^o}$ mixture which could <u>not</u> decay into $\pi\pi$.

The large cloud chamber handily discovered the K^o_L and almost discovered CP violation, ending up in 1958 with less than 200 K^o_L decays and a limit $\varepsilon < 10^{-2}$. In 1964 Cronin and Fitch found CP violation setting up two waves: (1) the prediction of a third generation (track 1a) via the KM matrix and (2) a mass movement to exploit the b-quark for illumination of the CP phenomenon (1b).

While we are on mass movements or full employment for physicists, we should glance at a track 2 which led from NEVIS muon physics to the 1957 parity violation (P and C were observed to fail in the $\pi \to \mu \to e$ chain). The concurrent discovery of how to make polarized muons (2a) led to precise measurements of the magnetic moment of the muon and two industries: the g-2 activity, which is one of the pillars of QED and is still a lively experiment at BNL, and the use of polarized muons as a tool to explore atomic and molecular magnetic fields in a technique (muon spin resonance) which holds annual international conferences of solid state physicists, materials scientists and chemists. Again at NEVIS, the weak process of negative muon capture from orbital states of muonic atoms culminated in the first measurement of the challenging process $\mu + p \to n + \nu$ (track 4). It was in fact this study (Weinrich's thesis) which was interrupted to discover parity violation in January of 1957.

Parity violation had also distracted the search for rare decay modes of the muon but by 1959, a new vision dominated T.D. Lee's blackboard at Columbia: why is $\mu \to e + \gamma$ forbidden? One suggestion was that the neutrino in $\pi \to \mu + \nu$ decays is different from Pauli's neutrino $n \to p + e + \nu$ and so grew the two neutrino experiment at the AGS (1961).

This has (at least) three major fallouts, recognized by the 1988 Nobel citation. The first (track 5a) is, of course, the introduction of flavor into physics, most directly leading to the prediction of charm and the creation of the Standard Model. "Standard Model physics" is, of course, one of the driving forces (pun?) of today's

HEP. The second (track 5b) is the mass movement to exploit neutrinos as a tool for weak interactions via deeply inelastic scattering to study quark structure functions and which today has merged into search for neutrino mass via neutrino oscillations. Neutrino beams at Brookhaven, CERN, Fermilab, and Serpukhov labored long and hard at this problem and oscillations have a big future at these laboratories.

The third line (5c) was the use of neutrinos to discover the W. These efforts failed but did stimulate a new tool (1969) for studying structure functions, namely $p+n \rightarrow \mu^+ + \mu^- +$ junk, the inclusive process that became known as the Drell-Yan process!

Two tracks then emerged, the ISR attempt to search for time-like resonances which was frustrated by the discovery of high P_T π^o's. Although this discovery was an indication of quark structure, the pay dirt was harvested by the use of the Drell-Yan process to find the J at Brookhaven.

At Fermilab we arrive at the merged experience and, with the awesome power of R. R. Wilson's 400 GeV protons, we break open the muon pair mass domain above 3 GeV and find upsilon near 10 GeV.

It is interesting that this history of frenetic activity was at least partially responsible for almost all of the now planned next decade of HEP research: I. B-physics (with two factories, HERA B, BTeV, LHCB, Cornell B, CDF and D0 B), II. Search for Standard Model pieces: Higgs and n_t, and III. neutrino oscillations.

Fuller accounts of these events and references can be found in "Observations in Particle Physics from Two Neutrinos to the Standard Model," Leon M. Lederman, *Science*, Vol. 244, 644–672, May 12, 1989.

The Discovery of the b Quark at Fermilab in 1977: The Experiment Coordinator's Story

John Yoh

Fermilab, Batavia, IL 60510

Abstract. I present the history of the discovery of the Upsilon (Υ) particle (the first member of the b-quark family to be observed) at Fermilab in 1977 by the CFS (Columbia-Fermilab-Stony Brook collaboration) E288 experiment headed by Leon Lederman. We found the first evidence of the Υ in November 1976 in an early phase of E288. The subsequent discovery in the spring of 1977 resulted from an upgraded E288 — the $\mu\mu$II phase, optimized for dimuons, with about 100 times the sensitivity of the previous investigatory dimuon phase (which had been optimized for dielectrons). The events leading to the discovery, the planning of $\mu\mu$II and the running, including a misadventure (the infamous Shunt Fire of May 1977), are described. Some discussions of the aftermath, a summary, and an acknowledgement list end this brief historical note.

OUTLINE

I. INTRODUCTION

II. THE 1968 BROOKHAVEN DIMUON EXPERIMENT, PRECURSOR TO E288/CFS

III. WHAT WERE YOU DOING WHEN THE J/PSI WAS DISCOVERED ... (NOVEMBER 1974 REVOLUTION)

IV. THE EARLY HISTORY OF E288/CFS

V. FIRST HINT OF UPSILON IN NOVEMBER 1976

VI. E288/CFS $\mu\mu$II PHASE — PLANNING AND IMPLEMENTATION

VII. THE DISCOVERY

VIII. AFTERMATH (IS THERE LIFE AFTER ...)

IX. SUMMARY AND OUTLOOK

X. ACKNOWLEDGEMENTS

I INTRODUCTION

The search for di-leptons (charged-charged or charged-neutral) in hadronic interactions has been one of the most rewarding strategies in High Energy Physics. Almost all important discoveries in hadron collisions in the last 25 years have been made in this mode (J, Υ, W, Z, ...), and many others have been made in modes involving leptons (much of bottom and charm physics, top, ...).

The Upsilon discovery in 1977 at Fermilab marked a major landmark in this progression. It initiated the beginning of precision muon searches (in contrast to the J discovery, based on precision electron searches), bringing muon physics to a parity with electrons.

The genesis of the muon searches in hadronic interactions began with ground-breaking experiments by groups headed by Leon Lederman, described in later sections. As a graduate student working at Brookhaven in the beam line next to Leon's BNL dimuon experiment, I remember thinking "Why would anyone be interested in that?" — I think that a lot of hadron-collision experimentalists shared that feeling at that time. (I prefer to forget that my thesis experiment turned out to be a Baryonium experiment — fortunately after-the-fact, so that I was not sucked into that quagmire!)[1]

II THE 1968 BROOKHAVEN DIMUON EXPERIMENT, PRECURSOR TO E288/CFS

In the mid-sixties, Leon Lederman and his collaborators initiated a series of experiments looking first at single muons, then dimuons. This came after Schwartz, Steinberger, and Leon's Nobel-prizewinning second neutrino experiment at Brookhaven (where they missed neutral currents, calling those events "Crapons" — Leon gets my vote for the physicist who missed the most discoveries, as well as one of the, or even THE physicist, after Einstein, making the most discoveries).

Leon was interested in finding the W and Z, at that time postulated particles which could have had masses as low as a few GeV, which would then be accessible at Brookhaven, with proton beam energy of 28 GeV. This was about a decade earlier than the establishment of the electro-weak theory.

The 1968 Brookhaven dimuon experimental setup was based on a novel idea — ranging. An intense extracted proton beam was steered into a Uranium beam dump, where all hadrons, electrons, and photons were absorbed. Only muons — directly produced or from decays — survived. By measuring the range and direction of each muon, one could reconstruct the mass of the dimuon, albeit with poor mass resolution (of order of 1 GeV at a mass

---

[1] "Never have so many HEP physicists toiled so hard for so little!"

of 3 GeV). Decay muons contributed over 90% of the dimuon spectrum, but could be subtracted using measurements of accidentals. Surprisingly, a large rate of direct dimuon production was found [1]. This led Drell and Yan to publish their famous virtual-photon paper [2], so that their names were added to the HEP lexicon (the "Drell-Yan" process). (Some pundits opined that the correct terminology should be Yamaguchi-Lederman-Drell-Yan, since Yamaguchi's paper inspired Leon's dimuon experiment, but that would be too much of a mouthful.)

A rather enigmatic feature of the direct-dimuon spectrum was a broad bump at 3 GeV (obviously the first evidence for J/ψ in hindsight). However, Leon and his collaborators were not sure what to make of this —

- Could this be just another ρ' resonance (since the bump could be either very narrow, or broad — up to 1 GeV, one could not rule out this hypothesis)?

- Some light-cone theorists claimed that they could reproduce this bump without resorting to resonance.

- Some collaborators were vehemently against making a big deal over the resonance interpretation.

Leon decided to pursue this physics further with proposals at the CERN ISR and the soon-to-be-built Fermilab machine. One of his collaborators, Peter Limon, proposed a follow-up dielectron experiment at Brookhaven using existing detectors from the Lindenbaum group, but that idea died from lack of interest. A year afterward, Sam Ting proposed his BNL dielectron experiment, and the rest was history.

I was witness to an aftermath in August of 1974. Sitting in the Fermilab cafeteria, I heard Mary K. Gaillard (see paper mentioned in Section III) tell Leon that his Brookhaven bump was charmonium. It's clear that Leon's BNL bump was ahead of its time — had his result come after the acceptance of the GIM hypothesis [3], it would have been natural to interpret it as a charmonium state!!!

III   WHAT WERE YOU DOING WHEN THE J/ψ WAS DISCOVERED ... (NOVEMBER 1974 REVOLUTION)

The series of experiments E70/E288/E494 was proposed by Leon and his collaborators on June 17, 1970. The co-authors included Taiji Yamanouchi and Jeff Appel; many other co-authors on the proposal to the then-National Accelerator Laboratory (since renamed Fermilab) either started new similar experiments (W. Lee, L. Read), or dropped out before the experiment was approved (J. Sculli, M. Tannenbaum, T. White). This is one prong of Leon's two-prong follow-up of the Brookhaven dimuon experiment discussed above. The Fermilab prong stressed the highest luminosity with the highest-energy accelerator. The CERN ISR prong (CERN-Columbia-Rockefeller collaboration in I1[2]) stressed the highest collision energy, using a two-arm dielectron non-magnetic spectrometer — another of Leon's experiments that missed the J/ψ, but discovered lead-glass darkening [4] and the copious production of high-p_t neutral pions, the first evidence for a power-law-vs.-p_t distribution, and indirectly for jets.

The goal of the E70/E288 CF (subsequently joined by S, for Stony Brook) collaboration at Fermilab was to do a complete survey of all leptons produced using the highest-intensity extracted proton beam from the new Fermilab Main Ring 300 GeV (eventually upgraded to 400 GeV) accelerator. The experiment would be performed in the Fermilab Proton Center hall, which was designed explicitly for the P70 single- and di-lepton experiments.

The first stage of the experiment would study how to do a single-arm electron-spectrometer experiment well. This would be followed by single muons, dielectrons, and dimuons. The electron spectrometer consisted of 1) a target box, with a small aperture whose position could be set at angles between 50 mrad and 100 mrad, 2) a sweeping magnet to sweep out all low-momentum particles and to bend the interesting electrons (along with charged hadrons) into the 3) detectors, which were placed outside the neutral-beam envelope (the boiling sea of photons and neutrons which would have swamped any detector). The detector consisted of scintillator-hodoscope arrays to measure the electron positions and bend angle (from which the momentum could be deduced), backed up by a lead-glass array to measure the electron energy and to differentiate between electrons and hadrons.

David Saxon and Maurice Bourquin arrived at Fermilab in 1972 to lay preparatory groundwork for the experiment. By 1973, Jeff Appel and many others arrived to set up the E70 single-arm electron spectrometer. Irwin Gaines and Hans Paar, the thesis students, Jean-Paul Repellin, Jean-Marc Gaillard, Bruce Brown, and myself arrived to join Leon, Jeff, Taiji, Dave, and Maurice.

By 1973, we started taking data with the single-arm electron spectrometer. Within six months, there were indications that direct electrons (*i.e.*, those not coming from photon conversions or Dalitz decays of the neutral pion) were observed, at a rate a few times 10^{-4} that of hadrons of equal p_t. We were thus diverted from the di-lepton phase to study these direct electrons in detail, taking data at various angles.

While we were on the direct-electron "kick," the November revolution happened. Sam Ting, in redoing Leon's Brookhaven dimuon experiment using dielectrons and the newly-available multiwire proportional chambers for much

---

[2]) ISR interaction region 1

better mass resolution, discovered the J, with preliminary indications in the late summer of 1974. Unfortunately, he did not publish until the SLAC Mark I experiment found the ψ (the same particle) at the electron-positron storage ring SPEAR in early November. Hence the double simultaneous publication in PRL [5], and eventually the double Nobel Prize.

The J/ψ particle was actually expected, at least by much of the theoretical-physics community. The charm hypothesis was originally a speculation of Bjorken and Glashow as early at 1964 [6]; however, it was not until 1970 that GIM (Glashow, Iliopoulos, Maiani [3]) provided a compelling motivation for charm — it handily explained one of the major mysteries of HEP at that time, the suppression of strangeness-changing-neutral-currents. But much of the experimental community was not impressed, and remained on the resonance kick (Argand diagrams, spin-parity analysis, X, Y, split A_2, ...).

An interesting anecdote is what Shelly Glashow preached to the experimental skeptics at their stronghold — the 4/26–27/1974 4th International Conference on Experimental Meson Spectroscopy, held at Northeastern University in Boston. His prediction for EMS 76, the next conference of the series held every 2 years, was:

"There are just three possibilities:

1. Charm is not found, and I eat my hat.

2. Charm is found by hadron spectroscopy, and we celebrate.

3. Charm is found by outlanders, and you eat your hats."

This was just six months before the November revolution!!! Shelly obviously got to keep his hat.

Another interesting aspect was the paper by Gaillard, Ben Lee, and Rosner [7] entitled "Search for Charm." The preprint was dated August 1974 (Fermilab-Pub-74/86-THY), but was published in Reviews of Modern physics only after the discovery (text unchanged, except for an appendix updating the discovery). All the physics of the charmonium (renamed J/ψ by the discoverers) and charm particles was expounded in glorious detail, and most was correct except for one glaring mis-prediction and one even-more-glaring omission. The mis-prediction was on the branching ratio of charm mesons to $K\pi$, where the paper predicted a BR 10 times higher than measured later — this would in 1976 make some people believe that J/ψ did not represent charmonium. The omission, which came from the experimental naïveté of theorists, was the statement that charmonium would not be discoverable in electron-positron collisions, since it was so narrow — they did not appreciate the radiative tail of the electron beam, which makes a significant fraction of the collisions occur not at twice the beam energy, but at lower energies. Thus, even when data-taking occurs at (2 × beam energy) = 3.2 GeV, enough collisions occur at 3.095 GeV to make the interaction rate 20% higher than normal, enough to make the puzzled Mark I people investigate this point more thoroughly, and discover the ψ.

IV THE EARLY HISTORY OF E288/CFS

Proposal # 288 to the Fermilab management, "A Study of Di-Lepton Production in Proton Collisions at NAL," was dated February of 1974. The text is a short one-page digest, stating the following goals:

"1. Observe and measure the spectrum of virtual photons emitted in p-nucleon collisions via the mass distribution of e^+e^- pairs.

2. Search for structures in the above spectrum, publish these and become famous.

3. [charged hadron pairs]

4. [dimuons]

5. [dimuon structures]

6. [neutral pion pairs via conversion]"

After the November revolution, we at CF/E70 realized that we had missed the boat. The dielectron phase started soon after, featuring newly-installed MWPCs built under the direction of Bruce Brown, and the J/ψ was observed at Fermilab within one year of its discovery. The dielectron phase involved a two-arm spectrometer. Again, the acceptance was small due to the need to place the detectors out of the neutral-beam envelope (at least a factor of 5 loss in acceptance). The incoming proton-beam intensity had to be scaled down since the charged-hadron rates were so high.

We took dielectron data until 1976. In the mean time, a proposal from the Stony Brook group headed by Bud Good implementing the dihadron part of P288 was accepted. This dihadron experiment would run simultaneously with our dielectron search, though with a separate experiment number — E494. The Stony Brook group built gas Cherenkov counters to differentiate among π's, K's and p's. Several of the Stony Brook physicists also joined E288 (hence the S in CFS).

By the middle of 1976, a substantial chunk of dielectron data had been taken. The first look revealed a clustering of events near 6 GeV; the probability of such a clustering anywhere in the plot was estimated conservatively at one chance in 50. We thus gave talks suggesting that this might be evidence for a new resonance. Jeff Weiss did an "availability search" of the Greek alphabet and found that the Greek letter Upsilon was not yet used (Iota was rejected since it resembles a question-mark — in hindsight, it would have been a better choice!). Walter Innes added that the name allowed us to make a Leon-type joke — Upsilon if the resonance is real, and the similar-sounding "Oops-Leon" if the resonance is false. Since our collaboration was a sucker for bad puns (considering our genealogy), we were taken in. Saner heads, such as Taiji Yamanouchi, were ignored. In our Phys. Rev. Letter [8], we backpedaled a little, by suggesting that the name Upsilon could be assigned either to the resonance (if real) or to the "onset of high-mass di-lepton physics."

In the spring of 1976, we took some data in the dimuon mode, using the detector setup optimized for dielectrons. This provided only a factor of 5 increase in sensitivity — but that was sufficient to show that the 6 GeV "resonance" was an "Oops-Leon" and not an "Upsilon."

V FIRST HINT OF UPSILON IN NOVEMBER 1976

As the "Oops-Leon" 6.0 GeV dielectron bump faded with the summer, I kept up with the data coming in, doing data reduction as well as a first look at the spectrum — within days of the data-taking. Soon after we reverted to dielectrons (E288/E494), I noticed another clustering and wrote an internal note dated 11/17/76, entitled "From the people who brought you the Υ, a bigger (but not necessary better) resonance." This note was triggered by two recent dielectron events at 9.51 and 9.67 GeV. When combined with other events from the ee spectrum and a cluster of 6 dimuon events near 9.5 GeV, these resulted in a cluster of 10 events within 300 MeV, compared to 7 events in adjacent bins 4 times wider (*i.e.*, a 1.75-event estimated background) — a probability of less than one in 200 or so, even accounting for possible clustering anywhere in the mass plot. As I was writing the memo, another event came in at 9.44 GeV, strengthening the clustering. The significance of this clustering was thus much stronger than the "Oops-Leon." Some collaborators even claimed that I underestimated the significance. My conclusion in that internal note was that "$\mu\mu$II," a phase then under planning and scheduled to run in the Spring of 1977, just 6 months away, "should settle this in 1 month [of running]." I also put a bottle of French champagne (Moët) with the written label "Υ 9.5" pasted on in the refrigerator at the experiment's trailer.

Thus the year 1976, which was so disastrous for CFS in mid-summer, ended on a hopeful note.

VI E288/CFS $\mu\mu$II PHASE — PLANNING AND IMPLEMENTATION

It is well known that by searching for muons in the final state in hadronic interactions, one could reach much higher sensitivity than by searching for electrons. This involved putting absorbers just downstream of the interaction point to absorb all the hadronic debris from the interactions, reducing the rates of particles in the detectors by orders of magnitude. Furthermore, there would be no "neutral envelope" to worry about, making possible the placement of detectors much closer to the bending magnet and giving a factor of 3–5 increase in acceptance. Thus, the sensitivity for dimuons could be about two orders of magnitude higher than for dielectrons.

Unfortunately, the absorbers traversed by the muons would result in multiple scattering, worsening the eventual di-lepton mass resolution (*e.g.*, Leon's Brookhaven dimuon spectrum, and his joke – if memory serves – that "anything that could flatten the $[J/\psi]$ skyscraper into the mound of rubble observed by us at BNL in 1968 should be proscribed by SALT [the anti-nuclear treaty]").

A detailed analysis of this problem was undertaken by Leon, Steve Herb, myself, and others. The trick is to put the densest absorber near the interaction point, and only low-Z absorber afterwards. This leads to a smearing of the production-angle measurement, but not to a large error in momentum determination. The resultant mass resolution would be about 2% near 10 GeV mass, in contrast to the 30% or so mass resolution for Leon's BNL dimuon spectrum.

Initial work on this optimized dimuon phase (called $\mu\mu$II, since $\mu\mu$I was the dimuon phase using the apparatus optimized for dielectrons modified with absorbers, but not optimized for dimuons) began with:

- Leon's 2/12/75 memo starting with the words "We propose to do dimuons without a movable filter using fixed beryllium filter to attenuate hadrons."

- The "Super 288 White Paper," signed by Leon and Taiji, dated January 28, 1976.

Memos flew by with increasing frequency. For example,

- I wrote a note dated 2/17/76 entitled "I: expected E288 I/II signal and backgrounds, II: options for improving E288 I/II signal/background," projecting the two orders of magnitude increase in sensitivity with $\mu\mu$II over dielectrons.

- Bruce Brown wrote a note dated 5/10/76 proposing "Muon momentum confirmation with a steel magnet" (remeasurement), a proposal that was adopted.

- A note by Leon, Walt, and Steve dated 6/22/76 on a proposed PWC system for $\mu\mu$II.

- Steve Herb, in a note dated 7/3/76, gave a detailed PWC proposal; the chambers are much closer to the magnet and the acceptance is much higher than in $\mu\mu$I.

- — etc.

Leon, Steve, and others thus worked hard to design a target box with mostly Be absorber in the aperture, but with an option to place interchangeable Be,

Cu, or W absorbers immediately downstreams of the target. Extreme care was taken to avoid cracks, and to angle the possible interfaces to avoid even hairline cracks pointing to the interaction point. This was the major innovation in the $\mu\mu$II phase of E288.

Many other aspects of the upgrade to $\mu\mu$II were worked on by other collaborators: Bruce Brown proposed a "remeasuring" iron magnet to confirm the momentum and provide rejection against backgrounds; Dan Kaplan worked on the on-line system; Walt Innes worked on the track reconstruction; Koji Ueno on the Monte Carlo; Chuck Brown on monitoring and alignment; Bob Kephart and Hans Sens on the Directional Drift Chamber; Steve on gas system and survey; Hans Jöstlein on measuring the iron-magnet field, etc.

The installation of the target box and rigging of the detector and shielding piles were undertaken in early 1977, led by Steve Herb and Karen Kephart, allowing us to take a short test run in April 1977. The 9.5 GeV resonance was alive and well, though not yet definitive.

VII   THE DISCOVERY

$\mu\mu$II data-taking commenced at 13:00 on May 13, 1977. I took the three or so data tapes generated each day to the Hi-Rise and submitted a batch job doing the data reduction and subsequent first-pass analysis. Thus, prelimilary results were available within two days of the data-taking.

However, the gods were not through toying with us yet. On May 20 just before 11 pm, barely 7 days after data-taking started, there was a magnet shunt that failed disastrously (rather than fail-safe!). It melted and started a fire in the cables in the adjacent cable tray. Chlorine- and fluorine-laden smoke filled the experiment pit and deposited acidic residue on the amplifier cards mounted on the wire chambers. This residue could possibly eat into the printed-circuit traces and electronic components, and thus increase the failure rate to an unacceptable level — we could be down every few hours replacing electronics!!! The problem was obvious — a finger rubbed gently on a circuit board picked up a sour-tasting coating. Data-taking was stopped for a week while we figured out how to recover.

Leon remembered a similar fire incident at CERN, and, more importantly, was able to find by 3 am (barely 4 hours after the fire) the phone number of a Dutch fire-salvage expert, and convince him to come immediately to Fermilab, bringing his "magic" liquids. However, his visa was a problem — it might take days to obtain. Leon got lucky again — he found a high official at the local embassy who was a Columbia alumnus. Being a Columbia professor, Leon was able to convince him to provide a visa speedily. The expert arrived the next day, and was busy telling us what to do. We (physicists, technicians, girl friends, *et al.*) worked 'round the clock to remove the electronic cards, dip them in the magic liquid, brush them, and dry them. It worked marvelously

— and the failure rate of the electronics was in fact lower than before!!!

By 6/4/77, barely one week after data-taking resumed, the 9.5 GeV-resonance significance was already more than 8σ. We spent the next weeks taking more data and doing studies on efficiency and systematics, to make sure that the effect was not an artifact. We took data with a different analysis-magnet current to make sure that there were no geometric aberrations; we compared the data before and after the fire. My analysis results were checked by many other people. Acceptances were calculated by Koji and Hans, *et al.* Many meetings were held to discuss the results, with many people (Leon, Steve, Walt, etc.) making suggestions of what to check and study. These studies were done by many of the collaborators. Finally, even Taiji was convinced that we had now finally discovered a new particle.

On June 30th, 1977, Steve Herb announced the discovery at Fermilab. The PRL paper [9] was submitted the next day, July 1, 1977. I gave a talk at Brookhaven and Walt at SLAC soon after. The HEP world finally took this seriously after Leon gave his talks at the Budapest EPS and the Hamburg Lepton-Photon meetings in July and August of 1977. Thus, this discovery was made in six weeks (minus one week lost due to the fire), by 16 authors.

The discovery of Υ (or bottomonium) was actually more unexpected than that of the J/ψ (charmonium). The Kobayashi-Maskawa paper [10] speculating on six quarks, though published in 1973, was totally unknown in the U.S., having been published in the obscure Japanese journal Progress of Theoretical Physics. The preliminary evidence for the τ from Mark I in 1975 was weak, and not established for a long time, becoming believable only after more data were collected by PLUTO and Mark I (some Europeans would argue that the first believable evidence for τ was actually that of PLUTO!). However, that did not stop Haim Harrari in the summer of 1975 from speculating that this third charged lepton must indicate a new pair of quarks, which he named bottom and top. This lepton-quark-universality hypothesis was much weaker than the charm hypothesis, since it had no other supporting evidence. Remember that these were the days of the notorious Cline-Mann-Rubbia high-y anomaly and singlet-b-quark evidence!! The third-generation hypothesis only became believable after the discovery of Υ and the b mesons and hadrons. Kobayashi and Maskawa got their belated recognition, and the KM matrix entered the HEP language (eventually the CKM matrix, recognizing Cabibbo's contributions).

VIII AFTERMATH (IS THERE LIFE AFTER ...)

The E288/CFS experiment and its offspring continued for many years. Many people, such as Al Ito, Chuck Brown, Dan Kaplan, *et al.*, worked on the analysis effort, and produced many measurements, such as $\sqrt{s}$ dependencies, p_t dependencies, target material (A) dependencies, etc., as well as many dihadron and other results. Others joined the collaboration and made major contributions.

In September of 1977, I began teaching at Columbia, followed one year later by Steve. Thus, our roles in CFS were reduced. On one of my increasingly rare visits to Fermilab in October of 1977, I read in the CERN courier an advertisement on the availability of the Cornell CESR North Area for a proposal for a small experiment to complement the large CLEO detector being built in the South experimental hall. Steve, Leon, and I discussed proposing an experiment for that area, and it seemed an obvious place to pursue bottom physics. We wrote a proposal, essentially detailing the eventual CUSB experiment, consisting of a 3π-solid-angle non-magnetic tracking system followed by NaI crystals, with lead glass to catch the energy leakage. The forward and backward directions were empty of detectors (except for luminosity monitors). We were asked at the first program-advisory-committee meeting in late 1977 to look for more collaborators, and asked the Franzinis to join. Unfortunately, soon after the experiment was approved, Leon was given an offer he couldn't refuse — directorship of Fermilab. Thus, Steve, the Franzinis, I, and our collaborators built the CUSB detector and discovered the $\Upsilon(4S, 5S)$ (along with CLEO), χ states through photon transitions, and the first evidence for B mesons (via lepton-spectrum cutoff at $\Upsilon(4S)$ mass divided by 4, not by 2, as well as evidence for B to D transitions, not to ρ or π's only). Subsequently I moved to Fermilab to work on CDF; Steve moved to Cornell and eventually DESY to work on the machine there.

While many of our E288 collaborators remained at Fermilab (Appel, Yamanouchi, Bruce and Chuck Brown, Kephart, Jöstlein, Ito), others left — Innes (to SLAC, now on BaBar), Kaplan (now at IIT), Hom (in New York City), Sens (now at CNRS), Snyder (now at Gallaudet College).

IX SUMMARY AND OUTLOOK

The discovery of the Upsilon,[3] coming just three years after the November 1974 revolution, continued the string of new quarks, culminating in the recent top discovery at Fermilab (the 3rd-generation lab?) by CDF and D0.

In some respects, the bottom quark has significance way beyond "just another quark." Due to the long lifetime and mixing, CP violation in the B-meson systems becomes the new "Holy Grail" of HEP. Several B factories are being built. A crude estimate would suggest that roughly 1/3 of current HEP experiments are either studying B physics or using B as tags (*e.g.*, the top discovery).

---

[3] I intend soon to put the history of the Upsilon discovery on the Web, with links (hopefully) from the official Fermilab web pages, and with scanned images of crucial memos, pictures of the apparatus and collaboration, links to other B-physics web sites, etc.

Other aspects of B also increase its significance: it's the heaviest quark that still has a real meson (as opposed to the virtual T meson, which lives too briefly to be a real physical meson); the b-quark is heavy enough that theoretical calculations for the various Υ and other bottomonium mass levels can be reliably calculated; ...

(Note that there are other papers covering the discovery of the Upsilon, such as Dan Kaplan's version [11].)

X   ACKNOWLEDGEMENTS

As the experimental coordinator of CFS/E288 during the period 1975 to 1977, I participated in one of the most exciting episodes in my physics career. Most of the credit for this discovery belongs with

- Leon Lederman, the "founding father" and "leading light" of the experiment, as well as the collaboration's spokesman. Many of the key concepts of the E70 and E288 experiment designs originated with him. In particular such crucial ideas for $\mu\mu$II as the Be absorber and many other issues (expected rates, detector arrangement, etc.) were first considered by him.

However, many others of the CFS/E288 collaboration played important roles in this discovery. (First, a caveat: the list below is based on my faulty 20-year-old memories, as well as internal notes from that period — fortunately, several are job lists I wrote that have people assigned to each item; nevertheless, some significant effort is likely to be missed in the list below, and I apologize in advance to those collaborators who might feel slighted — note that the contributions listed below are specifically for the months leading up to the discovery, not to any prior or subsequent contributions):

- Steve Herb — the $\mu\mu$II upgrade manager, the major architect with Leon of the target box, shielding pile, etc. Also worked on PWC placement, survey, and off-line work on fiducial cuts, position of detectors, monitoring.

- Walter Innes — the production event-reconstruction expert and architect; off-line work on resolution, etc.

- Bruce Brown — the wire-chamber expert, who also proposed the iron remeasurement magnet.

- Dan Kaplan — $\mu\mu$II thesis student from Stony Brook; on-line expert; off-line work on analysis, MMPWC, magnetic-field study, muon criteria.

- Koji Ueno — the Monte Carlo expert and main architect; worked on acceptances, energy loss, etc.

- Jeff Appel — experimental coordinator from 1973 to 1975, a "founding father," worked on the design of E70 and early phases of E288, $\mu\mu$II beam quality monitoring.
- Chuck Brown — monitoring, triggering, facilities support, off-line work on resolution, chamber alignment.
- Dave Hom — thesis student for previous dielectron data.
- Al Ito — collimator.
- Hans Jöstlein — iron-magnet field measurements.
- Bob Kephart — Cherenkov expert, "keeping John honest," Directional Drift-Tube intensity monitors.
- Hans Sens — target box, DDC (Directional Drift Chamber), Monte Carlo acceptance.
- Dave Snyder — thesis student for previous dielectron data.
- Taiji Yamanouchi — head of the Fermilab contingent; a "founding father;" the "voice of caution."
- I myself (JKY), besides being the experiment coordinator, wrote and ran single-handedly the first-phase data-reduction process, using a "quickie-track-reconstruction" algorithm I wrote that complemented the "full-track-reconstruction" algorithm written by Walt Innes that was used in the reconstruction phase. Also a host of off-line studies (event display, trigger studies, resolution using J/ψ, intensity dependences, etc.). I also did much of the physics analysis and studies for the discovery, along with efforts by Dan and others.

Of course, all collaborators worked on various other aspects of analysis, discussions of results and how to validate them, etc. — far too much to list.

- Our Fermilab-resident technicians — Karen Kephart, Frank Pearsall, and Jack Upton from Fermilab, Ken Grey from Columbia/Nevis, and Tom Regan from Stony Brook. They worked hard on chambers, target box, rigging, etc. and deserve much credit.
- The Nevis (Columbia Univ.) support people, especially Bill Sippach, whose pioneering electronic designs have had a major influence on all of HEP, Yin Au, mechanical engineer who participated in much of the apparatus design, and Art Timm (plastics expert), Herb Cunitz and Ed Taylor (electronics shop), *et al.*

And thanks especially to Fermilab —

- Bob Wilson, Director.

- the accelerator group (Helen Edwards *et al.*), which provided the beam without which we would not have made the discovery.

- the Research Division (John Peoples *et al.*).

- the Proton Department – Brad Cox, Ron Currier, Bill Thomas, Dave Eartly, Age Visser, Al Guthke, Bob Shovan, Ed Tilles, Fred Rittgarn, and the many mechanical and electrical tecnhicians there. In addition to providing support for our experimental area, they provided the bulk of the effort in implementing the all-important Be-filled target box.

- the beams group and other support groups.

- and especially to our funding agencies (ERDA, NSF) and the American taxpayers.

REFERENCES

1. J. H. Christenson *et al.*, Phys. Rev. Lett. **25**, 1523 (1970); Phys. Rev. D **8**, 2016 (1973).
2. S. Drell and T. M. Yan, Phys. Rev. Lett. **25**, 316 (1970); Ann. Phys. **66**, 578 (1971).
3. S. L. Glashow, J. Iliopoulos, and L. Maiani, Phys. Rev. D **2**, 1285 (1970).
4. J. S. Beale *et al.*, Nucl. Instrum. Meth. **117**, 501 (1974).
5. J. J. Aubert *et al.*, Phys. Rev. Lett. **33**, 1404 (1974); J. E. Augustin *et al.*, Phys. Rev. Lett. **33**, 1406 (1974).
6. J. D. Bjorken and S. L. Glashow, Phys. Lett. **11**, 255 (1964).
7. M. K. Gaillard, B. W. Lee, and J. L. Rosner, Rev. Mod. Phys. **47**, 277, (1975).
8. D. C. Hom *et al.*, Phys. Rev. Lett. **36**, 1236 (1976).
9. S. W. Herb *et al.*, Phys. Rev. Lett. **39**, 252 (1977).
10. M. Kobayashi and T. Maskawa, Prog. Theor. Phys. **49**, 652 (1973).
11. D. M. Kaplan, in **History of Original Ideas and Basic Discoveries in Particle Physics**, H. B. Newman and T. Ypsilantis, eds., NATO ASI Series B: Physics, vol. 352, Plenum, New York (1996), p. 359.

Observing $\Upsilon \to \mu\mu$

John P. Rutherfoord

University of Arizona, Tucson, Arizona 85721 USA

Abstract. Starting with the pioneering Brookhaven experiment by Leon Lederman's group, four successive detector techniques have been employed to measure inclusive dimuon production in a proton beam, each an improvement on the previous, particularly in dimuon invariant mass resolution. We describe these approaches along with their strengths and weaknesses.

INTRODUCTION

In a beam of protons the cross section for the inclusive production of a direct μ^+ and a μ^- (an electroweak process) of large invariant mass is quite small. Thus high luminosity is required to collect a significant number of events. Along with the luminosity comes a prodigious rate of ordinary, soft hadronic collisions producing hadrons which would flood the detectors. In order to withstand this onslaught the detector must have high rejection of such backgrounds while maintaining good efficiency for detecting the muons. In addition, good dimuon invariant mass resolution is desired to pick out any possible narrow resonances on top of the continuum production.

These requirements were appreciated from the beginning and were met with increasing success in a series of experiments, mostly by Lederman's group. Other talks at this conference will concentrate on Fermilab E288, the Υ discovery experiment, and its famous BNL predecessor so I will place a bit more emphasis on lesser known efforts.

Four distinct approaches have been brought to bear on the challenges of detecting dimuons and I will describe each, not necessarily in chronological order but rather in order of evolution of the technique.

THE RANGE TELESCOPE

For their exploratory experiment into uncharted waters [1] Lederman's group chose a range telescope which triggered on two muons. At the Brookhaven AGS where this experiment was performed, the produced muon energies were low enough that it was practical to range out the muons. Iron absorber served the combined functions of 1) momentum analysis via range, 2) muon identification via penetration, and 3) hadron rejection by absorption in the massive amount of iron. Hodoscope counters at various depths in the iron gave both momentum and angle determination. But the charge sign of the muons could not be determined so it was assumed that one was of opposite sign to the other. The massive iron and hodoscopes made for a simple experiment.

As is well known, the dimuon invariant mass resolution of $\Delta M_{\mu\mu}/M_{\mu\mu} \sim 11\%$ was insufficient to show the J/ψ as any more than a broad shoulder on top of the rapidly falling continuum.

Somewhat more elegant because of its simplicity was the range telescope of Eartly, Giacomelli, and Pretzl [2] at Fermilab, shown very schematically in Figure 1. Located in the Meson Lab tunnel upstream of the experimental floor, the geometry was set to detect symmetric dimuons produced near $y_{cm} = 0$. The small geometric acceptance was offset by the large flux of protons on their target. This experiment readily saw the J/ψ and evidence for yet another narrow dimuon resonance near 6 GeV as seen in the integral spectrum of Figure 2. I am not aware that this experimental artifact was ever explained.

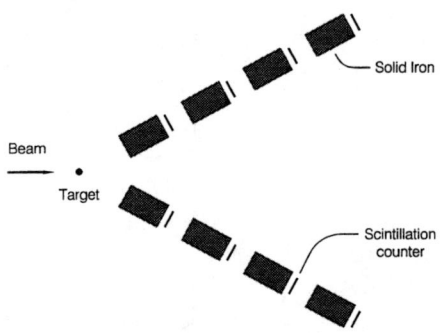
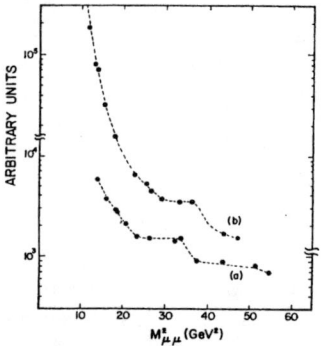

FIGURE 1. Dimuon detector using the range telescope technique. Acceptance is set for symmetric pairs near $y_{cm} = 0$.

FIGURE 2. Integral dimuon mass spectrum showing a narrow enhancement near $M_{\mu\mu} \sim 6$ GeV where (b) is data collected with less iron in front than in (a).

THE SOLID IRON MAGNET

It was clear the range telescope technique was reaching its limits at Fermilab energies. The depth of iron necessary to range out the higher energy muons

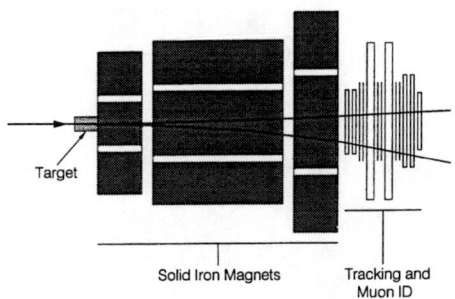

FIGURE 3. Dimuon detector using the solid iron magnet technique. Three such magnets are shown with the fiducial region in the middle and magnetic field direction out of the plane of the page. Return yokes are at top and bottom with gaps for the coils.

was impractically large. The next-generation technique employed iron in a much more compact configuration in Fermilab experiment E439 (see Figure 3). Rather than ranging out each muon, the iron was instead magnetized, and the muon bending gave a measure of its momentum. The magnetized iron then served the combined functions of 1) momentum analysis, 2) muon identification via penetration, and 3) hadron rejection by absorption in the massive amount of iron. Because the depth of iron was far less than required to range out the muons, relatively large geometric acceptance could be achieved. The muons exiting the magnets were measured by plastic hodoscopes and MWPC's with interspersed iron to guarantee that the measured particles were indeed energetic, thereby improving the muon identification and hadron rejection. For this experiment $\Delta M_{\mu\mu}/M_{\mu\mu} \sim 7\%$, limited by multiple scattering in the iron at lower dimuon masses, and lack of knowledge of the event origin at higher dimuon masses.

Convinced there was another quarkonium resonance above the J/ψ, E439 experiment spokesman David Garelick frantically pushed the experiment construction in a race with Lederman's group, who were coincidentally evolving the third-generation technique to be discussed in a moment. I was a member of the E439 team. A test run in late '76 with only hodoscopes showed the J/ψ peak growing in an on-line display in just a few accelerator pulses. While the resolution with only hodoscopes was comparable to Lederman's BNL experiment, the cross section at the higher energies at Fermilab transformed a shoulder into a "dramatic" peak as seen from this copy of a page from the log book [3] reproduced in Figure 4.

Our MWPC's were tied up in another experiment and became available only in Spring 1997. A short run was completed before the summer which showed a less than 3σ enhancement in the Υ region. But these data were available

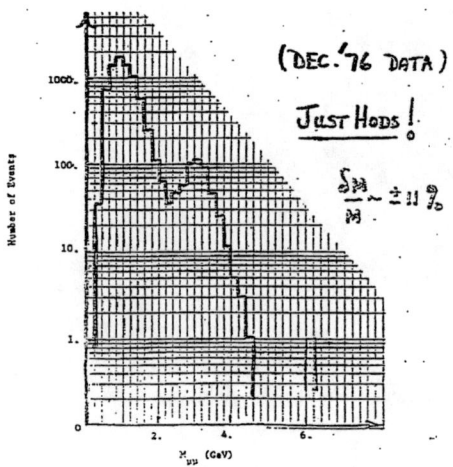 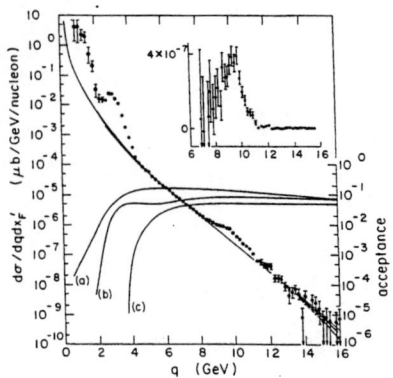

FIGURE 4. On-line J/ψ data from solid iron magnet technique from a test run using only hodoscopes.

FIGURE 5. Dimuon spectrum confirming the Υ. Inset shows the continuum-subtracted resonance.

only after Lederman's group had conclusively discovered the anticipated new quarkonium resonances. Our major run was completed in early 1988 showing a convincing enhancement (see Figure 5) over a wide range of rapidity [4–6]. Of the techniques discussed here this is the only one not used by Lederman's group.

THE SEPARATED-FUNCTION TECHNIQUE

A few months ahead of E439, Lederman's next-generation detector (E288) was evolving from a dielectron spectrometer. Major changes were made to meet the requirements for a competitive dimuon experiment. A low-Z (beryllium) absorber to stop hadrons was placed very close to the target as shown in Figure 6. Low Z gives optimum hadron rejection with minimal muon multiple scattering. Furthermore placing the absorber close to the target minimizes the effect of this multiple scattering on the dimuon invariant mass determination. A tungsten beam dump was buried in the midst of the beryllium. The muon momentum was determined with air- (actually helium-) gap magnets and MWPC's followed by solid-iron magnets to positively ID the muons (and additional iron absorber not shown in the figure.) Thus hadron rejection, momentum measurement, and muon ID were all determined by separate elements, in contrast to the first two techniques.

As is well described elsewhere in these proceedings (see e.g. the paper by

John Yoh) Lederman's group not only discovered the Υ, but the excellent dimuon invariant mass resolution of $\Delta M_{\mu\mu}/M_{\mu\mu} \sim 2\%$ showed that the Υ enhancement was actually three sub-threshold resonances [7].

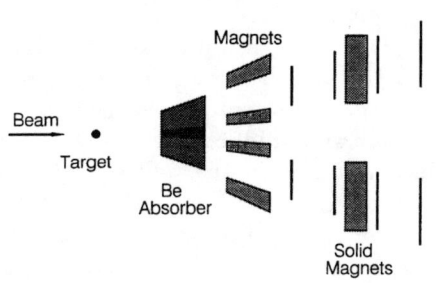

FIGURE 6. Dimuon detector using separated function technique.

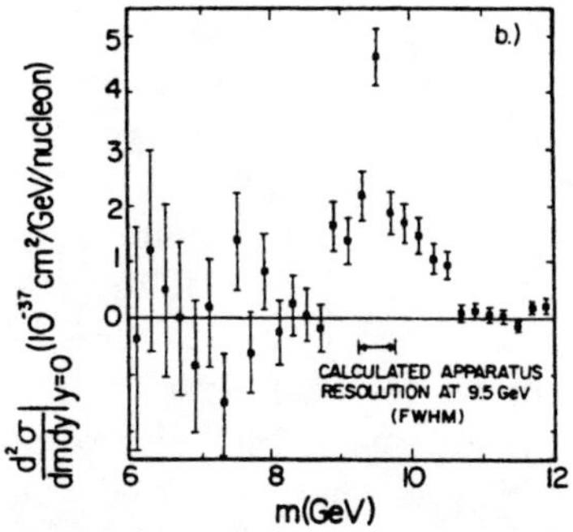

FIGURE 7. E288 continuum-subtracted Υ data.

THE MODIFIED SEPARATED FUNCTION TECHNIQUE

E605 started as a multi-purpose experiment. To permit detection of electrons and hadrons no absorber was placed in the path of the target-produced particles. By 1984 the original E605 goals were accomplished but only 200 dimuon (and 200 dielectron) events had been collected in the Υ region. So special configurations were being considered for a dedicated dimuon run. A straightforward modification would have been to stuff low-Z absorber close to the target as in E288. But Bob McCarthy challenged us to find a better way. I realized that we could achieve even better mass resolution with a high-Z absorber after the first magnet, but could not predict the rates. Chuck Brown and Dan Kaplan independently simplified my resolution calculation in the next day or two and within weeks we piled a lead-brick wall as a half-day test. The rates were low enough to push to the required luminosities. At this same time the Crystal Ball had discovered a dramatic resonance at about 8.3 GeV, and Glashow and Machacek [8] interpreted it as a Higgs with an appreciable branching fraction to dimuons. This would show up in our new configuration as a narrow resonance about the size of the Υ''. But the accelerator run was nearly over and we needed some improvements, in particular a high-rate drift chamber just downstream of the absorber wall, so we waited until the 1995 run. By this time the crystal ball effect had disappeared, and no other exciting resonances showed up. Nevertheless the technique was a success yielding a sample limited only by the beam (which we reluctantly had to share with other experimenters) with a mass resolution of $\Delta M_{\mu\mu}/M_{\mu\mu} \sim 0.3\%$.

The absorber wall at the downstream end of the first magnet (see Figure 8) was required to be as geometrically thin as possible and yet absorb all the hadronic debris. Only dense materials (lead) would do, hence high-Z. The scheme worked because a smaller downstream magnet already provided a reasonable estimate of the muon momentum. The chamber system allowed us to project the track back to a point at the absorber wall. If there were no multiple scattering at the wall then the muon could be tracked all the way back to the target yielding much better momentum and production-angle estimates. However the muon suffered considerable scattering in the wall and the true path deviated from this ideal. Both the momentum and angle deviated from the estimates because of this. However if the momentum was larger, then the angle was smaller and vice versa, so that p_T was preserved. To first order it is p_T for each muon which goes into the dimuon invariant mass calculation. So correlations in the errors due to multiple scattering were arranged to cancel out, and the mass resolution was nearly the same as we had in earlier runs with no absorber wall, but the rates were vastly increased. We were able to collect of order 10^4 events in the Υ mass region as shown in Figure 9, with excellent resolution [9].

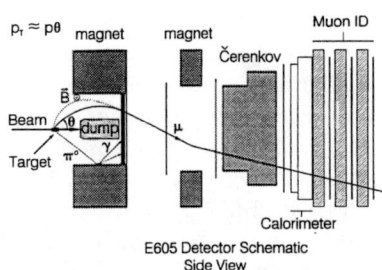

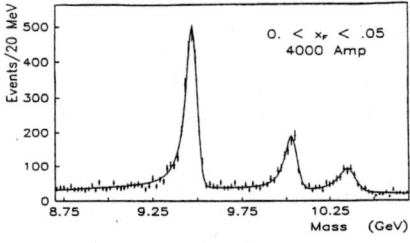

FIGURE 8. Dimuon detector using improved separated function technique. This compressed elevation shows the copper target, the beam dump, the momentum analyzing magnet with lead wall at the end, wire chamber planes and trigger hodoscopes indicated by vertical lines, momentum confirming magnet, the ring-imaging Čerenkov counter, calorimeters, and iron/proportional tube muon identifier.

FIGURE 9. E605 dimuon data showing the well resolved sub-threshold Υ resonances on top of the continuum without acceptance correction.

Many effects contribute to the tails on the Υ resonances on the low-mass side which are evident in Figure 9. One of the largest is due to the fluctuations to the high side of the muon energy loss in the wall, with contributions from pair production and bremsstrahlung.

REMARKS

In about 15 years and in four distinct steps experimenters were able to improve the dimuon mass resolution from $\Delta M_{\mu\mu} \sim 11\%$ to $\sim 0.3\%$ with roughly the same hadron rejection, allowing data samples with tens of thousands of events. The J/ψ (the discovery of the charm quark) *should have been* discovered by this technique, and the Υ (the discovery of the bottom quark) *was* discovered in such experiments. And it is interesting to note that with the best of these techniques, nothing of note was uncovered.

REFERENCES

1. J.H. Christenson *et al.*, Phys. Rev. Letters **25**, 1523 (1970); Phys. Rev. **D8**, 2016 (1973).
2. D. Eartly, G. Giacomelli, and K. Pretzl, Phys. Rev. Letters **36**, 1355 (1976).
3. Kindly provided by David Garelick.
4. D.A. Garelick *et al.*, Phys. Rev. **D18**, 945 (1978).
5. S.R. Smith *et al.*, Phys. Rev. Letters **46**, 1607 (1981).
6. S. Childress *et al.*, Phys. Rev. Letters **55**, 1962 (1985).
7. S.W. Herb *et al.*, Phys. Rev. Letters **39**, 252 (1977).
8. S. Glashow and M. Machacek, Phys. Lett. **145B**, 302 (1984).
9. G. Moreno *et al.*, Phys. Rev. **D43**, 2815 (1991).

b-Quark Physics at DORIS

Dietrich Wegener

Institut of Physics, University of Dortmund

Υ PHYSICS: THE EARLY YEARS 1977–1980

The high energy physics program ($E_{cms} \leq 8.6 \,\text{GeV}$) at DORIS was initiated by the PLUTO collaboration which sent its proposal to the Forschungskollegium June 30, 1977 [1]. The same day the observation of the $\Upsilon(9.46)$ resonance was announced to the public in a seminar at FNAL [2]. The physics program proposed by PLUTO included the measurement of σ_{tot} and the search for charm and τ. The search for a 3rd-generation quark is not mentioned in the proposal.

The news from FNAL spread fast. The first documented discussion at DESY between machine physicists and members of the PLUTO collaboration took place July 6, 1977 [3]. The energy upgrade of DORIS to $E_{cm} = 10 \,\text{GeV}$ at moderate cost within half a year turned out to be possible, if parts of the new PETRA cavities and power supplies were used. Moreover, minor changes of the DORIS lattice were envisaged to avoid strong saturation effects of the magnets. The proposal and its update [4] were discussed by the DESY Forschungskollegium at its meeting on July 15. The interest of measurements in the region of the new resonances was emphasized and the directorate was urged to consider an upgrade of DORIS to $E_{cm} = 10 \,\text{GeV}$ [5]. Note in this context that PETRA was under construction at this time and was scheduled to start running in late summer 1978.

The possible physics program at a 10 GeV machine was discussed at a DESY workshop in October 1977 [6]. J. Bürger and H. Schröder presented the physics program of the PLUTO and DASP II collaborations, the latter having just started to form. The "Physics Priorities at DORIS" from the theorists' point of view were discussed by T. Walsh. Astonishingly enough from today's point of view, mainly the physics of the 2nd generation was considered, with the $\Upsilon \to$ 3-gluon decay only briefly mentioned. Both experimental groups, on the other hand, discussed in detail the possibility of learning of a 3rd-generation-quark's properties in only a few days of running.

The steps leading from the 5 GeV double-ring DORIS to the 10 GeV single-ring DORIS I are collected in Table 1. The rapid energy upgrade of DORIS was unexpected. I remember a seminar given by A. de Rujula at CERN

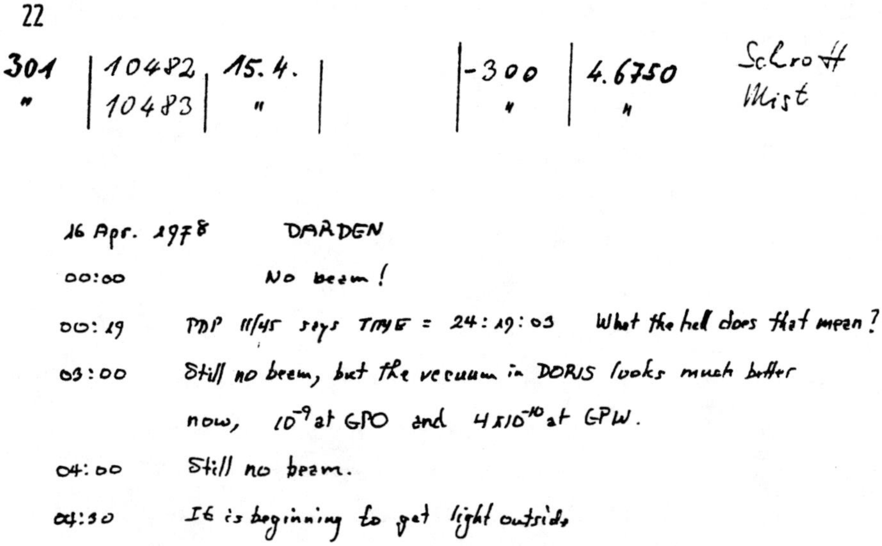

301	10482	15.4.		–300	4.6750	Schroff
"	10483	"		"	"	Mist

16 Apr. 1978   DARDEN

00:00   No beam!

00:19   PDP 11/45 says TIME = 24:19:03   What the hell does that mean?

03:00   Still no beam, but the vacuum in DORIS looks much better now, 10^{-9} at GPO and 4×10^{-10} at GPW.

04:00   Still no beam.

04:30   It is beginning to get light outside.

FIGURE 1. A page from the DASP II runbook (15/16.4.1977).

in March 1978 where he discussed Υ physics. According to him the first experimental results were to be expected from CLEO early in 1979. The scan in the $\Upsilon(1S)$ region started at DESY on April 15, 1978. Both the machine and the detectors had problems in the beginning (Fig. 1). A fluctuation observed first by DASP II, and less prominently by the PLUTO collaboration after applying sophisticated cuts, was convincing enough to motivate the DESY director to expend the first bottle of champagne. After a few days of running the peak vanished; its trace can still be found in the smaller step size of the scan around 9.38 GeV [9] in the published resonance curve. But finally, on April 30, the resonance signal was established. Why the Booze Up was delayed by 2 weeks (Fig. 2) cannot be reconstructed any more.

The results proved that the resonance was narrow, $\Gamma = (1.3 \pm 0.4)$ keV [9,10], and compatible with a $Q = -\frac{1}{3}$ charged quark. The mass of the resonant state was measured with high precision: $M(\Upsilon(1S)) = (9.46 \pm 0.01)$ GeV. These results established the $\Upsilon(9.46)$ resonance observed at FNAL [11] as a 3rd-generation quarkonium state. A few months later the DASP II and LENA collaborations – the latter replacing the PLUTO collaboration – with marginal statistics determined the parameters of the $\Upsilon(2S)$ state [12,13].

After the establishment of the quarkonium nature of the new resonances, the detailed study of the hadronic decays was of special interest, since the resonance was predicted to decay mainly via a 3-gluon final state [14]. Already the first study of the event topology by PLUTO revealed "a striking change in mean sphericity and thrust on the $\Upsilon(9.46)$ resonance" [15]. The PLUTO

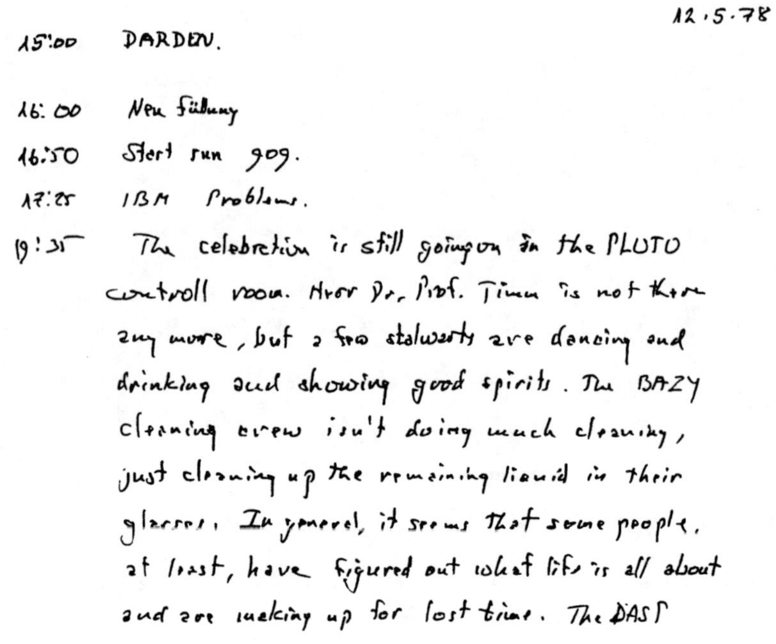
12.5.78

15:00 DARDEN.

16:00 Neu Füllung

16:50 Start run 909.

17:25 IBM Problems.

19:35 The celebration is still going on in the PLUTO controll room. Herr Dr. Prof. Timm is not there any more, but a few stalwarts are dancing and drinking and showing good spirits. The BAZY cleaning crew isn't doing much cleaning, just cleaning up the remaining liquid in their glasses. In general, it seems that some people, at least, have figured out what life is all about and are making up for lost time. The DASP

FIGURE 2. A page from the DASP II runbook.

collaboration addressed the problem of the 3-gluon final states in two further papers [16]. In the first paper, received by the publisher in December 1978, the authors concluded:

- The data are inconsistent with Υ decays into 2 light quarks (2-jet structure) and into multipion phase space.
- All quantities related to momentum phase-space configurations are found to be in agreement with the proposed 3-gluon decay mechanism. Vector gluons are consistent with the proposed 3-gluon decays but not proven.

Summarizing, one might say that vector gluons as the field quanta of the strong interaction were not *discovered* at DORIS I, but strong evidence for the decay of the $\Upsilon(1S)$ meson into three vector gluons *was* found [17]. This point is missed in some papers describing the discovery of the gluon [18].

The Crystal Ball (CB) [19] and ARGUS [20] collaborations later contributed further to our understanding of the $|b\bar{b}\rangle$ system.

DORIS II AND ITS DETECTORS

The major steps leading to the decision to upgrade DORIS I and to increase the machine energy to 11.2 GeV (DORIS II) are collected in Table 1.

TABLE 1. Milestones in the DORIS storage ring upgrade program.

6.7.1977	First discussion to upgrade DORIS to 2 × 5 GeV (DORIS I)
	Participants: Degele, Bürger, Criegee, Flügge
15.7.1977	Forschungskollegium: strong support for upgrade
16.12.1977	Proposal to upgrade DORIS to 2 × 5 GeV accepted
20.2.1978	Upgrade of DORIS starts → DORIS I
15.4.1978	Scan in $\Upsilon(1S)$ region starts
30.4.1978	$\Upsilon(1S)$ resonance observed
August 1978	$\Upsilon(2S)$ observed
July 1979	Low-beta insertion to increase luminosity proposed by K. Wille
March 1980	DORIS I stops running for high energy physics
February 1981	DORIS II (11.2 GeV machine) proposed
November 1981	DORIS II upgrade started
May 1982	DORIS II starts running
1991	DORIS II bypass upgrade for synchrotron radiation
October 1992	ARGUS stops data taking
May 1993	Tests to increase DORIS II luminosity fail
	High energy physics program at DORIS II ends

The driving forces were the growing interest in B physics and the possibility of upgrading DORIS I at moderate cost and manpower [21]. An essential criterion for the final choice of the DORIS II parameters was the requirement that the layout of the synchrotron-radiation beamlines be undisturbed. The essential changes of DORIS II with respect to DORIS I were the decrease of the gap width and the increase of the number of coil windings of the magnets, thus reducing saturation effects and power consumption. The injection was improved by installing separator plates and a faster kicker magnet. A major increase in the luminosity was achieved by mounting a special strong-focussing quadrupole at a small distance from the interaction point [22].

With these improvements DORIS II achieved a maximum integrated luminosity of 1.8 pb^{-1}/day and an average luminosity of 0.5 pb^{-1}/day.

The idea to build the ARGUS detector dates back to a dinner on September 14, 1977 [23]. Only one month later, at the DORIS workshop, the concept of "A New Detector at DORIS," including most of the features of the later ARGUS detector, was presented by W. Schmidt-Parzefall [6]:

- full coverage of the solid angle ($\approx 96\%$)

- good particle identification based on time-of-flight and dE/dx measurements

- shower counters inside the solenoidal coil to detect photons with energies $E_\gamma \geq 50$ MeV

- μ chambers to detect muons with a momentum $p \geq 0.9$ GeV/c.

TABLE 2. Milestones in the ARGUS detector program.

14.9.1977	First plans to build ARGUS
10.10.1977	Meeting of DORIS Experimenters
	Detector design study presented
October 1978	DESY proposal #146 : ARGUS – a new detector for DESY
July 1979	ARGUS proposal accepted by DESY directorate
April 1980	Interest in running ARGUS at 11.2 GeV emphasized
February 1981	B physics program at DORIS II discussed
6.10.1982	ARGUS starts running
September 1987	$B^0\overline{B}^0$ mixing observed
Autumn 1989	Observation of $b \to u$ transitions
8.10.1992	ARGUS stops taking data

TABLE 3. Integrated luminosity collected 1983–1986 at DORIS II.

	1983	1984	1985	1986
$\Upsilon(1S)$	9 pb^{-1}	23 pb^{-1}	-	31 pb^{-1}
$\Upsilon(2S)$	27 pb^{-1}	25 pb^{-1}	-	-
$\Upsilon(4S)$	6 pb^{-1}	14 pb^{-1}	45 pb^{-1}	44 pb^{-1}
Continuum	4 pb^{-1}	7 pb^{-1}	16 pb^{-1}	19 pb^{-1}

The ARGUS[1] proposal was presented to the Forschungskollegium in October 1978 and accepted in July 1979. The final design followed in many details the original idea, with only the layout of the drift chamber improved to account for the requirements of optimal pattern recognition. The physics benchmarks in the proposal were charm and τ physics. A detailed evaluation of a possible B-physics program was presented in April 1980 [24]. An expanded analysis of the possibilities of studying B physics with ARGUS followed in February 1981 [25] when it became clear that DORIS I could be upgraded to an energy of 11.2 GeV. The detector worked in a stable manner from 1982 through 1992.

During the DORIS workshop in February 1981 the idea arose to transfer the Crystal Ball (CB) detector from SLAC to DESY [19]. The proposal was soon presented and accepted in summer 1981. The CB detector was transported to DESY in spring 1982 and started data taking August 6, 1982, while ARGUS rolled in two months later. The competition between the two experiments delayed the B-physics program at DORIS for nearly 3 years because the CB collaboration preferred to run at the energy of the $\Upsilon(1S)$ and $\Upsilon(2S)$ resonances, since its detector was optimized for spectroscopic studies. As shown by Table 3, in the first years of DORIS II running, priority was given to the CB

---

[1] The official interpretation is **A**–**R**ussian–**G**erman–**US**–**S**wedish collaboration, indicating the nationalities of the original proponents of the experiment. The unofficial interpretation by one of the spouses knowing the senior members of the group too well reads **A**lle **R**ichtigen **G**enies **U**nter **S**ich.

physics program. The following facts may have contributed to the decision:

- CB was a running detector with a respectable record of discoveries.
- CB was an established and successful collaboration while the ARGUS senior members at that time were youngsters.
- CB observed an unexpected signal [26] and hopes were running high for a short time that a light Higgs had been discovered.[2] Unfortunately, the result turned out to be irreproducible [28].

Before discussing the most important ARGUS discovery a further obstacle met by ARGUS should be mentioned. As shown in Fig. 3, two major gaps in the data taking are manifest. They follow the most important ARGUS discoveries: in 1987 $B^0\overline{B}^0$ mixing was observed, and in 1989 $b \rightarrow u$ transitions were detected. One might wonder if the DESY directorate suspected ARGUS was not putting enough emphasis on analysis, and therefore wanted to give the collaboration a chance to improve in this respect. Note, however, that the official explanation is different: in 1987 HERA got priority and in 1990/1991 the DORIS bypass was built. From the latter "improvement" the machine never recovered for high energy running.

DISCOVERIES

The ARGUS collaboration for more than one decade substantially contributed to various fields of high energy physics. The results are summarized in [20]. In B physics the highlights are the following "firsts:"

- Observation of $B^0\overline{B}^0$ mixing [29].
- Observation of charmless B decays [30].
- Reconstruction of exclusive semileptonic B decays to D^* and D mesons [31].
- Reconstruction of exclusive hadronic B decays [32].
- Model-independent measurement of semileptonic B decays [33].
- Observation of charmed baryons in B decays [34].

Due to a lack of time, only the most important discovery is discussed in some detail.

---

[2] At this point it is appropriate to remind the reader of the guidelines for searches formulated 200 years ago: "One may notice that a shrewd intellect brings more artifice to bear the fewer data are available; indeed, to demonstrate his mastery he will select from all available data only those few favorable to his views; the remainder he will arrange so as not to obviously contradict his conclusions; and finally hostile data will be isolated, surrounded, and disarmed" [27].

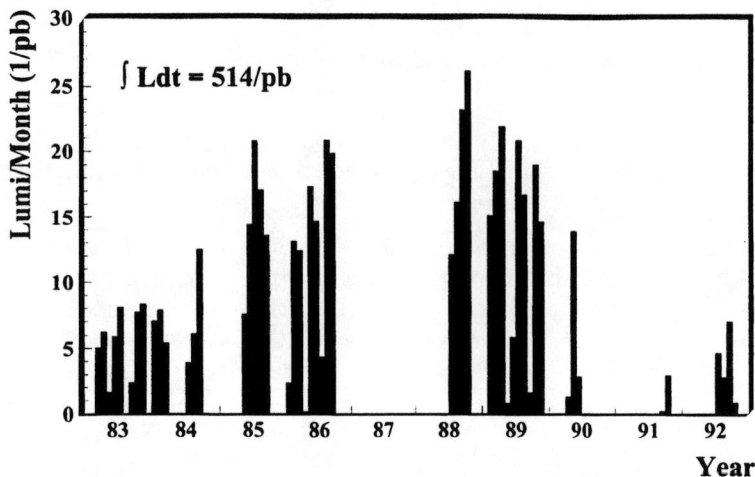

FIGURE 3. Luminosity collected by ARGUS 1982–1992.

$B^0\overline{B}^0$ Mixing

Present universal interest in B physics is largely due to the discovery of $B^0\overline{B}^0$ mixing by the ARGUS collaboration. As is well known [35] the process is mediated by box diagrams. The mixing parameter r_d derived from time-integrated measurements is given by the expression

$$r_d = \frac{N(B^0 \to \overline{B}^0)}{N(B^0 \to B^0)} = \frac{(\Delta M \tau_B)^2}{(2 + \Delta M \tau_B)^2} \sim m_{\text{top}}^4, \tag{1}$$

where

$$\Delta M = \frac{G_F^2}{6\pi^2} \, B_B f_B^2 m_B \mid V_{tb}^* V_{td} \mid^2 m_{\text{top}}^2 F\left(\frac{m_{\text{top}}^2}{m_W^2}\right) \eta_{QCD}, \tag{2}$$

i.e. mixing is dominated by virtual t-quark exchange. The experimental situation in 1986 was as follows: PETRA experiments did not observe a signal, *i.e.* $m_{\text{top}} \leq 23.3$ GeV, while UA1 claimed [36] a signal at $m_{\text{top}} \approx 40$ GeV. As a consequence a small mixing parameter $r_d \approx 0.01$ was expected. A scan of the literature by the author in September 1985, while preparing a memo to the DESY PRC, showed that under optimistic assumptions on f_B a mixing of $r_d \leq 0.05$ was predicted [37]. Mixing searches using b-quark jets by MARK II, MAC, and UA1 were not conclusive.

In summer 1986, for the first time ARGUS and CLEO had enough statistics to exploit the particularly clean conditions at the $\Upsilon(4S)$ to search for

$B^0\overline{B}^0$ mixing. The semileptonic decay $\overline{B}^0(\bar{b}d) \to l^+X$ served as a tag of the heavy flavor, i.e. $l^\pm l^\pm$ and l^+l^- events were used to measure the mixed and unmixed events respectively. At the Berkeley conference the groups presented their limits (90% CL): $r_d \leq 0.12$ (ARGUS [38]) and $r_d \leq 0.20$ (CLEO [39]). Immediately after the conference ARGUS prepared a publication which even got a DESY number (DESY 86-121). However, the distribution of the paper was stopped at the last moment by H. Schröder. He collected all preprints at the moment they left the printer's office. All copies were burned!

What observation led to this reaction? In August 1986 H. Schröder started an analysis of the $\overline{B}^0 \to D^{*+}l^-\bar{\nu}_l$ decay, which was of special interest, since a large branching ratio of $\sim 8\%$ was predicted but no measurements existed. Since the D^{*+} reconstruction capabilities of ARGUS were excellent and e and μ were identified with high efficiency, a high-statistics $\overline{B}^0 \to D^{*+}l^-\bar{\nu}_l$ sample out of ~ 25000 $B^0\overline{B}^0$ events was expected. However, a new method had to be developed to reconstruct these events containing an undetected ν_l, whose mass can be derived from the measurements:

$$m_\nu^2 = (E_B - E_{D^{*+}} - E_{l^-})^2 - (\vec{p}_B - (\vec{p}_{D^{*+}} + \vec{p}_{l^-}))^2. \tag{3}$$

From the first successful reconstruction of exclusive hadronic B decays [32] it was known that

$$2E_B = m(\Upsilon(4S)) \approx 2m_B. \tag{4}$$

Since $E_B = E_{beam}$, $|\vec{p}_B| = 0.33\,\text{GeV}/c$ and hence can be neglected in (3). Therefore,

$$m_\nu^2 \approx M_{rec}^2 = (E_{beam} - E_{D^{*+}} - E_{l^-})^2 - (\vec{p}_{D^{*+}} + \vec{p}_{l^-})^2. \tag{5}$$

As expected, a peak at $M_{rec}^2 \approx 0$ is observed with small, well-known background (Fig. 4). Though the application of this method was controversial at the time [40], in the following years it was applied in many analyses by the CLEO and ARGUS collaborations. In September 1986, 50 events with a reconstructed $B^0(\overline{B}^0)$ were available to tag the heavy flavor of the B^0. H. Schröder presented the first results of his analysis at the ARGUS group meeting September 25, 1986 (Fig. 5). He studied in detail the events with a fully reconstructed $B^0(\overline{B}^0)$ meson. The observed multiplicity and the number of kaons and leptons followed expectations, but he stumbled over a few events with wrong-sign kaons and leptons. In the data sample five candidates for mixed events were observed: 2 $\overline{B}^0 e^+$, 2 $\overline{B}^0 e^-$, and 1 $\overline{B}^0\mu^-$, along with 23 candidates for unmixed events. After background subtraction a mixing ratio of $r_d = 0.20 \pm 0.12$ was obtained.

The claim that $B^0\overline{B}^0$ mixing had indeed been observed was supported by the observation of one fully reconstructed event with 2 B^0 mesons in the final

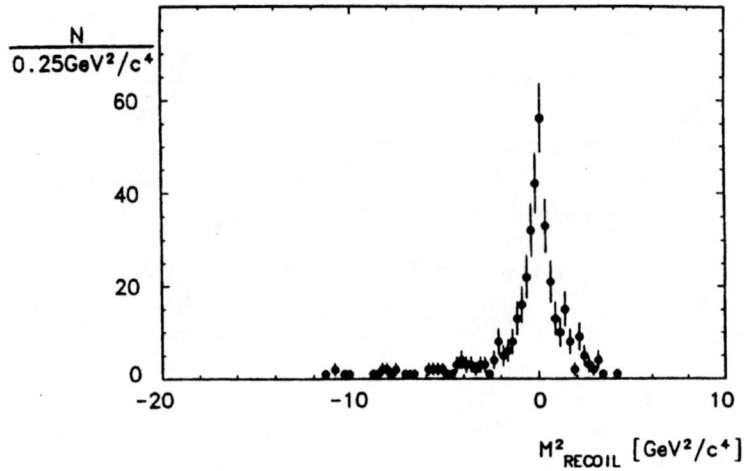

FIGURE 4. Measured recoil mass distribution.

state decaying via $B^0 \to D^{*-}\mu^+\nu_\mu$ (Fig. 6). The μ^+ and K^+ of both decays were uniquely identified. The observation of this event is a convincing example of the advantages of the ARGUS detector: precise momentum measurement, good particle identification, and hermeticity. The observation of D^*-lepton correlation therefore provided an extremely useful tool. This proved the existence of $B^0\bar{B}^0$ mixing with a large mixing parameter, totally unexpected at that time.

This result stimulated further activity. Y. Zaitsev and A. Golutvin repeated the same-sign lepton-pair analysis. A signal was observed in this sample as well. The major improvement compared to the previous analysis presented at Berkeley [38] was the increase in the collected luminosity of more than a factor of 2. Furthermore, the better understanding of the detector allowed improving the cuts applied in the analysis. The mixing parameter derived in this analysis was in good agreement with the result of the exclusive analysis. Combining the results ARGUS got

$$r_d = (0.21 \pm 0.08), \qquad (6)$$

in good agreement with the present world average [41].

To explain the large mixing parameter, ARGUS had to assume the top mass to be large, $m_{top} > 50$ GeV, 10 years ago an unconventional assumption in view of the UA1 claim [36]. The paper was published June 25, 1987, just 10 years after the discovery of the $\Upsilon(1S)$ resonance by Lederman and coworkers at FNAL. The large mixing in the B system raised hopes of observing CP violation in this system, a prospect attracting many scientists to the field.

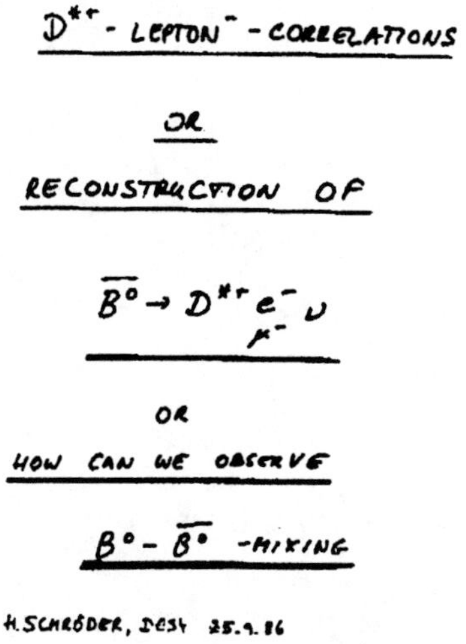

> D^{*+} - LEPTON$^-$ - CORRELATIONS
>
> OR
>
> RECONSTRUCTION OF
>
> $\overline{B^0} \to D^{*+} e^- \nu$
> $\phantom{\overline{B^0} \to D^{*+}} \mu^-$
>
> OR
>
> HOW CAN WE OBSERVE
>
> $B^0 - \overline{B^0}$ - MIXING
>
> H. SCHRÖDER, DESY 25.9.86

FIGURE 5. First page of H. Schröder's talk announcing the observation of $B^0\overline{B}^0$ mixing.

The experiments presently under construction [42] underline the importance of the seminal ARGUS result obtained 10 years ago.

SUMMARY

I will abstain from discussing in detail the other important contributions of ARGUS to B physics. I only want to address the question why the collaboration was so successful for nearly 10 years. The answer was given by David Cassel in his talk "The Impact of ARGUS on Experimental Heavy Flavor Physics" [43], where he discussed the lessons to be learned from ARGUS:

- Have a better detector that can see "all."
- Learn to use the hermeticity of the detector.
- Have excellent physics ideas and follow them.
- Have excellent physics analysis software.
- Have a little bit of luck.
- Do not underestimate the competition.

There are a bit too many *excellent*'s in this list but otherwise I have nothing to add. Hopefully the new generation of experiments will be as prolific as the 2nd generation, and the participants will have as much fun as the CLEO and ARGUS collaborations had.

FIGURE 6. The first fully reconstructed $B^0\overline{B}^0$ mixing event.

Acknowledgements

I thank the organizers for their generous hospitality and for a stimulating conference, where I learned many secrets of those experiments which established the field of B physics. Thanks for discussions, unpublished material, *etc.* to J. Bürger, H. Meyer, W. Schmidt-Parzefall, K. Wacker and K. Wille. Most helpful in reconstructing the steps leading to the discovery of $B^0\overline{B}^0$ mixing was the information I got from H. Schröder. This work was supported by the BMBF under contract number 6DO57I.

REFERENCES

1. C. Berger *et al.* (PLUTO), DESY proposal #144 (30.6.1977).
2. J. K. Yoh, these proceedings.
3. J. Bürger, G. Flügge, L. Criegee, D. Degele, Minutes of a meeting 6.7.1977.
4. Update to DESY proposal #144 (8.7.1977).
5. Minutes of 279. meeting of DESY Forschungskollegium (15.7.1977).
6. Meeting on DORIS Experiments 10/11.10.1977 DESY F15/01 (Nov. 1977).
7. C. W. Darden *et al.* (DASP II), DESY proposal #146 (24.2.1978).
8. Minutes of 284. meeting of the DESY Forschungskollegium (16.12.1977).
9. C. Berger *et al.* (PLUTO), Phys. Lett. **B76** (1978) 243.
10. C. W. Darden *et al.* (DASP II), Phys. Lett. **B76** (1978) 246.
11. S. W. Herb *et al.*, Phys. Rev. Lett. **39** (1977) 252.
12. C. W. Darden *et al.* (DASP II), Phys. Lett. **B78** (1978) 364.
13. J. K. Bienlein *et al.* (LENA), Phys. Lett. **B78** (1978) 360.
14. A. de Rujula *et al.*, Nucl. Phys. **B138** (1978) 387.
15. C. Berger *et al.* (PLUTO), Phys. Lett. **B78** (1978) 176.
16. C. Berger *et al.* (PLUTO), Phys. Lett. **B82** (1979), Z. Phys. **C8** (1981) 101.

17. D. Wegener in: A. Ali, P. Söding, High Energy e^+e^- Annihilation, p. 489, World Publishing 1989.
18. G. Jarlskog, Europhysics News **26** (1995) 88.
19. J. K. Bienlein, DESY F31-91-02.
20. H. Albrecht et al. (ARGUS), Phys. Rep. **276** (1996) 223.
21. K. Wille, DESY, 79–08 (Feb. 1978).
22. K. Wille, DESY 81–047 (Aug. 1981).
23. D. Wegener, UniReport 23 (1996) 17.
24. E. R. McCliment and H. Schröder, DESY F15 80/01 (April 1980).
25. K. Schubert in: Workshop on DORIS experiments, DESY Feb. 1981.
26. H. J. Trost et al. (Crystal Ball), Proc. XXII Int. Conf. High Energy Physics, Leipzig 1984, p. 201.
27. J. W. Goethe, Werke, Hamburger Ausgabe Vol. XIII, p. 10, Der Versuch als Vermittler von Objekt und Subjekt.
28. DESY Annual Report 1984, p. 123.
29. H. Albrecht et al. (ARGUS), Phys. Lett. **B192** (1987) 245.
30. H. Albrecht et al. (ARGUS), Phys. Lett. **B234** (1990) 409.
31. H. Albrecht et al. (ARGUS), Phys. Lett. **B197** (1987) 452.
32. H. Albrecht et al. (ARGUS), Phys. Lett. **B185** (1987) 218.
33. H. Albrecht et al. (ARGUS), Phys. Lett. **B318** (1993) 397.
34. H. Albrecht et al. (ARGUS), Phys. Lett. **B210** (1988) 263.
35. H. Schröder in: S. Stone, "B Decays," p. 282, World Publishing 1992.
36. G. Arnison et al. (UA1), Phys. Lett. **B147** (1987) 493.
37. ARGUS collaboration, Memorandum to the PRC: PRC 85/05 (20.9.1985).
38. H. Albrecht et al. (ARGUS), paper #9717 submitted to Berkeley conference 1986.
39. CLEO, paper submitted to Berkeley conference, CLNS 86–741.
40. S. Stone, Proc. Inter. Symp. Multiparticle Dynamics, Arles 1988, p. 13.
41. Particle Data Group, Phys. Rev. D **54** (1996) 1.
42. W. Schmidt-Parzefall, P. McBride, T. Nakada, W. R. Innes, R. Lipton, these proceedings.
43. D. Cassel, talk at the "ARGUS End of the Run Party," DESY 19.11.1993.

EXPERIMENTS AT e^+e^- COLLIDERS

CESR and CLEO—The Early Years

Albert Silverman

Nuclear Studies, Cornell University, Newman Lab, Ithaca, NY 14853-5001

Abstract. This narrative about the beginnings of CESR and CLEO covers mainly the period from 1974, when discussions about the possibility of building an e+e- collider at Cornell were begun, to 1981 when CLEO had been taking data for some eighteen months. Labeling this as a narrative is meant to emphasize that much of it is undocumented and relies on my memory.

INTRODUCTION

In 1974 the 10 GeV Cornell Synchrotron had been in operation for seven years and we started to think about a new machine. Two possibilities were considered. One, a 50 GeV electron synchrotron, and the other, an 8 GeV e+e- collider, using the 10 GeV synchrotron as the injector. We met perhaps a dozen times debating these alternatives (1). The collider was more exciting to us, but it had two problems. First, the synchrotron was a poor injector, requiring about an hour to fill the ring which was about the expected lifetime of the beams. Second, the physics seemed to be largely QED at short distances and we were more interested in hadron physics. There was information from Adone (2) at Frascati that the hadron cross section was larger than expected, but whether this signaled good hadron physics potential was not clear. Both of these reservations were removed in the winter of 1974 with the spectacular discovery of the ψ at SLAC and Maury Tigner's invention of a method to reduce the injection time to a few minutes.

CESR

I want to briefly describe the Tigner scheme for converting the synchrotron to an acceptable injector because it is very pretty and it was crucial to having a plausible proposal. CESR was designed with two interaction regions and two circulating bunches of electrons and positrons. In order to achieve the desired luminosity, one had to make the bunch currents in the collider large enough to reach the linear tune shift limit and this took an hour with the positron bunch current then produced by the synchrotron. Tigner invented a method for making an intense collider bunch by superimposing sixty synchrotron bunches. This was done by making the circumference of the collider bigger than that of the synchrotron by the distance between two synchrotron bunches. Sixty bunches were accelerated in the synchrotron and transferred to the collider. The second in the train of sixty collider bunches was sent back to the synchrotron, allowed to circulate for one turn and returned to the collider where it fell on the first bunch in the train. The third bunch was then returned to the synchrotron for two turns and put back into the collider also superimposed on the first turn, and so on for all sixty turns.

These developments persuaded the laboratory director, B.D. McDaniel, and most of the faculty to go for a collider and started an intense design period which resulted in a proposal to the National Science Foundation in May, 1975. The proposal was for an 8 GeV e$^+$e$^-$ collider with two interaction regions and a peak luminosity of 10^{32} cm^{-2}sec^{-1}. This energy filled a gap between SPEAR at 2.5 GeV and PEP/PETRA at 16 GeV. One might suppose that PEP/PETRA could operate at lower energies and so they could, but not very efficiently. For an e$^+$e$^-$ collider, the luminosity decreases at least as fast as the energy squared. However, the luminosity at the peak energy is more or less the same for all colliders, independently of the peak energy. The luminosity of CESR at 5 GeV was expected to be about a factor of five larger than that expected of PEP or PETRA. The National Science Foundation showed some interest in the proposal, and in 1976 and 1977 they provided funds for design work and prototypes. Full approval for the project came in 1977.

Maury Tigner was in charge of the construction. I will dispense with the usual parameter list. In general the CESR design, except for the Tigner injection scheme, was quite conventional. The radius was predetermined since it was to fit in the existing tunnel. Construction proceeded very rapidly. Electrons were first stored in April 1979. The first measurable colliding beams were observed on August 14 of that year, and by October, just two years after the project was approved, the first experimental data was taken. I will return to the experiment soon. But I want to spend a little time on the operation of the collider.

The most important measure of the success of a collider is, of course, the luminosity. Figure 1 is a record of the integrated luminosity per month from 1981 to 1996, and shows an increase of several orders of magnitude during this period. How was this steady increase achieved? Of course some of the luminosity increase resulted from continuous improvement in component reliability, from better theoretical understanding and the usual accidental discovery of better operating conditions. By far the greater part of the luminosity increase resulted from planned changes which I want to describe.

Assuming there are no other technical limitation, like insufficient RF or the inability to support the required current or single beam instability, the luminosity per bunch is limited by instability due to the interaction of the colliding beams. The strength of this interaction is characterized by the linear tune shift which is a measure of the spread in betatron frequencies introduced by the beam–beam interaction. It is proportional to the bunch density. In most e$^+$e$^-$ colliders, the maximum achievable tune shift is about 0.05. A convenient expression for the luminosity of a collider operating at the limiting linear tune shift, ξ, is given below.

Long Term Upgrade Plans

There is an active research program aimed at a Luminosity of 3×10^{34}. The plan calls for two rings with 180 bunches, 3 amperes, in each ring, and $\beta_v = 0.7$ cm. The decrease in β_v will be possible within the present upgrade plans. The two rings would cost about $35 million. There are many unsolved technical problems, but nothing that seems to be a show stopper. The conviction that a collider of this luminosity would do important B physics for many years provides ample motivation for solving these technical problems.

CLEO

CESR was to have two interaction regions. One, the "South Area" was large enough for a general purpose magnetic detector. A coalition was formed with faculty, research associates, and graduate students from Cornell, Harvard, Rochester, and Syracuse groups doing experiments on the 10 GeV synchrotron. The collaboration was soon joined by groups from Rutgers and Vanderbilt and adopted the name CLEO, which was not an acronym but seemed to go well with CESR. The "North Area" was much smaller, but could accommodate an experiment the size of the Crystal Ball at SPEAR. A competition was held and a group was chosen from Columbia and Stonybrook (CUSB), made up partly of the group whose discovery we are celebrating here today. CUSB was modeled after the Crystal Ball and did important work, particularly in Upsilon spectroscopy, about which you will hear more at this symposium (4).

After some preliminary work in 1976, the CLEO collaboration started intensive design studies at its first group meeting on March 4, 1977. The detector as it was in 1981 is shown in Figure 2. CLEO was strongly influenced by Mark II at SLAC, but there were two components that were unusual. The "thin" superconducting solenoid and the DE/DX proportional chambers. The choice of the magnet presented us with a problem. We wanted to have a field of at least 1 Tesla produced by a solenoid that was as thin as possible since most of the particle identification was to be outside the coil. Any warm coil would either require too much power or would be too thick in radiation lengths to achieve such a field. With several megawatts, a reasonable power requirement, and one radiation length one could achieve about 0.5 Tesla with a warm coil. At the time we started no "thin" superconducting solenoid (<1 r.l.) had been built. There was a design study at Lawrence Berkeley Lab for such a coil to be used at PEP. They were producing a full bore 1/4-length prototype to test whether such a coil could survive quenches without any active protection. We decided that one could proceed immediately with a coil based on the Lawrence Berkeley Lab design with active quench protection and did so. The design called for a field of 1.2 Tesla and 0.7 r.l. Since we were not confident that we could have the superconducting coil in the fall of 1979 when CESR was scheduled to begin experiments, we also decided to make an aluminum coil which would be 1.2 r.l. thick and would produce 0.4 Tesla with power supplies we had available. The coil cost

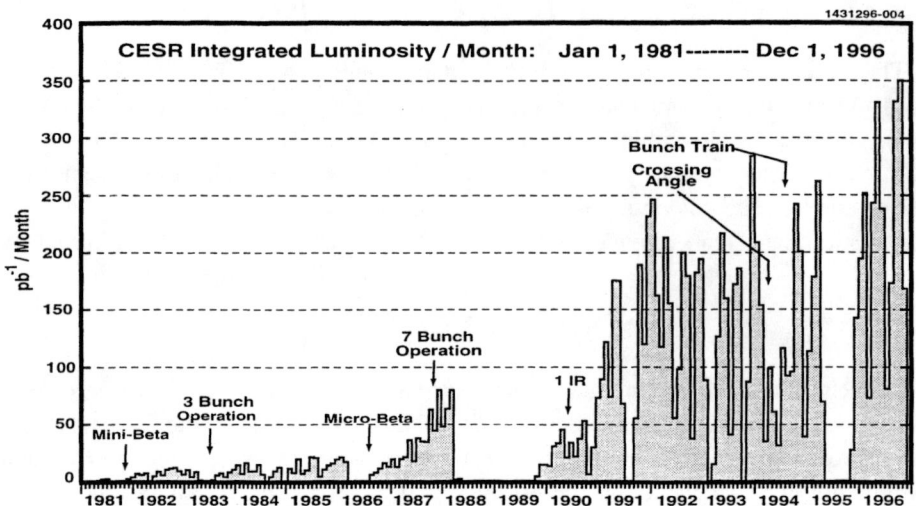

Figure 1. CESR integrated montly luminosity from January 1, 1981 through December 1, 1996.

$$L = 2.17 \times 10^{32} \, E(GeV) e N_e f \xi / \beta_v \tag{1}$$

β_v measures the strength of the focusing at the interaction region. At the tune shift limit the only parameters one can control are β_v and f which is the number of collisions per second (i.e., the number of bunches in the storage ring times the rotation frequency). Changes in these two accounted for most of the increase in luminosity. Initially, β_v was 10 cm. This was reduced twice, in 1981 to 3 cm and in 1986 to 1.8 cm, leading to something like a factor of five improvement. Another great leap forward was Raphael Littauer's invention of a method for having many bunches in the ring colliding at the interaction point but separated elsewhere. This made it possible to achieve the same limiting tune shift with many bunches. This "pretzel" scheme, as Littauer called it, enabled us to go initially to three, then to seven, and now to fourteen bunches in seven trains of two bunches. The best luminosity now is 4×10^{32}, about a factor of five better than any other e^+e^- collider. In the next two years, we plan to collide fortyfive bunches (nine trains of five) and to control a troublesome longitudinal instability which limits the present current. This should increase the luminosity to about 1.5×10^{33}.

So far we are talking about flat beams. We have been studying the behavior of round beams and have found that one can easily achieve a linear tune shift of 0.08, about twice what we do with flat beams. Dick Talman has invented a particular method for obtaining round beams which he calls the Mobius accelerator (3) and which seems to have some advantages over the usual method of coupling horizontal and vertical oscillations to produce round beams. Sometime this summer we will begin testing the Talman scheme.

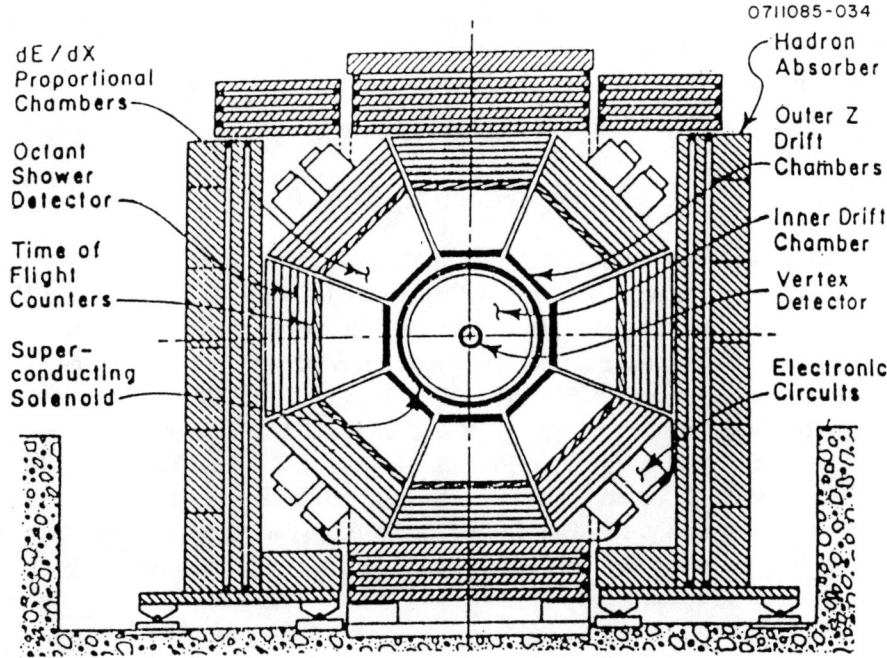

Figure 2. CLEO detector as of 1981.

$20,000. This precaution turned out to be wise. The superconducting coil was not ready until the summer of 1981, about 18 months after the start of the experimental program. I believe it was the first "thin" superconducting solenoid to be used in a collider experiment.

The other unusual component was the DE/DX proportional chambers. They consisted of octants of 117 cell multi-wire proportional chambers which measured the mean ionization of charged particles with an accuracy of 6%. This, together with the time-of-flight information from the scintillation counters (Fig. 2), provided reasonable particle identification up to 800 MeV/c, which included most of the particles produced in decay of the Upsilon states. Initially we planned to provide some particle identification for relativistic particles and for this purpose half the octants contained threshold gas Cerenkov counters. But we soon discovered that there were very few high momentum particles in these events (a fact we could have predicted) and replaced the Cerenkov counters with DE/DX chambers. It is interesting that as the emphasis has shifted to the rare two body decays of the B meson, identifying relativistic particles has become very important and CLEO is now building, under the leadership of Sheldon Stone, ring imagining Cerenkov counters for this purpose.

Inside the coil was a tracking drift chamber with seventeen layers, eight stereo. Outside the coil, the octants contained, in addition to those components described, an electromagnetic calorimeter consisting of layers of lead plate interleaved with proportional counters. Finally, the muon detector, with iron absorbers separated by drift chambers.

Experimental Results

The first measurement was the hadronic cross section versus energy. The three Upsilon bound states were observed by Christmas 1979 and reported in CLEO's first publication (5). Shortly thereafter we observed the wider $\Upsilon(4S)$ (6). The state of the hadronic cross section measurements in the summer of 1981 is shown in the Figure 3, reproduced from a report by CLEO to the 1981 Lepton Photon Conference in Bonn (7).

From the widths, it seemed clear the $\Upsilon(1S)$, $(2S)$, and $(3S)$ are bound $b\bar{b}$ states and the $\Upsilon(4S)$ was likely to be a state above the threshold for $B\bar{B}$ decay. The earliest measurements established the broad feature of the bound state decay modes. Here we had very good guidance from the Psi data. In fact, in 1976, a year before the discovery of the Upsilon, Kurt Gottfried, assuming clairvoyantly a 5 GeV b-quark, accurately predicted the bound state $b\bar{b}$ spectroscopy (8). The main result of the bound state measurements is that the quark forces are flavor independent. Many of the photon decays of the bound states, crucial for establishing the spectroscopy, were first observed by CUSB (4).

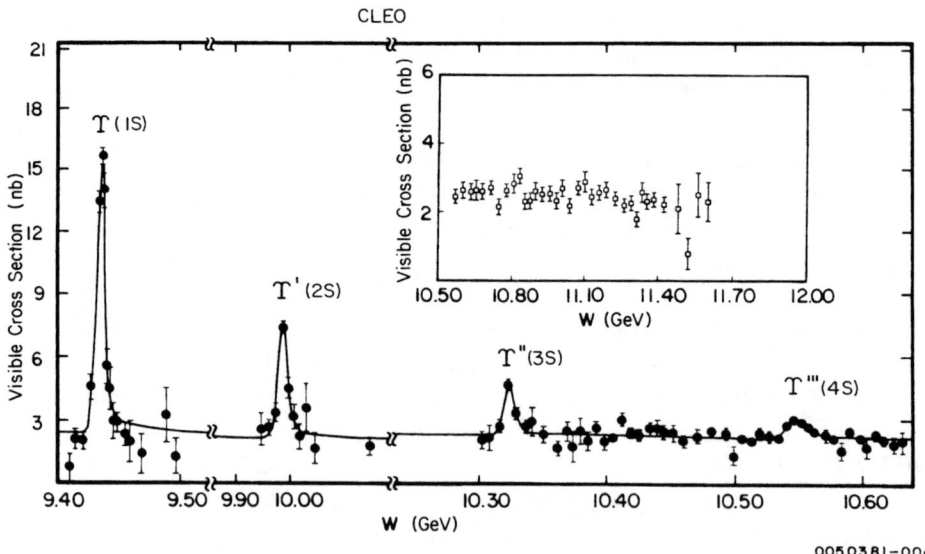

Figure 3. Total cross section for $e^+e^- \to$ hadrons (CLEO).

Attention then turned to the properties of the B meson, from the presumed decay $\Upsilon(4s)$ → $B\bar{B}$, the study of which has been the main focus of CLEO. By 1981, many of the general features of $\Upsilon(4S)$ decays had been determined. These were summarized in the CLEO report to the conference in Bonn (7) as follows.

> We find that B-meson decay is well described by Kobayashi-Maskawa. Diverse measurements have shown that the B-meson decays primarily by b → c, but a contribution of b → u of 25% or less cannot be ruled out.
>
> We have tested a variety of exotic models, including a broad class with no top quark and find them in conflict with the data.
>
> We have evidence from hadronic decays of the $\Upsilon(2S)$ and $\Upsilon(3S)$ that soft gluon processes are well described by the multipole expansion formalism. A new determination of the QCD scale parameter, $\Lambda_{\overline{ms}}$, yields a value of about 100 MeV, considerably lower than the value of 500 MeV popular a year ago.

The measurements which lead to these conclusions are also listed in this report.
- Inclusive properties of B-meson decays.
- Semileptonic branching ratios.
- Dileptons. Limits on flavor changing neutral currents,
  K^0 and $K^\pm$ yields from B-decay
- Search for exotic decays

The inclusive properties showed that the $\Upsilon(4S)$ decays were spherical (Fig. 4). The mean multiplicity of charged particles was large and the hadron momenta spectra soft (Fig. 5). These characteristics were expected for the decay $\Upsilon(4S)$ → $B\bar{B}$, with the B-mesons produced nearly at rest, and was in sharp contrast to the jet-like continuum events. The large lepton yields provided evidence for the weak decay of a new flavored quark. Assuming the leptons came from B meson decay, the semileptonic branching ratio and the lepton spectrum showed that the decay was primarily b → c as expected (Fig. 6). This conclusion was strengthened by the large yield of kaons per decay at the $\Upsilon(4S)$, about twice the number observed in continuum events. At this time one could set

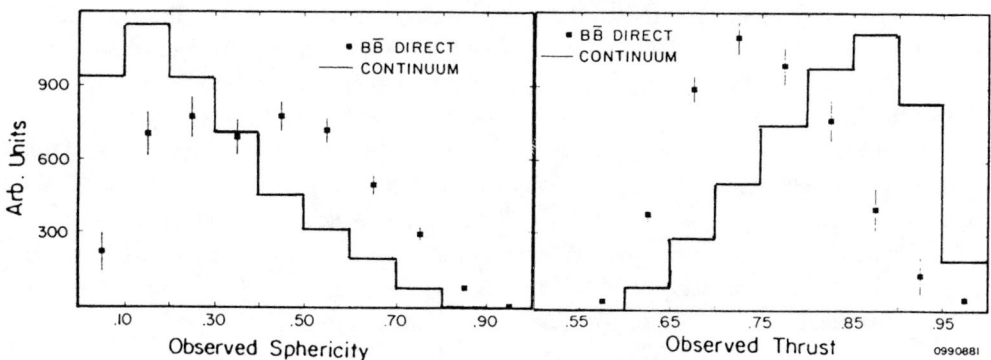

Figure 4. Shericity and thrust distribution for $B\bar{B}$ and continuum events.

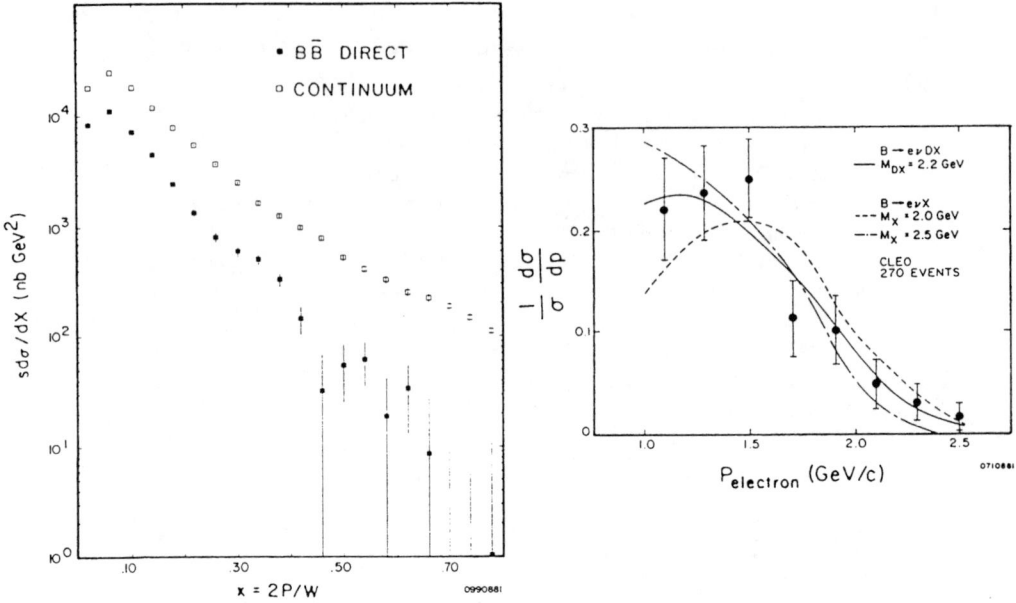

Figure 5 (left). Fractional momentum distribution of charged particles for continuum of $B\bar{B}$. **Figure 6 (right).** Momentum distribution of electrons from B decay. The curves are Monte Carlo calculations for various assumptions about the mass of the hadrons in the decay.

a limit of about 25% to the b → u branching ratio. It took another several years to show that B(b → u) is finite and about 1% (9). However, this early data established that the CKM (then the KM) quark mixing matrix describes the main features of the weak decay of the b-quark.

An interesting result reported to the same conference was an upper limit to flavor changing neutral current (FCNC) decays, B(B → $\ell^+\ell^-$x) < 0.74%. Several theoretical calculations showed that if the b is a singlet (no top quark partner) and decays through the W or Z, the ratio, r = (B → $\ell^+\ell^-$x)/(B → ℓvx), must be greater than 0.12 (10, 11). The measured ratio ruled out a singlet b-quark decaying weakly via W and Z. This somewhat indirect evidence that there was a top quark was comforting since the determined search for the top quark at DESY (12) had been unsuccessful and had set a lower limit for its mass of 20 GeV. The first hint of just how massive the top quark is came from the important discovery by ARGUS of the large $B\bar{B}$ mixing (13). The mixing result provided strong evidence that the top quark mass is greater than 40 GeV.

CLEO Upgrades

The early work with CLEO revealed several weaknesses. It was clear that we needed better tracking, particle identification, and photon detection for the more demanding experiments to come.

In 1984 the tracking was improved by adding, at small radius, ten layers of straw drift chambers with spatial resolution of about 100 microns. This made it possible to measure charm and tau lifetimes. In 1986 we replaced the 17-layer drift chamber with a chamber containing 51 layers which improved the tracking and momentum resolution very considerably. It also improved particle identification because the DE/DX resolution was as good as the octant DE/DX measurements and, importantly, did not suffer from the problems caused by interactions in the coil.

The difficult problem was improving the photon detection. Two proposals were considered. One was a lead glass shower counter replacing the DE/DX counters which had largely outlived their usefulness. This would have provided a substantial improvement but still left the problem of conversions in the coil. The other more ambitious plan was to put 7000 CsI crystals inside the coil and replace the superconducting coil with a larger one at 1.5 Tesla. Fortunately, the more daring plan, carried out under the leadership of Bernie Gittelman and Sheldon Stone, was adopted, and in 1989 CLEO became the first collider detector with excellent charged particle and photon detection.

The improved detector, combined with the large improvements in CESR luminosity, have made CESR/CLEO one of the important players in heavy quark and tau physics. At the moment, CLEO has accumulated a total luminosity of some 5 fb^{-1}, 10^7 B decays, and 5×10^6 charmed and tau decays. Some of the more recent CLEO results will be reported in these proceedings (14).

REFERENCES

1. For a more detailed history of CESR and CLEO, see Berkelman, K. "A Personal History of CESR and CLEO," CBX 95-87, Nuclear Studies, Cornell Univ., Ithaca, NY.
2. Bernardini, C., "Results on e^{+-}Reactions at Adone," in *Proceedings of the Conference on Electron, Photon Interactions*, 1971, Cornell Univ., Ithaca, NY.
3. Talman, R., "A Proposed Mobius Accelerator," *Phys. Rev. Lett.* **74**, 1590, 1995.
4. Lee-Franzini, J., "Hidden and Open Beauty in CUSB," presented at this conference.
5. Andrews, D. et al. (CLEO Collaboration), "Observation of Three Upsilon States," *Phys. Rev. Lett.* **44**, 1108, 1980.
6. Andrews, D. et al. (CLEO Collaboration), "Observation of a Fourth Upsilon State in e^+e^- Annihilations," *Phys. Rev., Lett.* **45**, 219, 1980.
7. Silverman, A., "Recent Results from CESR," in *Proceedings of the 1981 International Symposium on Lepton and Photon Interactions at High Energies*, pp. 138–163, Univ. Bonn Physics Institute, Bonn, Germany.
8. Gottried, K., "Heavy Quark Spectroscopy Before the Discovery of the Upsilon," presented at this conference.

9. Behrends, S. et al. (CLEO Collaboration), "$\Gamma(b \to ul\nu)/\Gamma(b \to cl\nu)$ from the End Point of the Lepton Spectrum in Semileptonic B Decay," *Phys. Rev. Lett.* **59**, 407, 1987.
10. Barger, V. and Pakvasa, S., "Weak Isospin of the b-Quark and Possible Non-Existence of a t-Quark," *Phys. Lett.,* **81B**, 195, 1979.
11. Kane, G. L. and Peskin, M. E., "A Constraint from B Decay of Models with no t-Quark," UMHE 81, 1981.
12. Burger, J., "Results from Petra: Search for New Particles," in *Proceedings of the 1981 International Symposium on Lepton and Photon Interactions at High Energies,* pp. 115–137, Univ. Bonn Physics Institute, Bonn, Germany.
13. Albrecht, H. et al. (Argus Collaboration), *Phys. Lett.* **192B**, 245, 1987.
14. Poling, R. A., "B Physics at CLEO," presented at this conference; Honscheid, K., "Beauty and the Beast: What Hadronic B Decays Tell Us," presented at this conference.

The Discovery of the B Mesons

Sheldon Stone

Physics Department, 201 Physics Building
Syracuse University, Syracuse, NY 13244-1130

Abstract. In 1983 the CLEO collaboration first fully reconstructed B meson decays. I describe here what was done and how these proceedures have evolved.

INTRODUCTION

In the early 1980's experiments at CESR discovered the $\Upsilon(4S)$ resonance, and measured its width to be about 20 MeV, far wider than the narrow $\Upsilon(1S)$, (2S) and (3S) resonances [1]. The conjecture at the time, which proved to be true, was that the $\Upsilon(4S)$ decayed into either B^-B^+ or $\overline{B}^o B^o$. Further evidence came from the observation of lepton production from the (4S) with a momentum spectrum that ended at approximately half the (4S) mass [2].

Reconstructed B decays, however, had not been seen in 1982. Much physics could be learned from them. For example, specific decay modes could supply interesting information on the decay mechanisms. Furthermore, measurements of V_{cb} would be difficult not knowing the B mass.

The successful search described here used the CLEO I detector [3]. The inner part of the detector consisted of a series of concentric shells, used to perform the charged particle tracking, surrounded by a superconducting magnet coil, 0.7 radiation lengths thick, producing a 1.0 T field. The innermost component at a radius of 7.5 cm from the beam line was a 3 layer multiwire proportional-chamber placed just outside a 2 mm thick aluminum beam-pipe. This device was useful for pattern recognition but not momentum resolution. It extended out to a radius of 17.25 cm where a 17 layer drift chamber began. The drift chamber ended at a radius of 95 cm where the magnet coil started. Beyond the coil were high pressure proportional chambers used to identify charged hadrons by dE/dx, a time-of-flight system, an electromagnetic calorimeter and a muon detector.

Identification of charged kaons and pions was done only in the momentum range between 300-800 MeV/c. The dE/dx and TOF systems were redundant, but the dE/dx efficiency was substantially higher, mainly because many of the

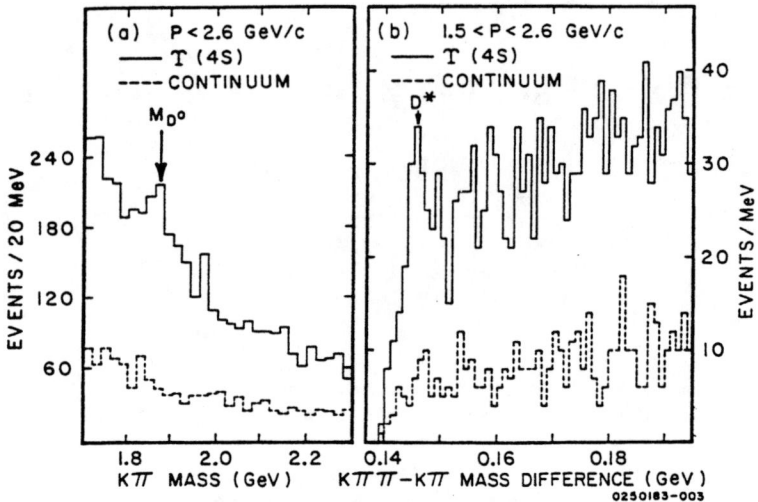

FIGURE 1. $K^\pm\pi^\mp$ invariant mass spectrum (a) and the mass difference spectrum for $K^+\pi^-\pi^+$-$K^+\pi^-$ (and the charge conjugate) (b) from 40 pb^{-1} of CLEO I data.

kaons decayed before reaching the TOF. The dE/dx system was important in this analysis in another way. Three out of the four individuals directly involved in the B discovery worked on its construction, maintenance and software. It was a difficult and time consuming task; the chambers ran pressurized at 3 atm. and took 4500 V to achieve adequate gain. That was our day job; our night job was to reconstruct B mesons.

ANALYSIS PROCEDURE

Our plan was to look for "simple" decay modes with a D^o or D^{*+} and one or two charged pions. (The CLEO I detector could not usefully detect neutral pions from B decay.) Fig. 1 shows the invariant $K^\pm\pi^\mp$ mass (a) and mass difference distribution $K\pi\pi$-$K\pi$ after a cut around the D^o mass [4]. Small signals for D^o production in (a) and D^{*+} production in (b) are visible in this 40 pb^{-1} data sample ($\sim$88,000 B mesons). Fig. 2 shows similar distributions from a 3.1 fb^{-1} sample from the CLEO II detector. The quality of the signals is produced not only by more data, but by a much more sophisticated detector now surrounding a thin beryllium beam pipe and utilizing almost sixty layers of drift chamber tracking.

First I will describe the "modern" method of reconstructing B's which is simpler than what we did in 1983 because of the superior quality of the newer

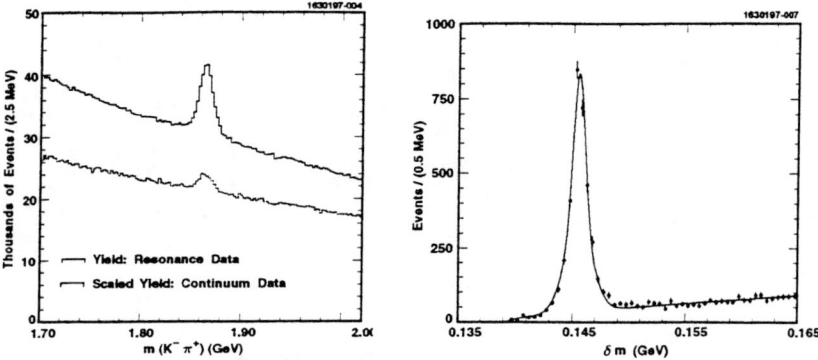

FIGURE 2. The $K^{\pm}\pi^{\mp}$ invariant mass spectrum (a) and the mass difference spectrum for $K^+\pi^-\pi^+$-$K^+\pi^-$ (and the charge conjugate) (b) from 3.1 fb^{-1} of CLEO II data.

data. Then I will contrast it to what we did earlier. Let me refer specifically to the $B^- \to D^0\pi^-$ decay [5]. Generalization to other modes should be obvious. The first step is to select D^0 and π^- candidates. The next is to calculate the energy difference (ΔE) between the beam energy and the sum of the energy of the D^0 and the π^-. Since the B production reaction is

$$e^+e^- \to \Upsilon(4S) \to \overline{B}B, \qquad (1)$$

the energy of each B must be equal to the beam energy. The invariant mass of B meson candidates can then be calculated as:

$$M_B^2 = E_{beam}^2 - (\vec{p}_D + \vec{p}_\pi)^2. \qquad (2)$$

B meson signals appear as Gaussian shaped bumps in both ΔE and M_B. Usually the data is cut on ΔE and the projection on M_B is shown. For current CLEO II data the resolution on ΔE is 25 MeV r.m.s., considerably less than a pion mass, so there isn't a problem with background from B decays with one additional pion. The ΔE resolution is shown in Fig. 3 from the data [6]. Fig. 4 shows recent M_B distributions. The resolution on M_B is ≈ 2.3 MeV r.m.s. and arises almost entirely from the energy spread in the beam.

Now let us review the technique circa 1983 when reconstructing B's was far more arcane. One serious problem was that the ΔE resolution was of the order of a pion mass. In order to overcome that difficulty we chose to kinematically fit the decay products of the D^0 to the known D^0 mass and thus improve the momentum resolution [7]. (The primary constraint was that the candidate B energy had to equal the beam energy.) The optimal solutions was to use an overall two constraint fit utilizing the old (even at that time) bubble chamber fitting program "SQUAW" [8].

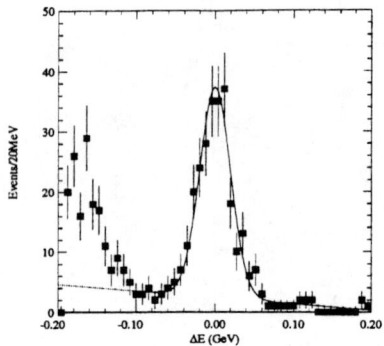

FIGURE 3. The difference between the measured energy of D^o plus π^- candidates (ΔE) when the combination is consistent with the B mass, from CLEO II data.

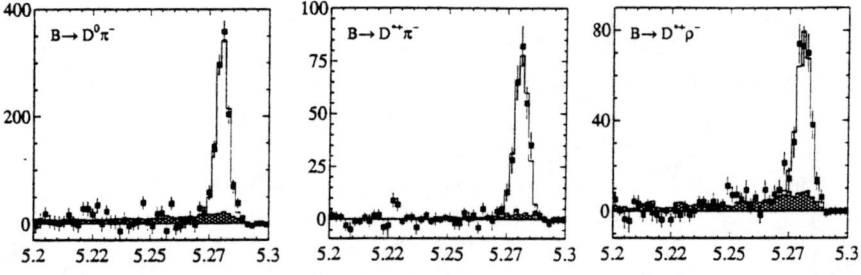

beam-constrained mass (GeV/c^2)

FIGURE 4. Some sample M_B distributions from CLEO II data. (The data have been continuum subtracted.) The hatched histograms show the estimated backgrounds from other B decays.

The quality of the fit to a particular hypothesis is represented by χ^2. Since we were not sure that the track errors were properly evaluated we investigated a sample of well understood events $\Upsilon(3S) \to \pi^+\pi^-\Upsilon(1S)$, where $\Upsilon(1S) \to \ell^+\ell^-$. Fig. 5 shows the expected χ^2 distribution from Monte Carlo (a), the experimental distribution (b) and the result of embedding six additional tracks in the event to more accurately simulate the environment of a single B (c). The χ^2 distribution clearly widens as more tracks are added causing an inefficiency. On the basis of plots such as these the χ^2 cut was chosen.

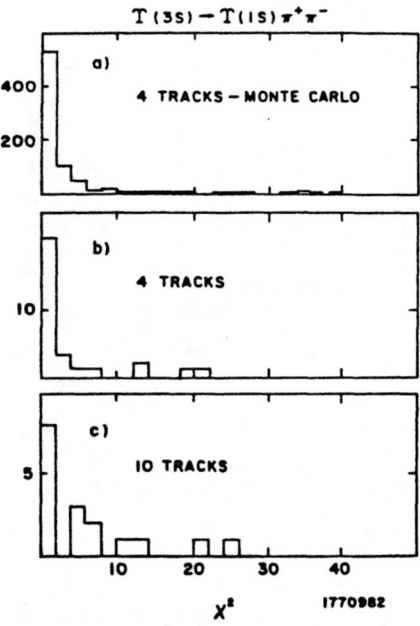

FIGURE 5. The χ^2 distributions for $\Upsilon(3S) \to \pi^+\pi^-\Upsilon(1S)$ where the $\Upsilon(1S)$ subsequently decayed to e^+e^- or $\mu^+\mu^-$. The errors on the electrons or muons have been increased to simulate a two-constraint fit. (a) Monte Carlo simulated events. (b) $\Upsilon(3S) \to \pi^+\pi^-\Upsilon(1S)$ real events. (c) $\Upsilon(3S) \to \pi^+\pi^-\Upsilon(1S)$ with six tracks added.

Three background samples were also analyzed at the same time. The first used data taken in the "continuum," 60 MeV in center-of-mass energy below the $\Upsilon(4S)$. The second used fake D^o candidates, where the mass range was selected either above or below the D^o mass peak; this sample was called "D^o sidebands." The third was B meson decays which cannot exist with comparable branching ratios to the sought after modes. For example $B^- \to \overline{D}^o\pi^-$ was looked for as a background to $B^- \to D^o\pi^-$; these were called "wrong sign modes."

In the days of yesteryear simply doing a computer analysis of the data

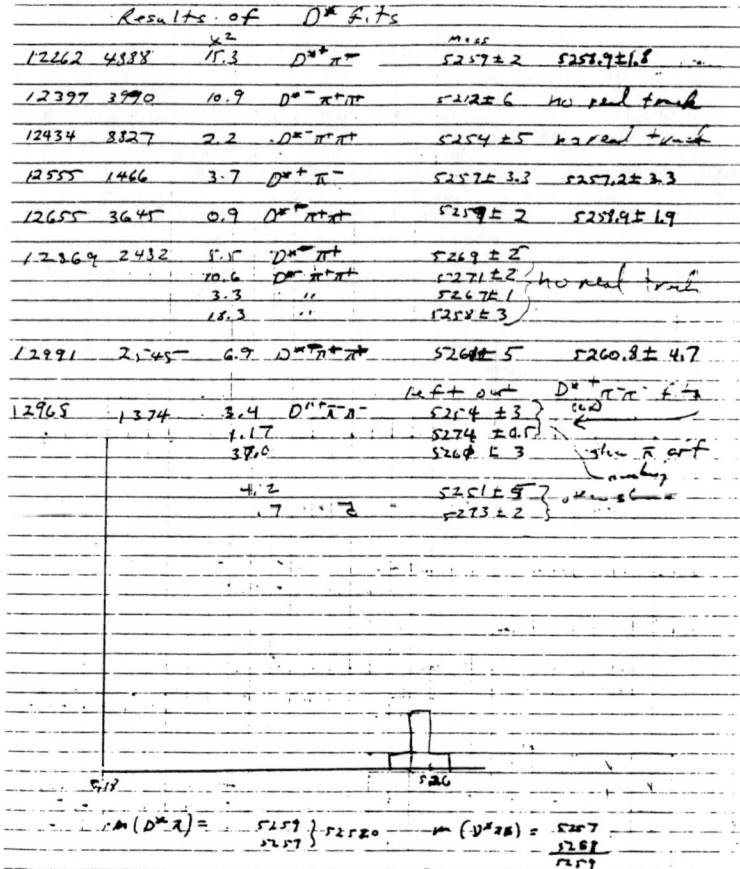

FIGURE 6. One page in the log book showing the results of a scan. The plot shows a peak near 5260 MeV. The low mass occurs because the orginal beam energy estimate was about 15 MeV lower than present estimates.

was insufficient. We were dealing with a handful of candidate events and it behooved us to view the track images in these events. This "scanning" is a time consuming effort which is fraught with possible systematic biases. To remove these possible biases we scanned both signal and background samples together. A reproduction of one page of our log book is shown in Fig. 6. Note the many references to "no real tracks." This comment referred usually to a collection of hits which reconstructed to a track but was obviously false. In Fig. 7, I show the display of one real event containing a $\overline{B}^0 \to D^{*+}\pi^-$ candidate.

FIGURE 7. The viewing of a $\overline{B}^o \to D^{*+}\pi^-$ event candidate by (right to left) Sheldon Stone, David Kreinick, Rainer Wilcke and Yuichi Kubota.

RESULTS

The results are shown in Fig. 8 [4]. The left side shows the mass peak in the signal region for B meson candidates, while the right side shows the result of analyzing the two background samples. There is a signal near 5275 MeV which does not appear in the background samples. Quantitative results are shown in Table 1. Not shown is the result obtained by analyzing the continuum sample. With about half the integrated luminosity as on the $\Upsilon(4S)$ we found 2 background events in the signal region.

Inspection of Fig. 8 does not show a signal in the $D^o\pi^+\pi^-$ mode, yet we included it and even quoted a branching ratio. This was done because we had decided that the search would be performed for modes with a D^o or D^{*+} and we included the results of all modes in the search. (This was probably a very naive thing to have done.) Furthermore all branching ratios were quoted without background subtraction, though the backgrounds are substantial. In the Table, I have estimated the background from the D^o sidebands and wrong charged event samples shown in Fig. 8. (This was not in the original paper.) However, the published paper does state *"If a B decay contains a low-energy π which escapes detection, the remaining particles from that B may still be consistent with the beam-energy constraint and give an acceptable fit."* This

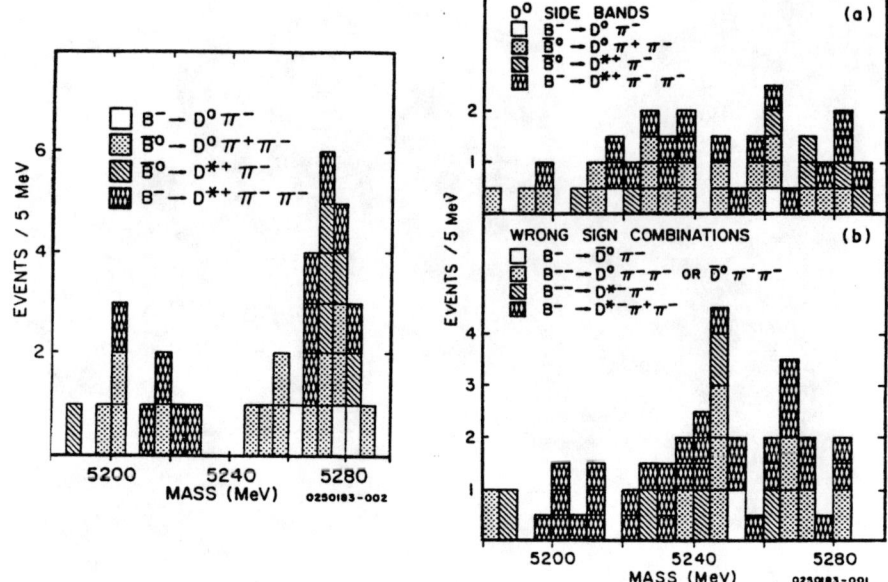

FIGURE 8. The left side shows the mass distribution of B meson candidates. Charge conjugate modes are also included. The right side shows the mass distribution of two separate background estimates. In (a) D^o's are chosen from sidebands. The events shown are plotted with a weight of 1/2 an event since there are twice as many events in the sidebands as in the D region. In (b) we view wrong charge combinations, which have also been scaled to account for the number of allowed combinations.

TABLE 1. The first B exclusive decay results

Final State	$D^o\pi^-$	$D^o\pi^+\pi^-$	$D^{*+}\pi^-$	$D^{*+}\pi^-\pi^-$
# of events (S+B)	2	5	5	6
Branching Ratio (%)[a]	4.2±4.2	13±9[b]	2.6±1.9	4.8±3.0
Bkgrnd estimate (new)	~0.2	~2	~1	~3
Present projected #	3	0	1	1/3

[a] Backgrounds were not subtracted.
[b] There is no mass peak in this mode. The branching ratio should not have been quoted.

applies especially to the $D^* \to \pi D$ modes feeding the D^o modes but also to $D^{*+}\rho^-$ feeding $D^{*+}\pi^-\pi^-$ by missing a π^o from the ρ^- decay and adding a π^-. It was unknown in 1983 that the ρ^- in this final state is almost fully polarized [9], making it easy for a slow π^o to be replaced by a slow π^- from the other B in the event.

I have also included in the last line of the Table an estimate of the number of events we should have seen based on current branching fractions. The average B meson mass we measured was $5272.3 \pm 1.5 \pm 2.0$ MeV compared with the mass now known to be 5280 ± 2.0 MeV [9].

CONCLUSIONS

In 1983 CLEO successfully reconstructed B mesons and measured their mass though we were low by two standard deviations. In the two-body decay modes $D^o\pi^-$ and $D^{*+}\pi^-$, at least some of the events were signal and not background.

In the next 15 years we vastly improved the detectors (ARGUS and CLEO II) and learned how to do this analysis better. Perhaps most importantly we showed that B mesons could be reconstructed, a possibility that had been discounted by many [10], and this led to a plethora of reconstructed final states. Fully reconstructed B's have become one of the most important ways to learn about weak decays and they will be an important way for us to learn more about CP violation.

ACKNOWLEDGEMENTS

I thank Dave Kreinick, Yuichi Kubota and Rainer Wilcke for past collegiality and useful comments on this manuscript. This work was supported by the National Science Foundation.

REFERENCES

1. D. Andrews et al., *Phys. Rev. Lett.* **45**, 219 (1980); G. Finocchiaro et al., *Phys. Rev. Lett.* **45**, 223 (1980).
2. C. Bebek et al., *Phys. Rev. Lett.* **46**, 84 (1981); K. Chadwick et al., *Phys. Rev. Lett.* **46**, 88 (1981).
3. The detector is described in Al Silverman's talk at this conference, and in more detail in D. Andrews et al., *Nucl. Instrum. Methods* **211**, 47 (1983).
4. Behrends, S. et al., *Phys. Rev. Lett.* **50**, 881 (1983).
5. The first decay mode we tried was $B^- \to \psi K^-$, but the effective branching ratio is now known to be an order of magnitude lower than $B^- \to D^o\pi^-$.
6. B. Barish, et al., (CLEO Collaboration), "Exclusive Reconstruction of $B \to D^{(*)}(n\pi)^-$ Decays," CLEO CONF 97-01 (1997).

7. This procedure is still employed for some final states such as ψK, $\psi \to \ell^+\ell^-$, where fitting the $\ell^+\ell^-$ to the ψ mass improves the ΔE resolution considerably.
8. O. I. Dahl, T. B. Day, F. T. Slomitz and N. L. Gould, "SQUAW Kinematic Fitting Program," Group A Programming Note No. P-126, Univ. of Cal. LRL, Berkeley, Cal. (1968).
9. M. S. Alam et al., *Phys. Rev.* **D50**, 43 (1994).
10. S. Herb private communication; see also S. Herb, "Prospects for the Study of B Meson Decays," in *Proceedings Of The 1982 DPF Summer Study On Elementary Particle Physics And Future Facilities*, ed. by R. Donaldson, R. Gustafson and F. Paige, American Physical Society, NY (1983).

Hidden and Open Beauty in CUSB

Juliet Lee-Franzini

Laboratori di Frascati dell'INFN
CP-13, Via Enrico Fermi 40, I-00044, Frascati Roma and
Physics Department, SUNY at Stony Brook, N.Y. 11794, U.S.A.

Abstract. We present a brief history and a summary of the physics results of CUSB at CESR.

HISTORY

Since the discovery of the Υ system at FNAL, as told at the begining of this volume, the CUSB (Columbia University-Stony Brook) collaboration at CESR (Cornell Electron Storage Ring) has played a pivotal role in the study of the upsilon system, potential models, QCD, and weak interactions of the b quark [1-9].

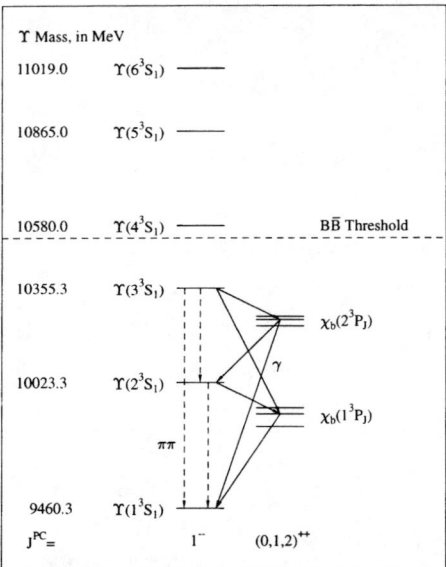

Figure 1. Υ's Level Diagram, CUSB observed and studied all levels and transitions between levels indicated above. Open beauty was discovered in the decay $\Upsilon(4S) \rightarrow$ high energy electrons. The B^*'s were found in the continuum above the $\Upsilon(4S)$.

In retrospect, things happened very fast. CUSB's formal proposal was submitted to Cornell on December 2nd, 1977 and was approved by February 15th, 1978! On August 16th, 1979 we paid our first visit to the North Area (NA) underground *cul de sac* which became the home for our electronics. In May 1979 we installed the first NaI qradrant at the NA interaction region. In September of 1979 a caravan of two trailers containing electronics, escorted by a truck and passenger cars driven by and containing a dozen physicists, arrived at the Northern edge of the Cornell Alumni Atheletic Fields. The trailers functioned as the CUSB data acquisition center, until May 1991 when they were honorably retired, their outer skins showing various indentations resulting from seasonal sport projectiles. On October 18th, 1979, with half of the sodium iodide, NAI, detector, 8.3 radiation lengths (X_0) deep, centered upon the NA interaction area, we observed the first Bhabha ever seen at CESR.

Figure 2. In this form, *i.e.* just one half of the NaI crystal array, CUSB observed the three Υ's below threshold and on December 1979 had evidence for the $\Upsilon(4S)$, the first level above threshold.

From then until January 1984, in between collecting 950,000 events, we completed CUSB-I. The CUSB-I detector comprised 320 rectangular NaI crystals in a square geometry, surrounded by an array of lead glass cubes ($15 \times 15 \times 17.5$ cm^3, 7.7 X_0) for shower leakage containment. These in turn are surrounded by scintillator muon counters [10]. In this period we did a whole series of exploratory experiments, and one of us (J. L-F) also had a wonderful time re-learning quantum mechanics from "guru" K. Gottfried. We resolved the first three Υ resonances to the chime of the CUSB "R" meter (a scalar counting a restrictive trigger) [11]. We saw the first $\Upsilon(4S)$ signal in December 1979 [12]. We observed the $\pi\pi$ transitions between the Υ bound states [13]. We discovered the χ_b states from inclusive and exclusive E1 transitions [14]. We even foretold the existence of χ_b transitions from studying event shapes [15], studied K^0 production at all the resonances [16], and made the first determination of α_s from study of γgg on the Υ resonances [17].

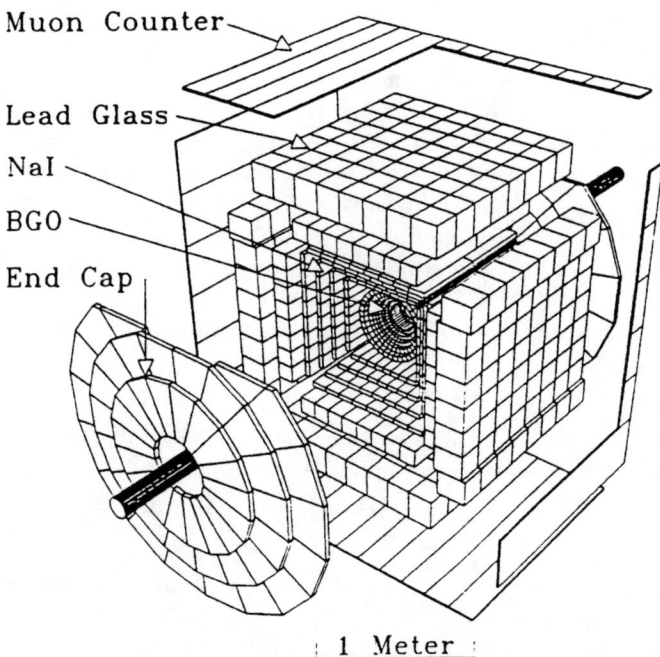

Figure 3. The complete CUSB detector. This configuration was only reached in 1989. The BGO calorimeter and the endcap trigger counters were not present at the beginning. Because of the high cost of NaI, lead glass was used as a catcher for the electromagnetic showers. The lead glass was used later for muon identification.

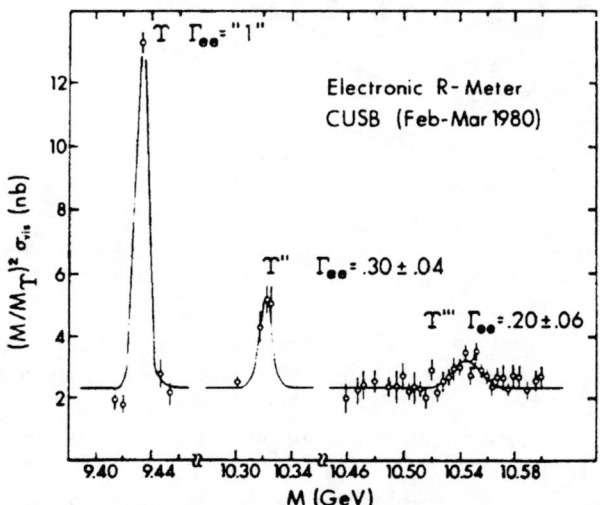

Figure 4. The $\Upsilon(4S)$ was observed by CUSB by counting on scalers small angle Bhabha and hadronic Υ+continuum events. Thus the fourth upsilon was discovered without analysis of the collected events. The latter confirmed the former and gave information on event shape and on the production of B mesons.

From the observation of high energy electron production in 4S events we inferred the B-mass, plotted out the B meson semileptonic decay spectra [18], and had long discussions with phenomenologists [19]. We searched for the B^* repeatedly on the Υ(4S), in vain, finally observing it as a broadish bump in the continuum above (due to the Doppler shift of the parent B's) [20]. This last signal inspired us to improve our photon resolution by using BGO crystals in addition to NaI. We saw a rise in ΔR of $\simeq 1/3$ due to the production of $b\bar{b}$-quarks [21], and did a complete coupled channel analysis to obtain the parameters of the higher Υ resonances [22]. In July 20, 1984 we inserted a BGO quadrant inside the CUSB-I NaI array. With this set up, together with the Cornell Polarization group, we made a precision Υ mass measurement [23]. We also continued to establish limits for non existent particles such as axions, short lived particles, light gluinos, ζ (8.3), light Higgs, and light squarks [24].

In December 1985 we installed the complete BGO array, the heart of CUSB-II, inside CUSB-I. The array is a cylinder composed of 360 trapezoidal-cross-sectioned BGO crystals. The 5 layers's thickness is twelve radiation lengths.

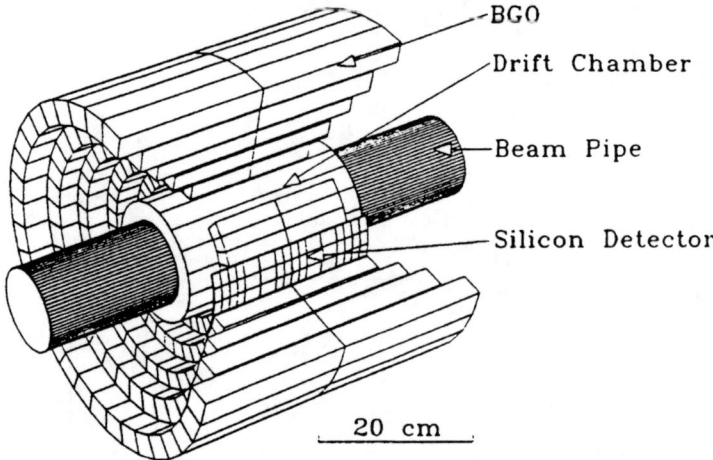

Figure 5. The "BGO cylinder" improved the CUSB energy resolution for low energy photons by a factor of two. In its final configuration, silicon pads were between BGO crystals to obtain better angular resolution for photons. A small drift chamber inside the BGO array detected charged particles.

We proved by study of $\approx$5 GeV electrons from Bhabha scatterings that the new CUSB-II detector had achieved an improvement of a factor of two in resolution [25]. The CUSB-II physics results obtained in this configuration include accurate measurements of the branching ratios $B_{\mu\mu}$(3S) and $B_{\mu\mu}$(1S) [26], stringent limits on direct photon production from the Υ(4S) [27], precise determinations of hyperfine splittings in the B-meson system, evidence for B_s-meson production on the Υ(5S) resonance [28] and semileptonic branching ratios from B_u-mesons and B_s-mesons semileptonic decays at the Υ(4S) and Υ(5S) [29].

Table I. The data taking history of CUSB is presented. For each sample the integrated luminosity and the number of collected events is listed. Some major physics topic studied are also listed.

Run 394-5513	Oct. 18, 1979 – Jan 25, 1984		CUSB
	$\mathcal{L}$ (pb^{-1})	Events	Physics
1S	7.1	111,405	Resonances first seen in "R-meter" (1979).
2S	29.6	224,163	($\pi\pi$, E1) transitions to 1S, discovery of χ_b.
3S	17.7	89,096	3S (1979), ($\pi\pi$, E1) to 2S,1S, discovery of χ_b'.
4S	63.1	190,280	1st seen in Dec. 1979, B semileptonic decay, no B^*.
cont	32.7	72,865	$\Delta R \simeq 1/3$.
>4S	113.8	260,443	Coupled channel $\Upsilon(5S, 6S, 7S)$, observe B^*'s.
Total	263.7	948,252	K^0 production measured, no axion seen.

Run 5515-7284	Jul. 20, 1984 – Aug. 2, 1985		CUSB-1.5
	$\mathcal{L}$ (pb^{-1})	Events	Physics
1S	28.9	529,537	Precision (1S) mass (10ppm), γgg
4S	60.3	243,916	Semileptonic $B \to \mu, e, \nu$ spectra, no $b \to u$.
cont.	23.9	70,283	No B^*'s here (below 4S), nor on 4S.
Total	113.3	844,599	No $\zeta(8.3)$.

Run 10391-12210	Dec. 16, 1985 – Apr. 23, 1988		CUSB-II
	$\mathcal{L}$ (pb^{-1})	Events	Physics
1S	24.4	458,492	$B_{\mu\mu}(1S)$.
3S	144.3	1,001,329	$B_{\mu\mu}(3S)$, no η_b seen, no h_b seen.
4S	273.7	1,163,213	Precise (4S) semileptonic decay studies.
cont	122.3	395,013	No B^*'s below 4S.
> 4S	140.2	467,729	HFS, B^*, B_s, B_s^* masses, $B \to e, \nu$ at 5S.
Total	704.9	3,090,763	Limits: Higgs, squark, gluino, short τ particles.

Run 20001-22542	Oct. 20 1989 – Oct. 8, 1990		CUSB-II+
	$\mathcal{L}$ (pb^{-1})	Events	Physics
1S	18.5	332,413	$\alpha_s(1S)$, hadronic width.
3S	143.6	969,8903	$\alpha_s(3S)$, hadronic width, precision FS, $\pi\pi$ spectra.
4S	36.9	150,750	No direct photons from 4S, no $\pi\pi$ to (2S), (1S).
cont	32.4	99,608	
BB^*	63.1	204,356	Precision B^* mass.
Total	294.5	1,757,617	$\Lambda_{\overline{MS}}$.

Run 394-22542	Oct. 18 1979 – Oct. 8, 1990		CUSB to CUSB-II+
	$\mathcal{L}$ (pb^{-1})	Events	Physics
All	1376.4	6,641,231	Scalar nature of confining potential.

From April 23, 1988 to October, 1989, during CESR's shutdown for completing CLEO-II, we constructed and installed the remaining portions of CUSB-II, including new BGO read-out electronics, a shower centroid detector, and a forward-backward lepton end-cap trigger. Data obtained from the last CUSB run, October 20, 1989 to October 8, 1990, resulted in precision measurements of the relative contributions of the spin-orbit and tensor interactions to the fine structure of the 2P state and we determined that the long-range confining potential transforms as a Lorentz scalar [30]. We made precise studies of the sequential decays of the $\Upsilon(3S)$, measured the hadronic widths of the χ_b states and observed the rare decay transition from the $\Upsilon(3S)$ to the χ_b [31]. We measured the $\pi\pi$ mass spectra of the dipions from $\Upsilon(3S)$ hadronic transitions to (2S) and (1S) and determined the beam energy window where single B^*B production is dominant [32].

The total running period for CUSB-I, CUSB I.5, and CUSB-II span from October 1979 to October 8, 1990. The CUSB collaboration had always been a small one, with at most two dozen members at any one time during the CUSB-I stage [33] and no more than one dozen at any one time, including students, technicians and senior physicists, for CUSB-II [34]. Table I gives a summary of the run history and physics highlights of each run period.

SELECTED PHYSICS RESULTS

$\Lambda_{\overline{MS}}$ from Υ's $\to \mu\mu$

From our measurements of the branching ratio for $\Upsilon \to \mu^+\mu^-$, $B_{\mu\mu}$, and the other measured parameters of the Υ's we obtain the branching ratio for $\Upsilon \to gg$ and thus we determine $\alpha_s(m_b)$ and $\Lambda_{\overline{MS}}$, Table II.

Table II. Measurements of $B_{\mu\mu}$ and the derived $\alpha_s(m_b)$ values.

Resonance	$B_{\mu\mu}$ (%)	Γ (keV)	$\alpha_s(m_b)$	$\Lambda_{\overline{MS}}$
$\Upsilon(1S)$	2.61±0.09	51.1±3.2	0.174±0.004	150±13
$\Upsilon(2S)$	1.38±0.25	42.3±9.2	0.176±0.016	167±58
$\Upsilon(3S)$	1.73±0.15	27.7±3.7	0.173±0.008	154±29
Average	—	—	0.1736±0.0033±0.017	157±12±60

For the average values of $\alpha_s(m_b)$ and $\Lambda_{\overline{MS}}$ we have included a reasonable guess of the theoretical uncertainty in fixing the energy scale in the systematical error. The α_s and $\Lambda_{\overline{MS}}$ obtained by us are in excellent agreement with those obtained using a number of other processes, proving that the Upsilon system provides an independent probe of QCD.

Hyperfine Splitting of B and B_s Mesons

We have studied the inclusive photon spectrum from 2.9×10^4 $\Upsilon(5S)$ decays. We observe a strong signal due to $B^* \to B\gamma$ decays, both from inclusive hadronic events a) and from electron tagged events b). From a detailed analysis we obtain: i) the average B^*-B mass difference, (46.7 ± 0.4) MeV, ii) the photon yield per $\Upsilon(5S)$ decay, $\langle \gamma/\Upsilon(5S) \rangle = 1.09 \pm 0.06$ and iii) the average velocity of the B^*'s, $\langle \beta \rangle = 0.156 \pm 0.010$, for a mix of non-strange (B) and strange (B_s) B^*-mesons from $\Upsilon(5S)$ decays. From the shape of the photon line, we find that significant production of B_s is required implying nearly equal values for the hyperfine splitting of the B and B_s meson systems.

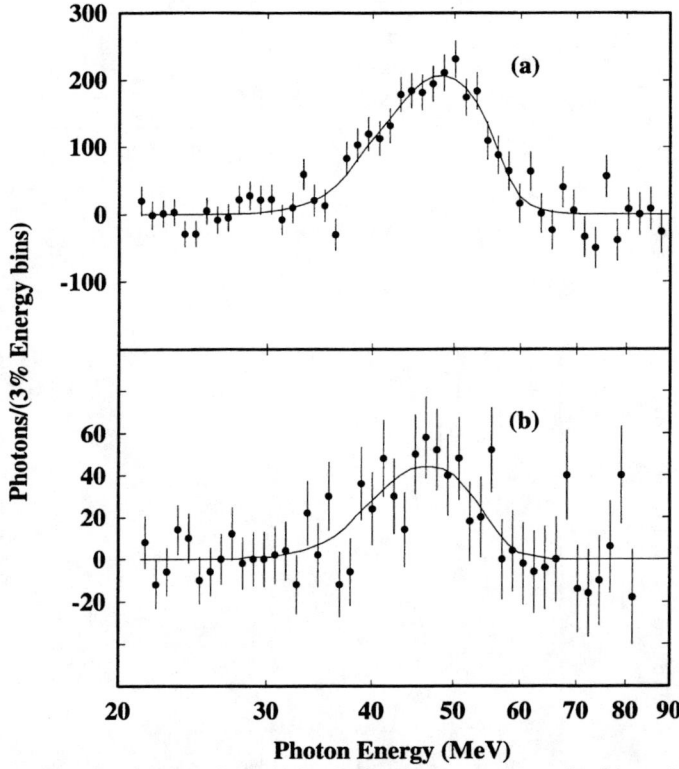

Figure 6. Photon spectrum in inclusive hadronic events above the flavor threshold. (a) The line at ~48 MeV is due to the decay $B^* \to B+\gamma$, proving the existence of the B^* meson and measuring its mass. (b) The line is present also selecting events with a high energy electron, confirming the presence of a B meson decaying semileptonically.

B Semileptonic Decays at the $\Upsilon(4S)$ and the $\Upsilon(5S)$

B meson semileptonic decay spectra have been obtained at the $\Upsilon(4S)$ and at the $\Upsilon(5S)$. The branching ratio for $B \to e\nu X$ at the $\Upsilon(4S)$ is found to be $(10.0 \pm 0.5)\%$. The electron spectrum of $B \to e\nu X$ at the $\Upsilon(5S)$ is observed for the first time and the average branching ratio for $B, B_s \to e\nu X$ is consistent with that for B's from $\Upsilon(4S)$ decays. The shape of the electron spectrum at the $\Upsilon(5S)$ indicates production of B mesons which are heavier than non-strange B's, presumably B_s's.

$b\bar{b}$ Spectroscopy from the $\Upsilon(3S)$ State

We have made a detailed study of electric dipole (E1) and hadronic transitions between the Υ and χ_b states. The amplitude for these transitions are depicted in figure 7.

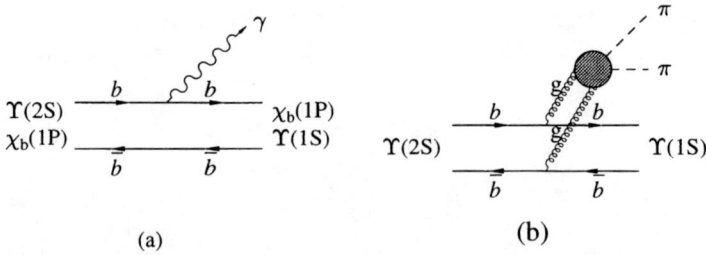

Figure 7. Amplitudes for: (a) electric dipole transitions (E1) $^3S(^3P) \to {}^3P(^3S)+\gamma$ and (b) double *color-electric* dipole transitions $n\,^3S \to (n-1)\,^3S + gg(2\pi)$.

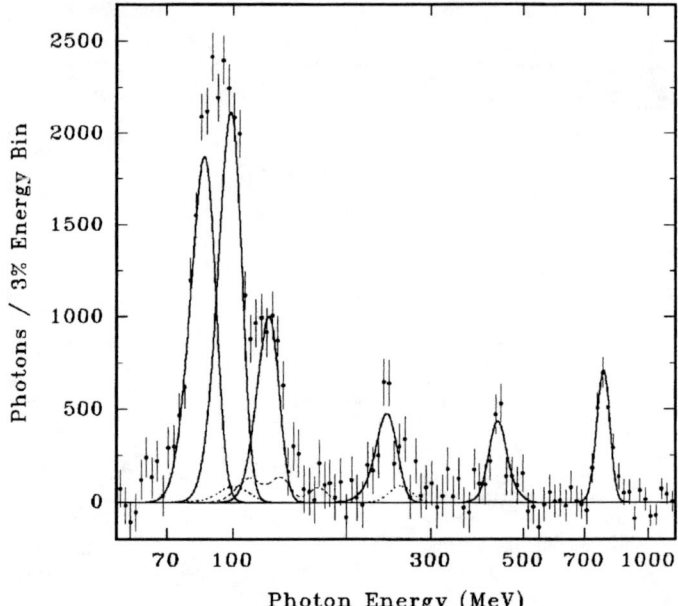

Figure 8. The inclusive photon spectrum at the $\Upsilon(3S)$, after background subtraction. Both lines from direct $\Upsilon(3S)$ decays as well as from decays of daughter $b\bar{b}$ states, see figure 1, are observed.

The data were collected both from exclusive and inclusive channels. An inclusive photon spectrum at the 3S is shown in figure 8. We have determined their branching ratios: BR($\Upsilon'' \to \chi_b(2P_{2,1,0})\gamma$)=(11.1±0.5±0.4)%, (11.5±0.5±0.5)%, (6.0±0.4±0.6)%; BR($\Upsilon'' \to \chi_b(2P_{2,1,0})\gamma \to \Upsilon'\gamma\gamma$)=(4.2±0.6±0.5)%; BR($\Upsilon'' \to \chi_b(2P_{2,1,0})\gamma \to \Upsilon\gamma\gamma$)=(2.0±0.2±0.2)%. We have measured the center of gravity of the $\chi_b(2P)$ states to be (10259.5±0.4±1.0) MeV. We have made precision measurements of the electric dipole transition rates from Υ'' to χ'_b, which are in excellent agreement with theory.

The fine structure splittings obtained using all data are $M(\chi'_2)-M(\chi'_1)$= 13.5±0.5 MeV and $M(\chi'_1)-M(\chi'_0)$=23.2±1.0 MeV which determines the spin orbit contribution as a=9.5±0.22 MeV and that of the tensor interaction as

$b=2.3\pm0.14$ MeV. The resolution of the fine strucure in terms of spin orbit and tensor interactions is illustrated in figure 9.

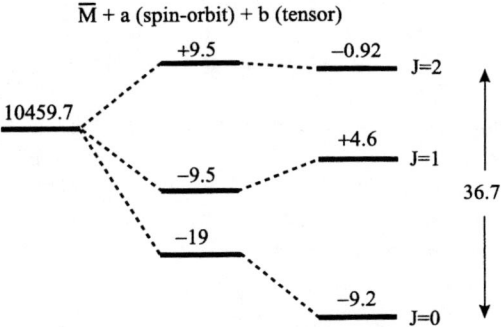

Figure 9. Fine structure of the χ'_b state, resolved in their spin-orbit and tensor contributions as determined by CUSB. Energies are in MeV.

This is an important results which confirms expectactions that the long range, confining part of the quark anti-quark potential is due to the exchange of an effective (multi-gluon) scalar state, while the short range, coulombic part is due to single gluon (vector) exchange. The long range potential has the form kr, with k the so called string tension. We define the scalar fraction f as the part of the kr contribution due to exchange of a Lorentz scalar. Then the 'vector' and 'scalar' potentials are respectively $V_v = -(4/3)(\alpha_s/r) + (1-f)kr$ and $V_s = fkr$. The fine structure of the χ_b states or, equivalently, the values of the spin orbit and tensor contributions to the splitting, are sensitive to f

The value of k is presently known to be between 0.14 and 0.18 GeV^2. Combining this with our measurements of the fine structure of the $\chi_b(2P)$ states results in $f \sim 1$ with an error of about 5% from our measurement and another similar uncertainty due to the abovementioned uncertainty in the value of k. The value of f determined from the fine structure of the $\chi_b(1P)$ states is consistent with the same value, however no precise measurments have been performed for this case.

Unlike in the charmonium system, the hadronic width of the $b\bar{b}$ P-wave states, which are expected to be of the order of tens to hundreds of keV, are too narrow to be measured directly. We have pioneered the method of deriving them from our measured branching ratios of $\chi_b \to \Upsilon\gamma$ and calculated widths of electric dipole transitions, the latter had been well verified from our measurements. The resultant hadronic widths are in excellent agreement with QCD predictions, and can be used to extract α_s. For example, the values obtained from the $\chi_b(2P)$ states are 0.16 ± 0.01, 0.21 ± 0.02, and 0.14 ± 0.03 from the $J=2,1,0$ states respectively. These values and those from the $\chi_b(1P)$ states are in good agreement with the results of measuring α_s in the $b\bar{b}$ system. We have also observed the suppressed transition $\Upsilon(3S) \to \chi_b(1P)\gamma$. The measured branching ratio suggests that potential models which include relativistic effects are correct.

Study of $\pi^+\pi^-$ Transitions from the $\Upsilon(3S)$ State

We have investigated the decay $\Upsilon(3S)\to\Upsilon(1S, 2S)\pi^+\pi^-(\pi^0\pi^0)$, where the final state $\Upsilon(1S, 2S)$ decays to a pair of leptons. We found $\sim$ 390 events of the type $\Upsilon(3S)\to\Upsilon(1S)\pi^+\pi^-$ and $\sim$ 140 events of the type $\Upsilon(3S)\to\Upsilon(2S)\pi^+\pi^-$. The corresponding branching ratios are $(3.27 \pm 0.30)\%$ and $(3.59 \pm 0.49)\%$ respectively. We have also studied the $\pi\pi$ invariant mass spectrum where earlier data already indicated a deviant behavior from theoretical expectations. Both $\pi^+\pi^-$ and $\pi^0\pi^0$ invariant mass spectra, albeit the latter has less statistics, exhibits the same behavior. In short, we see an unusual double-peak behavior on the dipion mass spectrum from $\Upsilon(3S)\to\Upsilon(1S)\pi^+\pi^-(\pi^0\pi^0)$, figure 10 is the one from the $\pi^+\pi^-$ transitions. It is quite different from the spectra for $\pi\pi$ decays from other Υ and ψ states and disagrees with theoretical predictions of a single peak in the high mass region of the distribution. We have compared our spectrum with several current theoretical modifications to various models and found that none of them could successfully explain the observed shape of the double-hump spectrum[35].

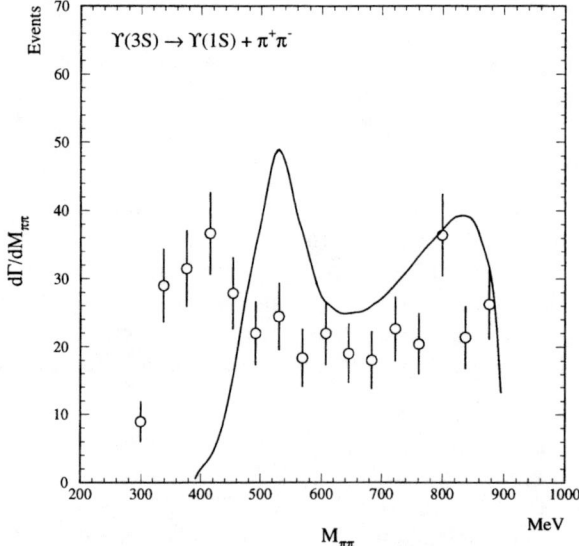

Figure 10. Dipion mass spectrum observed in $\Upsilon(3S)\to\Upsilon(+S)\pi^+\pi^-$. The curve is from the model ref. 35.

REGRETS

While we recall our venture of CUSB at CESR with great affection and sense of accomplishment, there are regrets that we could not have stayed, run on the $\Upsilon(3S)$ longer, on the $\Upsilon(2S)$ especially and found a few more known-to-exist states, such as the η_b and h_b [6, 7]. We could have solved a few more puzzles such as the hadronic widths of the χ_b states [8] and whether there is a pseudoscalar field in the $Q\bar{Q}$ potential [7]. There are, however, other problems which no amount of additional CUSB running would have helped. For example, there are no light Higgs particles nor sparticles, and after we established that there are neither direct photons nor excessive direct pions from the $\Upsilon(4S)$, there is still no explanations why the the B-semileptonic decay branching ratios do not agree with theoretical expectations [27,9].

ACKNOWLEDGEMENTS

The author wishes to thank Dan Kaplan and the other organizers for both an extremely enjoyable and a very scholarly reunion of the beauty physics community. She thanks Dean Schamberger and especially Paolo Franzini for retrieving archaic CUSB figures so they can have a new life here and PF for editing the manuscript.

REFERENCES

1. Lee-Franzini, J. "Quark Spectroscopy with a new flavor", Survey in High Energy Physics **2**, 3 (1981).
2. Franzini, P. and Lee-Franzini, J., "Upsilon Physics at CESR", Physics Report **81C**, 3 (1982).
3. Franzini, P. and Lee-Franzini, J., "Upsilon Resonances", Ann. Rev. Nuc. Part. Sci., **33**, 1 (1983).
4. Lee-Franzini, J., Ono, S., Tornqvist, N. A. and Sanda, A.I., "Where Are the $B\bar{B}$ Mixing Effects Observable in the Υ's Region?", Phys. Rev. Lett. **55**, 2938 (1985).
5. Franzini, P. and Lee-Franzini, J., "10 Years of Υ Spectroscopy", *Present and Future of Collider Physics*, ed. C. Bacci et al., Italian Phys. Soc. Conf. Proc., Editrice Compositori, Bologna, 255 (1991).
6. Franzini, Paula J., "Fine Structure of the P States in Quarkonium Fine Structure and the nature of the Spin-dependent Potential", Phys. Lett. **B296**, 199 (1992).
7. Franzini, Paula J., "A Pseudoscalar Field in the $q\bar{q}$ Interaction in Heavy Quarkonium?", Phys. Lett. **B296**, 195 (1992).
8. Lee-Franzini, J. and Franzini, Paula J., "Hadronic Width of the χ_b States", *Proc. of the Third Workshop on the Tau-Charm Factory*, Marbella, Spain, 457 (1993).
9. Wu, Q. W., "Study of $\pi^+\pi^-$ transitions from Upsilons", thesis, Columbia University, Nevis Labs. (1993).
10. Schamberger, R. D. et al., NIM **A309**, 450 (1991).
11. Yoh, J. K., et al., High Energy Physics-1980, Durand, Pondrum Eds., AIP, NY 700 (1981).
12. Finocchiaro, G. et al., Phys. Rev. Lett. **45**, 222 (1980).
13. Mageras, G. et al., Phys. Rev. Lett. **46**, 1115 (1981); Phys. Lett. **118B**, 453 (1982).
14. Han, K. et al., Phys. Rev. Lett. **49**, 453 (1982); Klopfenstein, C. et al., Phys. Rev. Lett. **51**, 160 (1983); Eigen, G. et al., Phys. Rev. Lett. **49**, 1616 (1982); Pauss, F. et al., Phys. Lett. **130B**, 1439 (1983).
15. Peterson, D. et al., Phys. Lett. **114B**, 277 (1982).

16. Gianinni, G. et al. Nuc. Phys. **B206**, 1 (1982).
17. Schamberger, R. D., Phys. Lett. **138B**, 225 (1984).
18. Spencer, L. et al., Phys. Rev. Lett. **47**, 771 (1981); Klopfenstein, C. et al., Phys. Lett. **130B**, 444 (1983); Levman, G. et al., Phys. Lett. **141B**, 271 (1984).
19. Altarelli, G. et al., Nuc. Phys. B **208**, 365 (1982).
20. Schamberger, R. D. et al., Phys. Rev. D **26**, 720 (1982); Phys. Rev. **D30**, 1985 (1984).
21. Rice, E. et al., Phys. Rev. Lett. **48**, 906 (1982).
22. Lovelock D. M. J. et al., Phys. Rev. Lett. **54**, 377 (1985).
23. Mackay, W. W. et al. and the CUSB Collaboration, Phys. Rev. D **29**, 2483 (1984).
24. Sivertz, M. et al., Phys. Rev. D **26**, 717 (1982); Mageras, G. et al., Phys. Rev. Lett. **56**, 2672 (1986); Tuts, P. M. et al., Proc. of the DPF-APS 1984 Meeting, M. Nieto and T. Goldman Eds., AIP, New York (1985); Phys. Lett. **B186**, 233 (1987); Franzini, P. et al., Phys. Rev. D **35**, 2883 (1987).
25. Lee-Franzini, J., NIM **A263**, 35 (1988).
26. Kaarsberg, T. et al., Phys. Rev. D **35**, 2265 (1987); Phys. Rev. Lett. **62**, 2077 (1989)
27. Narain, M. et al., Phys. Rev. Lett. **65**, 2749 (1990).
28. Lee-Franzini, J. et al., Phys. Rev. Lett. **65**, 2947 (1990).
29. Yanagisawa, C. et al., Phys. Rev. Lett. **66**, 2463 (1991).
30. Narain, M. et al., Phys. Rev. Lett. **66**, 3113 (1991).
31. Heintz, U. et al., Phys. Rev. Lett. **66**, 1563 (1991); Phys. Rev. D **46**, 1928 (1992).
32. Wu, Q. W. et al., Phys. Lett. **B301** 307 (1993); Phys. Lett. **273**, 177 (1991).
33. CUSB-I began with members from Columbia University which in total included: T. Böhringer, F. Costantini, J. Dobbins, P. Franzini (spokesperson), K. Han, S. W. Herb, D. M. Kaplan, L. M. Lederman, G. Mageras, D. Peterson, E. Rice, D. Son, J. K. Yoh, S. Youssef, and T. Zhao, and similarly, from SUNY at Stony Brook: G. Finocchiaro, G. Gianinni, J. E. Horstkotte, D. M. J. Lovelock, C. Klopfenstein, J. Lee-Franzini, R. D. Schamberger, M. Sivertz, L. J. Spencer, and P. M. Tuts. We were joined in 1981 by R. Imlay, G. Levman, W. Metcalf and V. Sreedhar of Louisiana State Univesity and G. Blanar, H. Dietl, G. Eigen, V. Fonseca, E. Lorenz, F. Pauss, L. Romero and H. Vogel of MPI Munich. Louisiana State and MPI Munich left in 1984.
34. CUSB-I.5, CUSB-II: P. Franzini (co-spokesperson), U. Heintz, J. Horskotte, T. Kaarsberg, S. Kanekal, J. Lee-Franzini (co-spokesperson), D. M. J. Lovelock, M. Narain, R. D. Schamberger, S. Sontz, P. M. Tuts, J. Willins, Q.W. Wu, C. Yanagisawa, S. Youssef, and T. Zhao.
35. Belanger, P., DeGrand, T. and Moxhay, P., Phys. Rev. D **39**, 257 (1988).

Upsilon Physics at MD-1 in Novosibirsk

S.E.Baru, A.E.Blinov, V.E.Blinov, A.E.Bondar,
A.D.Bukin, V.R.Groshev, Yu.I.Eidelman, V.A.Kiselev,
S.G.Klimenko, G.M.Kolachev, S.I.Mishnev, <u>A.P.Onuchin</u>,
V.S.Panin, V.V.Petrov, I.Ya.Protopopov, A.G.Shamov,
V.A.Sidorov, Yu.I.Skovpen, A.N.Skrinsky, V.A.Tayursky,
V.I.Telnov, Yu.A.Tikhonov, G.M.Tumaikin, A.E.Undrus,
A.I.Vorobiov, V.N.Zhilich

Budker Institute of Nuclear Physics, 630090, Novosibirsk, Russia

Abstract. Experiments with the MD-1 detector at the VEPP-4 collider in the energy region of Υ mesons were carried out in 1980-1985. Some of the MD-1's results are still among the most accurate: precise measurement of $\Upsilon(1S)$, $\Upsilon(2S)$ and $\Upsilon(3S)$ meson masses; leptonic widths of $\Upsilon(1S)$ and $\Upsilon(2S)$; R ratio in the energy region $\sqrt{s}$=7.2-10.3 GeV; upper limits on the $\Upsilon(1S)$ decays to $p\bar{p}$, K^+K^-, $\pi^+\pi^-$; upper limit on the electron width of new resonances in the energy region 7.2-10.3 GeV. New values of $\Upsilon(1S)$, $\Upsilon(2S)$, and $\Upsilon(3S)$ masses corrected to change of the electron mass are given.

MD-1 detector and VEPP-4 collider. The physical project of the MD-1 detector at the VEPP-4 collider [1] was approved in 1974. In 1980-1985, experiments with the MD-1 on study of upsilon mesons and two-photon processes were carried out in the energy region $\sqrt{s}$=7.2-10.3 GeV [2]. The integrated luminosity of 30 pb^{-1} has been collected including 10^5 Υ mesons.

A project of the MD-1 detector (Fig. 1) had two peculiar features. 1) For the first time a coil was moved out beyond the e.m. calorimeter. This feature improves an energy resolution of the calorimeter and gives an opportunity to make the coil thicker. Later, almost all detectors were made in that way. 2) The magnetic field of MD-1 was transverse to the beam orbit plane. This provided a high efficiency and accuracy of the scattered electrons energy measurements in two-photon processes. Besides, synchrotron radiation inside the

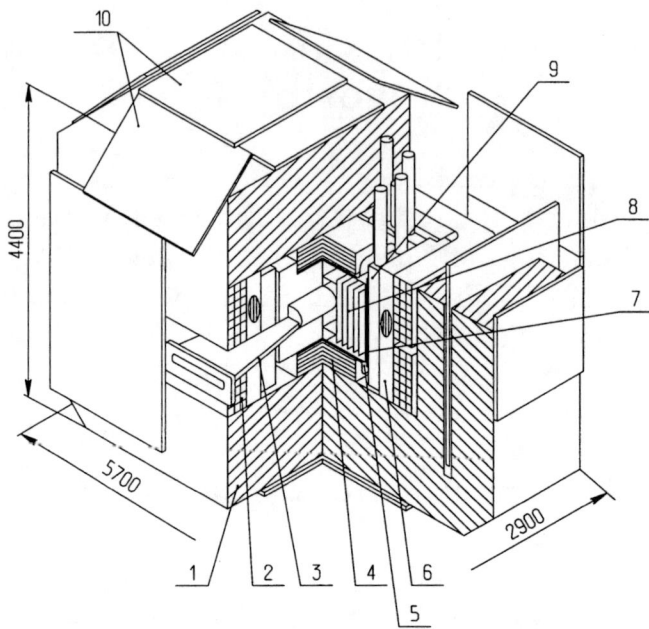

FIGURE 1. Layout of the MD-1 detector. 1 – magnet yoke, 2 – copper coil, 3 – beam pipe, 4,6 – shower-range chambers, 5,7 – scintillation counters, 8 – coordinate chambers, 9 – gas Cherenkov counters, 10 – muon chambers.

MD-1 was used for measurement of the beam polarization used for the absolute energy calibration.

The magnetic field in MD-1 had volume of about 10 m³ and was equal to 12 kG at $\sqrt{s}$=10 GeV. The detector's subsystems were based on multiwire proportional chambers, scintillation counters and gas Cherenkov counters.

Precise measurement of masses of Υ mesons. The masses of the Υ mesons were determined from the dependence of the measured hadronic cross section on the beam energy. The energies of the beams were measured by the method of resonant beam depolarization developed in Novosibirsk [3]. The method is based on the fact that the spin precession frequency Ω depends on the beam electron energy E:

$$\Omega = \omega_s(1 + \frac{\mu'}{\mu_0}\gamma), \qquad (1)$$

where ω_s is the beam revolution frequency, μ'/μ_0 is the ratio of the anomalous and normal parts of the electron magnetic moment, $\gamma = E/mc^2$ is the Lorentz factor of electrons. The frequency Ω is measured with the polarized electron

TABLE 1. MD-1 experiments on measurement of masses of Υ mesons.

Particle (exp. year)	Duration (days)	$\int Ldt$ (nb^{-1})	Energy calibr.	Scanning series	Mass (MeV)
$\Upsilon(1S)$ (1982)	25	70	100	2	9459.5±0.6
$\Upsilon(1S)$ (1983)	4	130	20	1	9461.0±0.5
$\Upsilon(1S)$ (1984)	30	2250	90	4	9460.60±0.09±0.05
$\Upsilon(2S)$ (1983)	40	600	180	4	10023.6±0.5
$\Upsilon(3S)$ (1983)	110	1250	370	5	10355.3±0.5

beam using a depolarizer with the frequency ajusted to the Ω. The method's accuracy of E measurement is better than 10^{-5}.

The beams in VEPP-4 were transversely polarized due to spontaneous synchrotron radiation, the polarization time was about 1 hour. The polarization degree was determined by the angular asymmetry of synchrotron photons scattered on the opposite beam. This method was developed in Novosibirsk [4]. For this experiment the transverse magnetic field of the detector was used. The energies of the electron and positron beams were measured simultaneuosly.

Three independent measurements of the $\Upsilon(1S)$ mass were performed with MD-1 (in 1982, published in 1982; in 1983, published in 1984; in 1984, published in 1986 and in 1992) [5-8]. In 1983, the masses of $\Upsilon(2S)$ and $\Upsilon(3S)$ were measured [6,9]. The results are presented in Table 1.

In Table 2 we present the PDG data prior to our experiments, our results, as well as the data of experiments at DESY [10] and Cornell [11] performed with the method of resonant depolarization. Our result on $\Upsilon(1S)$ mass differs from the result of Cornell [11] by 3.8σ. The reason is not clear.

TABLE 2. Υ meson masses (MeV) after 1981.

Particle	Mass PDG (1982)	Mass	Laboratory	Publ.year
$\Upsilon(1S)$	9456±10	9460.60±0.09±0.05	Novosibirsk	82,84,86,92
	9456±10	9459.97±0.11±0.07	Cornell	84
$\Upsilon(2S)$	10016±10	10023.6±0.5	Novosibirsk	84,86
	10016±10	10023.1±0.4	DESY	84
$\Upsilon(3S)$	10347±10	10355.3±0.5	Novosibirsk	84,86

TABLE 3. New (corrected due to the new electron mass) Υ meson masses from the MD-1 experiments.

Particle	Mass (MeV)
$\Upsilon(1S)$	9460.51±0.09±0.04
$\Upsilon(2S)$	10023.5±0.5
$\Upsilon(3S)$	10355.2±0.5

In 1987 [12] and in 1995 [13], the value of the electron mass was changed (PDG 1996). Since in the resonant depolarization method the quantity $\gamma = E/mc^2$ is measured, this change of the electron mass caused a change of the Υ masses (Table 3).

Leptonic widths of $\Upsilon(1S)$ and $\Upsilon(2S)$. Leptonic width of resonance is determined experimentally from the dependence of $\sigma(e^+e^- \to resonance \to hadrons)$ on the energy. The very important feature of these measurements is the correct account of radiative corrections and the detection efficiency.

The most accurate theoretical calculation of the radiative corrections ($\sim 0.1\%$) has been done by Kuraev and Fadin [14]. In 1984 we used their formula in data processing and showed that the Γ_{ee} of $\Upsilon(1S)$ is shifted by

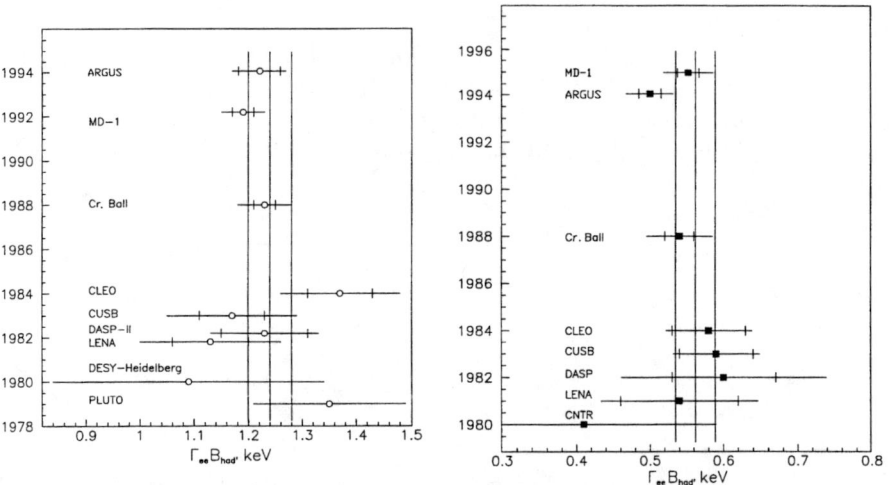

FIGURE 2. Compilation of $\Gamma_{ee} \cdot B_{had}$ values for $\Upsilon(1S)$ (left figure) and $\Upsilon(2S)$ (right figure). The full error bars represent total errors, the small ones show statistical errors separately. The vertical lines illustrate the world average and its error (without MD-1 and ARGUS experiments).

10%, and the mass is shifted by 0.1 MeV [15].

To minimize the detection efficiency uncertainty, the trigger was chosen to be as efficient as possible. In our experiment, the trigger efficiency was about 98%, the efficiency of selection was 90%, and the systematic uncertainty of efficiency knowledge was about 1%.

The experiment on measurement of $\Gamma_{ee}(\Upsilon(1S))$ was carried out during one month, 4 scans were performed with 90 energy calibrations, the integrated luminosity of 2.4 pb^{-1} was collected, 0.5 million events were recorded, including $1.5 \cdot 10^4$ of $\Upsilon(1S)$ mesons. The result of this experiment [8] is $\Gamma_{ee} \cdot B_{had} = 1.187 \pm 0.023 \pm 0.031$ keV. A comparison with other measurements is shown in Fig. 2.

The same technique was employed for determination of the $\Upsilon(2S)$ leptonic width [2]. The data collected during the experiment on the R-ratio measurement were used for the analysis. The integrated luminosity of 0.4 pb^{-1} was recorded for 2E=10010 ÷ 10040 MeV in 3 scans. Averaging of results gives $\Gamma_{ee} \cdot B_{had} = 0.552 \pm 0.031 \pm 0.017$ keV. The results of experiments on $\Upsilon(2S)$ leptonic width measurement are collected in Fig. 2.

Search for narrow resonances. The energy region $2E = 7.2$–10.3 GeV (excluding Υ-mesons) in e^+e^- collisions has not been studied much before our experiment [16]. The scan was done in steps of $\Delta(2E) = 4 - 5$ MeV (this value is close to the c.m.s. energy spread in VEPP-4). The data in each energy point were taken in several runs. Each energy region was scanned 2–4 times. The integrated luminosity of 16 pb^{-1} was taken. The experimental data were

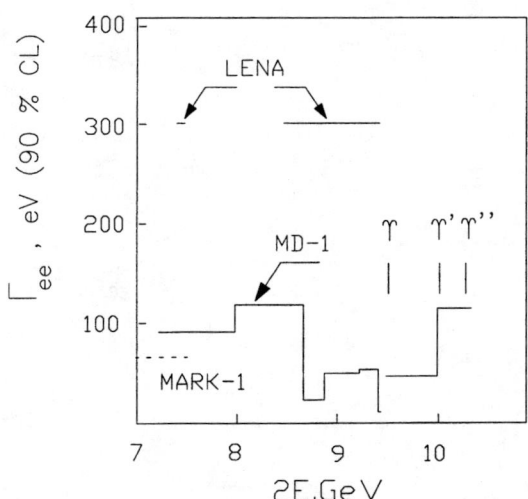

FIGURE 3. Upper limits on Γ_{ee} for narrow resonances (90% CL). Data from MARK-1 [17] (dashed line), LENA [18], and MD-1 [16].

fitted by two parameters: $\sigma_{con} \cdot \epsilon_{con}$ and $\sigma_{res} \cdot \epsilon_{res}$, where σ_{con} and σ_{res} are the hadron production cross sections for the continuum and resonance, and ϵ_{con}, ϵ_{res} are the corresponding efficiencies. The visible width of a narrow resonance is determined by the c.m.s. energy spread of the colliding beams. No new resonances were found. The upper limits on the leptonic width Γ_{ee} at a 90% confidence level are presented in Fig. 3.

Measurement of R. R is defined as $\sigma(e^+e^- \to hadrons)/\sigma(e^+e^- \to \mu^+\mu^-)$, where the non-resonant hadronic cross section does not include QED corrections and τ-pair decays; for μ-pair the Born cross section is used. From the visible cross section after background subtraction, we obtained the R-value using the expression:

$$R = \frac{\sigma_{vis}}{\varepsilon(1+\delta)} \cdot \frac{1}{\sigma_{\mu\mu}}, \qquad (2)$$

where ε is the detection efficiency for the multihadronic events with radiative effects included and $(1+\delta)$ is the radiative correction factor due to QED processes up to order α^3.

For the R measurement, the data with 16 pb^{-1} luminosity integral in the energy region 7.25-10.34 GeV were used [19]. The data are fitted very well by

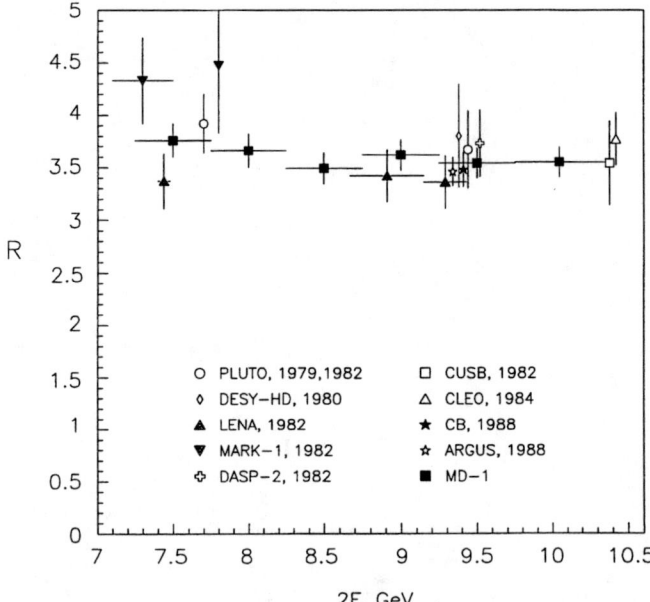

FIGURE 4. Compilation of the results for R in the energy region 7.0÷10.5 GeV. The statistical and systematic errors are summed in quadrature.

a constant ($\chi^2 = 28$ for 30 degrees of freedom). The systematic error in R is determined by several factors and is estimated to be 3.9%. The average value of R in the energy region $7.25 \div 10.34$ GeV is $\overline{R} = 3.578 \pm 0.021 \pm 0.140$.

In Fig. 4 our data on R are compared with the previous experiments in the region between 7.0 and 10.5 GeV. Our results are the most precise (only ARGUS measurement at 9.36 GeV has the same systematic error). The R value had not been measured before in the energy intervals of 7.8–8.7 GeV and 9.5–10.34 GeV. Averaging of the obtained results gives the value of $\overline{R}$ in the energy region $7.0 \div 10.5$ GeV: $\overline{R} = 3.579 \pm 0.066$ (all errors were summed in quadrature). This value is in a good agreement with the QCD prediction $R_{QCD} = 3.602 \pm 0.014$ for the average scaling parameter $\Lambda = 260^{+54}_{-46}$ MeV. From these data we obtained also: $\alpha_s = 0.174 \pm 0.039$ ($2E \approx 8.9\,GeV$). Although the average value of $\overline{R}$ has a good accuracy (1.8%), the corresponding error of α_s is about $2 \div 3$ times worse than the one obtained from study of the direct QCD effects.

Measurement of $B(\Upsilon(1S) \to \mu^+\mu^-)$. For the $B_{\mu\mu}$ measurement, the total integrated luminosity of 20.4 pb^{-1} in the mass region $\sqrt{s} = 7.6 - 10$ GeV was used, including 5.1 pb^{-1} in the $\Upsilon(1S)$ region. The number of $\Upsilon(1S) \to hadrons$ events is about 10^5.

For the first time, we took into account a systematic error in determination of $B_{\mu\mu} = (\Upsilon \to \mu\mu)/(\Upsilon \to hadr)$ due to the collider energy uncertainty. The ratio of the muon and hadron yields in the Υ decays gives the branching ratio $B_{\mu\mu}$ in the case if the center-of-mass energy W is equal to the resonance mass M. If W is not equal to M, the interference between the resonance and the continuum should be taken into account. The energy calibration by the resonant depolarization technique was carried out in this experiment. The obtained probability of the process $\Upsilon(1S) \to \mu^+\mu^-$ decay is [20]: $B_{\mu\mu} = (2.12 \pm 0.20 \pm 0.10)\%$.

Search for exclusive decays of $\Upsilon(1S)$ meson. The recorded 10^5 Υ mesons were used for search of the following exclusive decay modes: $\Upsilon \to p\bar{p}$ [2,20]; $\Upsilon \to \pi^+\pi^-, K^+K^-$ [20]; $\Upsilon \to \rho^0\pi^0$ [22]; $\Upsilon \to \xi(2.2)\gamma$ [21]; $\Upsilon \to X(2.2)\gamma$ [21].

For the $B_{p\bar{p}}$ measurement, the gas Cherenkov counters with $\gamma_{th}=5.16$ have been used. The γ-factor of p and $\bar{p}$ from the $\Upsilon(1S)$ decay is less than γ_{th}, and therefore the Cherenkov counters were very effective in background suppression. We have obtained the upper limit $B(\Upsilon(1S) \to p\bar{p}) < 5 \cdot 10^{-4}$ at 90% CL. Theoretical estimations give $B_{p\bar{p}} \sim 10^{-7}$. Upper limits obtained for other decays are listed in Conclusions.

Measurement of inclusive Λ production. One of interesting experimental facts in Υ physics is that the baryon yield in direct Υ decays is enhanced by a factor of larger than 2.5 in comparison with the continuum. We measured the inclusive Λ production in direct $\Upsilon(1S)$ decays and in the continuum [23]. The integrated luminosity of 5.6 pb^{-1} in the $\Upsilon(1S)$ region and 16.6 pb^{-1} in the continuum was used. Our result on the Λ yield per one event in the direct $\Upsilon(1S)$ decays is presented in Conclusions. In the continuum for two different cms energy intervals we obtained: $<n_\Lambda(cont)> = 0.076 \pm 0.018 \pm 0.015$ at $\sqrt{s} = 7.2 - 10.0$ GeV and $<n_\Lambda(cont)> = 0.070 \pm 0.027 \pm 0.020$ at $\sqrt{s} = 7.2 - 9.4$ GeV. Results of our measurement agree with the data from CLEO [24] and ARGUS [25]. In the range $\sqrt{s} = 7.2 \div 9.4$ GeV, the Λ yield was measured for the first time.

Measurement of Bose-Einstein correlations. An enhancement in the production of pion pairs of the same charge and similar momenta is attributed to Bose-Einstein (B-E) statistics. Our result on measurement of B-E correlation parameters [26] is presented in Conclusions.

Conclusions. The main results obtained with the MD-1 detector at the VEPP-4 collider on study of the Υ mesons and the nearby continuum are presented below:

1. $M(\Upsilon(1S)) = 9460.51 \pm 0.09 \pm 0.04$ MeV (corrected) [5-8];
2. $M(\Upsilon(2S)) = 10023.5 \pm 0.5$ MeV (corrected) [6,9];
3. $M(\Upsilon(3S)) = 10355.2 \pm 0.5$ MeV (corrected) [6,9];
4. $\Gamma_{ee}(\Upsilon(1S)) \cdot B_{had} = 1.187 \pm 0.023 \pm 0.031$ keV [8];
5. $\Gamma_{ee}(\Upsilon(2S)) \cdot B_{had} = 0.552 \pm 0.031 \pm 0.017$ keV [2];
6. $\Gamma_{ee}(M_X = 7.2 - 10$ GeV$) < (0.015 - 0.12)$ keV [16];
7. $\overline{R}(2E = 7.25 - 10.34$ GeV$) = 3.578 \pm 0.021 \pm 0.140$ [19];
8. $B(\Upsilon(1S) \to \mu\mu) = 2.12 \pm 0.20 \pm 0.10$ % [20];
9. $B(\Upsilon(1S) \to \xi(2.2)) \cdot B(\xi(2.2) \to K^+K^-) < 2 \times 10^{-4}$ [21];
10. $B(\Upsilon(1S) \to X(2.2)) \cdot B(X(2.2) \to \Phi\Phi) < 3 \times 10^{-3}$ [21];
11. $B(\Upsilon(1S) \to \rho^0\pi^0) < 3.3 \times 10^{-4}$ [22];
12. $B(\Upsilon(1S) \to \pi^+\pi^-) < 5 \times 10^{-4}$ [20];
13. $B(\Upsilon(1S) \to K^+K^-) < 5 \times 10^{-4}$ [20];
14. $B(\Upsilon(1S) \to p\bar{p}) < 5 \times 10^{-4}$ [20,2];
15. $<n_\Lambda(\Upsilon(1S)_{dir})> = 0.19 \pm 0.02 \pm 0.02$ [23];
16. $<n_\Xi(\Upsilon(1S)_{dir})> = 0.04 \pm 0.02$ [23];
17. $\pi\pi$ correlations. At the direct $\Upsilon(1S)$ decays: $r_0 = 0.73 \pm 0.11$ fm, $\lambda = 0.71 \pm 0.16$. In the continuum at $2E = 7.2$–10.3 GeV: $r_0 = 0.83 \pm 0.23$ fm, $\lambda = 0.51 \pm 0.19$ [26].

Note that even today the results 1–7, 12–14 are either comparable in accuracy with the best results from other detectors or surpass them.

Summary of MD-1 results including study of two-photon processes is published in our final report [2].

REFERENCES

1. Baru S.E. et al. (MD-1 collaboration), *Preprint INP* **83–39**. Novosibirsk 1983.
2. Baru S.E. et al. (MD-1 collaboration), *Physics Reports* **267**(1996)71.
3. Derbenev Ya.S. et al., *Part.Acc.* **10**(1980)177.
4. Blinov A.E. et al., *Nucl. Instr. and Meth.* **A241**(1985)80.
5. Artamonov A.S. et al. (MD-1 collaboration), *Phys. Lett.* **118B**(1982)225.
6. Artamonov A.S. et al. (MD-1 collaboration), *Phys. Lett.* **137B**(1984)272.
7. Baru S.E. et al. (MD-1 collaboration), *Z. Phys.* **C30**(1986)551.
8. Baru S.E. et al. (MD-1 collaboration), *Z. Phys.* **C56**(1992)547.
9. Baru S.E. et al. (MD-1 collaboration), *Z. Phys.* **C32**(1986)622.
10. Barber D.P. et al., *Phys.Lett.* **135B**(1984)498.
11. MacKay W.W. et al. (CUSB collaboration), *Phys.Rev.* **D29**(1984)2483.
12. Cohen E.Richard et al., *Rev.Mod.Phys.* **59**(1987)1121.
13. Farnham Dean L. et al., *Phys.Rev.Lett.* **75**(1995)3598.
14. Kuraev E.A., Fadin V.S., *Yadernaya Fizika* **41**(1985)733.
15. Baru S.E. et al. (MD-1 collaboration), *Preprint INP* **84-97**, Novosibirsk 1984. Submitted to the 22-th Int. Conf. on High Energy Physics, Leipzig, 1984.
16. Blinov A.E. et al. (MD-1 collaboration), *Z. Phys.* **C49**(1991)239.
17. Siegrist J. at al. (MARK-1 collaboration), *Phys.Rev.* **D26**(1982)969.
18. Niczyporuk B. et al. (LENA collaboration), *Z.Phys.* **C15**(1982)299.
19. Blinov A.E. et al. (MD-1 collaboration), *Z. Phys.* **C70**(1996)31.
20. Baru S.E. et al. (MD-1 collaboration), *Z. Phys.* **C54**(1992)229.
21. Baru S.E. et al. (MD-1 collaboration), *Z. Phys.* **C42**(1989)505.
22. Blinov A.E. et al. (MD-1 collaboration), *Phys. Lett.* **B254**(1990)311.
23. Blinov A.E. et al. (MD-1 collaboration), *Z. Phys.* **C62**(1994)367.
24. Behrends S. et al. (CLEO collaboration), *Phys.Rev.* **D31**(1985)2161.
25. Albrecht H. et al. (ARGUS collaboration), *Z.Phys.* **C39**(1988)177.
26. Blinov A.E. et al. (MD-1 collaboration), *Z.Phys.* **C69**(1996)215.

First Measurements of the b Lifetime*

John A. Jaros

Stanford Linear Accelerator Center, Stanford, CA 94309, USA

Abstract. This paper reviews the first measurements of the b lifetime, the theoretical and experimental climate in which they were made, and their considerable impact on knowledge of the CKM matrix and b phenomenology.

INTRODUCTION

The first measurements of the b lifetime came just six years after the discovery of the Upsilon, and only three years after the first compelling evidence for the production of hadrons with open b flavor. The average b-hadron lifetime was measured to be about 1.5 ps, surprisingly long by the theoretical standards of the day and remarkably close to today's accepted value. Despite the surprise, the fact that the b is long-lived was accepted almost immediately and was confirmed within a year by other experiments. The implications of the long b lifetime were clear even before the first experimental results were in print, and they were far-reaching. Significant b-meson mixing, a heavy top quark mass, and appreciable CP violation in the b system were among these expectations.

This recollection will review the early theoretical landscape and experimental limits, discuss the first collider lifetime measurements, and examine the impact the first measurements of the b lifetime had on our knowledge of the CKM matrix.

PREDICTIONS AND EARLY LIMITS

The spectator model for heavy-quark decays had been sufficiently developed for charm decays that its extrapolation to b decays was straightforward by the time of the Upsilon discovery [1]. Several authors [2] related the b lifetime to the strength of the $b \to c$ and $b \to u$ couplings, accounting for the phase-space

---

*) Work supported by the U.S. Department of Energy under contract DE-Aco3-765F00515

differences of the final states:

$$\tau_b = \frac{(M_\mu/M_b)^5 \tau_\mu}{2.75|V_{bc}|^2 + 7.7|V_{bu}|^2}.$$

If the mixing between the third and second generations were like that between the second and the first, the b lifetime would be very short, $\tau_b \sim 3 \times 10^{-14}$ s. Predictions of the lifetime used existing constraints on the CKM elements to limit $|V_{bu}|$ and $|V_{bc}|$. The ε parameter in K^0 decays, expressed in terms of CKM parameters from the box-diagram analysis, provided the basis of Harari's estimate [3]: 10^{-14} s $< \tau_b < 10^{-11}$ s.

A more aggressive limit was derived by Barger, Pakvasa, and Long [4], which included constraints from the $K_L^0 - K_S^0$ mass difference. They concluded 10^{-14} s $< \tau_b < 10^{-13}$ s, and this became the prevailing theoretical opinion before the first lifetime measurements: the b lifetime is short [5].

Unconventional ideas from Cahn and Fritzsch [6] suggested the b might be nearly stable and motivated two Fermilab searches [7] for the production of meta-stable b hadrons. Both experiments looked for massive 5 GeV/c^2 particles produced in 400 GeV p-Be collisions. They used existing secondary beamlines to select momentum, Cherenkov counters to reject backgrounds, and long-baseline time-of-flight techniques to measure velocity and so infer mass. Neither saw candidate events, establishing that $\tau_b < 5 \times 10^{-8}$ s. The JADE experiment at PETRA established a better limit [8], $\tau_b < 2 \times 10^{-8}$ s, by excluding the existence of charged tracks with anomalously high dE/dx in 30 GeV e^+e^- annihilations.

LIFETIME MEASUREMENT TOOLS

The 30 GeV e^+e^- storage rings PETRA at DESY and PEP at SLAC became operational soon after the Upsilon discovery. They were ideal laboratories for measuring the b lifetime because the $b\bar{b}$ production cross-section was known, clean b identification was possible, luminosities were adequate, and picosecond lifetimes were boosted into millimeter decay lengths, which were readily measurable.

Semi-leptonic b decays, with their distinctive high-transverse-momentum leptons, provided a clean b tag. CLEO and CUSB [9] at Cornell first measured the semi-leptonic branching ratio to be about 12% in 1981. Semi-leptonic b decays and the b fragmentation function were measured at PEP by the Mark II and MAC experiments [10] in 1982. Good agreement with the Cornell results established that b tagging was quantitatively understood.

The Mark II Collaboration pioneered vertex detection in the collider environment with their 1980 proposal [11] to add a precision drift chamber close to the interaction point. The physics motivations for the device included the measurement of the tau lifetime, measurement of charm-particle lifetimes, and the search for a finite b lifetime. A new technique was proposed to measure the tau lifetime by measuring the distance between the interaction point and the $\tau \to \nu 3\pi X$ decay vertex. This method was soon exploited by Mark II, MAC, and Cello [12] to provide the first indications that the tau lifetime is finite. The first results from the Mark II vertex detector were reported in 1982 [13]. The detector's superior impact parameter resolution dramatically reduced the tau lifetime measurement errors and showed exponential tails in distributions that had been broad, slightly offset Gaussians. Techniques for measuring beam positions, optimal decay lengths, resolutions, and systematic errors were developed at this time. The measured value of the tau lifetime [14] was in good agreement with theory, lending credibility to these new techniques.

THE LEPTON IMPACT-PARAMETER METHOD

The JADE experiment at PETRA reported a technique suitable for measuring the b lifetime in Spring 1982. It involved measuring the signed impact parameter of a lepton track, which presumably originated from the b decay. Signing the impact parameter positive (if the track appeared to come from a positively-displaced vertex) or negative meant one would see lifetime effects by a slight offset of the mean of the resolution function. Monte Carlo techniques were used to relate the amount of offset to the lifetime. Since the lepton spectrum in b decays was known and the fragmentation function hard, model dependence was manageable. JADE used this method to measure the average impact parameter of 27 high-momentum muons coming from a 10 pb^{-1} dataset. The result was consistent with zero, and was used to establish a much improved limit, $\tau_b < 1.4 \times 10^{-12}$ s [15]. The MAC and Mark II experiments quickly adopted similar strategies. MAC reported [16] $\tau_b = 1.7 \pm 1.0$ ps at the 1982 Paris Conference. The value was insignificantly positive, but tantalizing. Mark II also learned that it was hard to improve on JADE's limit when there are hints of finite lifetime in the data. We chose to keep mum.

FIRST MEASUREMENTS AT PEP

The 1982–83 year at PEP provided record luminosities to the MAC and Mark II experiments. PETRA meantime was trading high luminosity for high energy in its quest for a 20 GeV top quark. By year's end, MAC had accumulated and speedily processed 100 pb^{-1} of data, giving a sample of 270 electrons and muons measured with 600 μm impact parameter resolution. Mark II had 80 pb^{-1}, only 104 leptons, but 200 μm resolution. Both experiments were relying on the semi-leptonic b tag for their event identification, and lepton impact parameter as a measure of the lifetime. We in Mark II were convinced by early

Spring '83 that the mean impact parameter was positive, but struggled to implement a full maximum-likelihood fit to exploit our good resolution. MAC saw effects late in the spring when the full data set, electrons and muons, was available. MAC measured the mean of the lepton impact parameter, weighted by the inverse square of the impact parameter error. They beat Mark II to press and announced early in the summer that $\tau_b = 1.8 \pm 0.6 \pm 0.4$ ps [17]. Mark II reported its results one month later at the SLAC Summer Institute [17]: $\tau_b = 1.20\,^{+0.45}_{-0.36} \pm 0.3$ ps. The data are shown in Fig. 1.

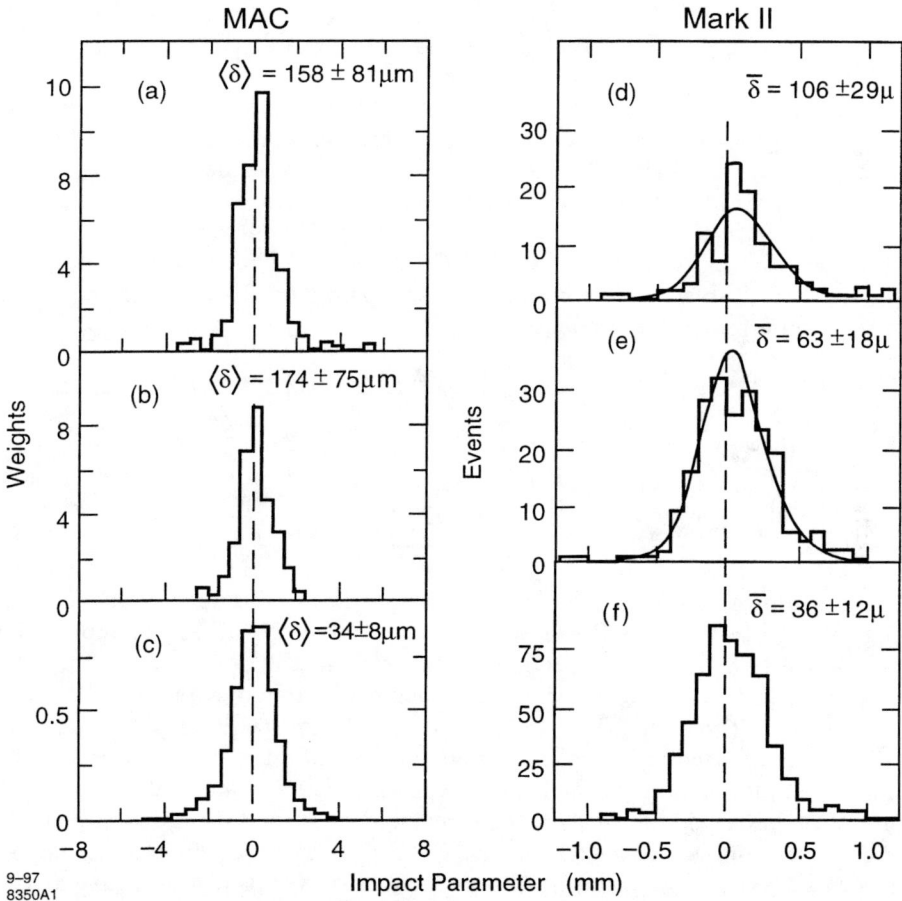

FIGURE 1. Impact parameter distributions from the two PEP experiments. MAC's results are shown for (a) muons, (b) electrons, and (c) hadrons. MARK II's results are shown for (d) "b leptons" with $p_t > 1$ GeV/c, (e) "c leptons" with $p_t < 1$ GeV/c, and (f) hadrons.

Bill Reay reviewed the lifetime results [18] at the Lepton-Photon Symposium that year: "My conclusion is that the three standard deviation effect seen by two experiments for the impact parameter is a strong indication that the b lifetime is of order 10^{-12} seconds." The result was widely accepted. There were after all two independent experiments, seeing effects in electrons and muons, checking that average hadronic impact parameters were very small as expected, and cross-checking that the charm lifetime was as expected. It was a strong case experimentally.

DELCO at PEP and TASSO and JADE at PETRA confirmed the result in 1984 [19]. The early results were quite consistent with values accepted today.

THEORETICAL IMPACT

The first b lifetime measurements provided the necessary final ingredient to fix the magnitudes of the CKM matrix elements. CLEO and CUSB had established in early 1983 that $|V_{bu}|^2/|V_{bc}|^2 < 0.04$ [20]. Consequently, the b lifetime is essentially a direct measure of $|V_{bc}|$, the $b \to u$ term being inconsequential. The first lifetime measurements established that $|V_{bc}| \sim 0.05$. With this input and the assumption of unitarity, the CKM matrix-element magnitudes were established. Table 1 shows how our knowledge of the CKM matrix advanced after the lifetime measurements [21].

TABLE 1. The CKM Matrix Before and After b Lifetime Measurements

	1982				1983		
	d	s	b		d	s	b
u	$.973 \pm .024$	$.224 \pm .006$	$.05 \pm .05$	u	$.973 \pm .024$	$.224 \pm .006$	$.007 \pm .007$
c	$.22 \pm .02$	$.89 \pm .09$	$.31 \pm .26$	c	$.22 \pm .02$	$.972 \pm .002$	$.053 \pm .017$
t	$.06 \pm .06$	$.28 \pm .28$	$.90 \pm .09$	t	$.007 \pm .007$	$.053 \pm .017$	$.998 \pm .001$

This improved knowledge of the CKM matrix had interesting phenomenological implications. As Paschos, Stech, and Turke [22] observed, appreciable mixing and CP-violation effects were expected in the b-meson system. This is a consequence of first-order b decays being so strongly suppressed that the second-order (box) diagrams were relatively significant. Ginsparg, Glashow, and Wise [23] used the new information on $|V_{bc}|$ and the ε parameter to infer that top must be heavy, which at the time meant $m_t > 45$ GeV/c^2. The failure of the prediction for a short b lifetime indicated that short-distance effects did not dominate the description of the $K_L^0 - K_S^0$ mass difference. Lastly, the smallness of $|V_{bc}|$ could not be understood in terms of the simple ansatz relating masses and mixing angles that was popular before the measurements.

EXPERIMENTAL IMPACT

The long b lifetime has made it possible to identify b hadrons by virtue of their decay topology. Early attempts to do so at PEP and PETRA had tagging efficiencies around five-percent and purities in the 60-70% range [24]. The art has developed rapidly since the introduction of high-precision silicon vertex detectors. The CCD vertex detector in SLD [25] tags b jets at the Z with 50% efficiency and nearly 99% purity. Efficient lifetime tags have made it possible to identify the top quark, measure heavy-quark electroweak parameters to high precision, and extend searches for the Higgs. This physics has underscored the importance and spurred the development of high precision vertex detectors. The new detectors, in turn, are sharpening our view of the underlying vertex structure of high energy interactions, and the wealth of physics implicit in these structures.

ACKNOWLEDGEMENTS

It is a pleasure to thank my MAC competitor, Bill Ford, and my Mark II colleague, Nigel Lockyer, for sharing their reminiscences of these early measurements with me.

REFERENCES

1. L. Maiani, in *Proceedings of the Eighth International Symposium on Lepton and Photon Interactions at High Energies*, Hamburg, 1977, edited by F. Gutbrod (DESY, Hamburg, Germany, 1977).
2. See for example M. Gaillard and L. Maiani, in *Proceedings of the 1979 Cargese Summer Institute on Quarks and Leptons*, edited by M. Levy et al (Plenum, New York, 1979), p. 433.
3. H. Harari, SLAC Report 2234, Nov. 1978.
4. V. Barger, W. F. Long, and S. Pakvasa, J. Phys. **G5**, L147 (1979).
5. G. Kalmus, J. Phys. (Paris), Colloq. **43**, C3–431 (1982).
6. Robert Cahn, Phys. Rev. Lett. **40**, 80 (1978); Harald Fritzsch, Phys. Lett. **78B**, 611 (1978).
7. D. Cutts *et al.*, Phys. Rev. Lett. **41**, 363 (1978); R. Vidal *et al.*, Phys. Lett. **77B**, 344 (1978).
8. W. Bartel *et al.*, Z. Physik **C6**, 295 (1980).
9. C. Bebek *et al.*, Phys. Rev. Lett. **46**, 84 (1981); K. Chadwick *et al.*, Phys. Rev. Lett. **46**, 88 (1981); L. J. Spencer *et al.*, Phys. Rev. Lett. **47**, 771 (1981).
10. M. E. Nelson *et al.*, Phys. Rev. Lett. **50**, 1542 (1983); E. Fernandez *et al.*, Phys. Rev. Lett. **50**, 2054 (1983).
11. Mark II Collaboration, PEP-5, Supplement B, *Proposal to Add a Secondary Vertex Detector to the Mark II Detector*, July, 1980.
12. G. J. Feldman *et al.*, Phys. Rev. Lett. **48**, 66 (1982); W. T. Ford *et al.*, Phys. Rev. Lett. **49**, 106 (1982); H. J. Behrend *et al.*, Nucl. Phys. **B211**, 369 (1983).
13. J. Jaros, J. Phys. (Paris), Colloq. **43**, C3–106 (1982).
14. J. Jaros *et al.*, Phys. Rev. Lett. **51**, 955 (1983).

15. W. Bartel et al., Phys. Lett. **114B**, 71 (1982).
16. D. M. Ritson, J. Phys. (Paris), Colloq. **43**, C3–52 (1982).
17. E. Fernandez et al., Phys. Rev. Lett. **51**, 1022 (1983); N. S. Lockyer et al., Phys. Rev. Lett. **51**, 1316 (1983).
18. Bill Reay, in *Proceedings of the 1983 International Symposium on Lepton and Photon Interactions at High Energy*, Cornell, 1983, edited by D. Cassel and D. Kreinik (Ithaca, 1983).
19. D. E. Klem et al., Phys. Rev. Lett. **53**, 1873 (1984); M. Althoff et al., Phys. Lett. **149B**, 524 (1984); W. Bartel et al., Z. Phys. **C31**, 349 (1986).
20. C. Klopfenstein et al., Phys. Lett. **130B**, 444 (1983); A. Chen et al., Phys. Rev. Lett. **52**, 1084 (1984).
21. S. Pakvasa, in *Proceedings of the 21st International Conference on High Energy Physics*, Paris, 1982, edited by P. Petiau and M. Porneuf, Editions de Physique, 1982. (Journal de Physique, v. 43, Colloque C-3, suppl. 12, Dec. 1982).
22. E. Paschos, B. Stech, and U. Turke, Phys. Lett. **128B**, 240 (1983).
23. P. Ginsparg, S. Glashow, and M. Wise, Phys. Rev. Lett. **50**, 1415 (1983).
24. Paul Weber, Ph.D. Thesis, University of Colorado, 1990; W. Braunschweig et al., Z. Phys. **C42**, 17 (1989).
25. See Su Dong's talk at this Symposium.

B Physics with CESR and CLEO

Ronald A. Poling

School of Physics and Astronomy
University of Minnesota
Minneapolis, Minnesota 55455

Abstract. Since 1980 the CLEO collaboration has studied the production and decay of b-flavored hadrons produced in the Cornell Electron Storage Ring. This talk recounts some of the milestones of this program, including recent developments in semileptonic and rare B-meson decays.

INTRODUCTION

This year we mark the twentieth anniversary of the discovery of hidden beauty at Fermilab, and the seventeenth year since the exposure of naked beauty with the CLEO detector at the Cornell Electron Storage Ring (CESR). The b20 symposium is a celebration of the glorious past, the exciting present and the beautiful future of b physics. My contribution to this includes personal reflections on the early days of CLEO, and an overview of selected milestones of our research program, including recent results in semileptonic and rare B-meson decays.

HISTORICAL REFLECTIONS

As the first graduate student to write a thesis based on CLEO data [1] I had a somewhat different perspective from the other CLEO contributors to b20. While Al Silverman was masterfully managing the final stages of CLEO installation in January, 1980, I was crawling under 400 tons of flux-return iron installing muon chambers. When that was done Ed Thorndike and I put finishing touches on muon-chamber analysis software and joined with our colleagues in the first serious CLEO run at the $\Upsilon(4S)$ resonance in February and March. Within weeks we had indications of excess muon production at the $4S$, the first evidence ever of bare b decays, and confirmation that the upsilons were in fact the $b\bar{b}$ states we believed them to be. It was a wonderfully exciting time, and we all understood that it was just the beginning.

Figure 1 shows the state of CLEO's inclusive muon and electron signals as of

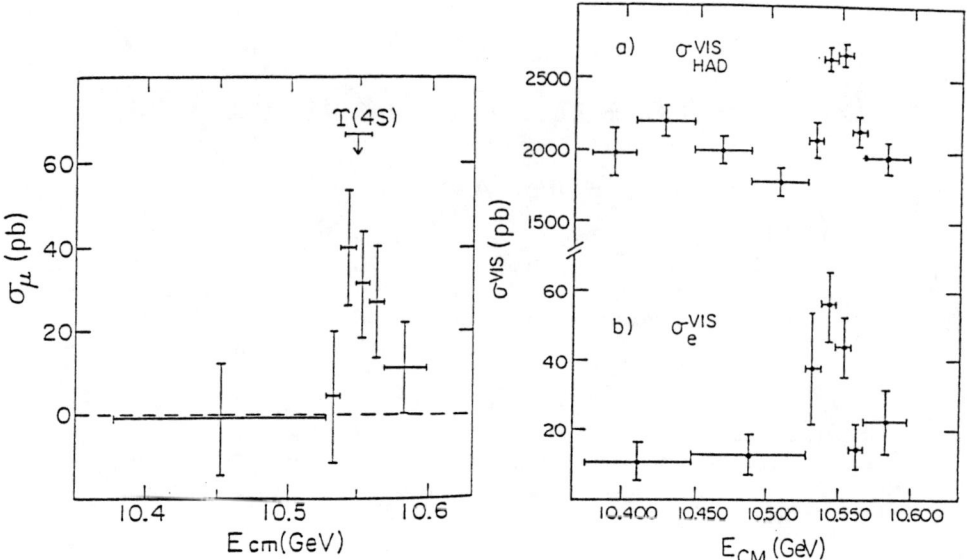

FIGURE 1. At left, the visible cross section for inclusive muon production, and at right the hadronic cross section (a) and the electron cross section (b) as a function of total energy in 1980 CLEO data in the vicinity of the $\Upsilon(4S)$ resonance.

summer 1980. These data were presented by Ed Thorndike at the Rochester conference in Madison, and attracted considerable attention. "Beauty is Found; Where is Truth?" proclaimed the title of an article in Newsweek's August 11, 1980 issue.

CLEO had been constructed by what for the time was a very large group. The early CLEO publications were authored by about 66 physicists belonging to six university groups. For comparison, the author lists for the most recent CLEO papers number 24 groups and ~ 220 authors. Among these are five of the original six universities and 23 of the original authors.

With bare b convincingly established, our attention came to be focused on detailed studies of B decays, where it has remained ever since. The data came quite slowly at first. The first year of B studies with CLEO produced about 700 $B\bar{B}$ events, representing less than an hour of current CESR running, and mere minutes of full-luminosity running with any of the three B factories. A succession of innovative upgrades to the storage ring and associated accelerators, as well as the laboratory's total dedication to efficient CESR running and integrated luminosity, have resulted in a remarkable record of productivity (Figure 2). CESR holds world records for essentially every measure of storage ring performance, and the prospects for continuing improvement are excellent.

Almost all CESR running has been at the peak of the $\Upsilon(4S)$ resonance, where we obtain one $B\bar{B}$ for every three events of continuum $q\bar{q}$, and at energies just below B threshold to determine the continuum background. The result is a recorded

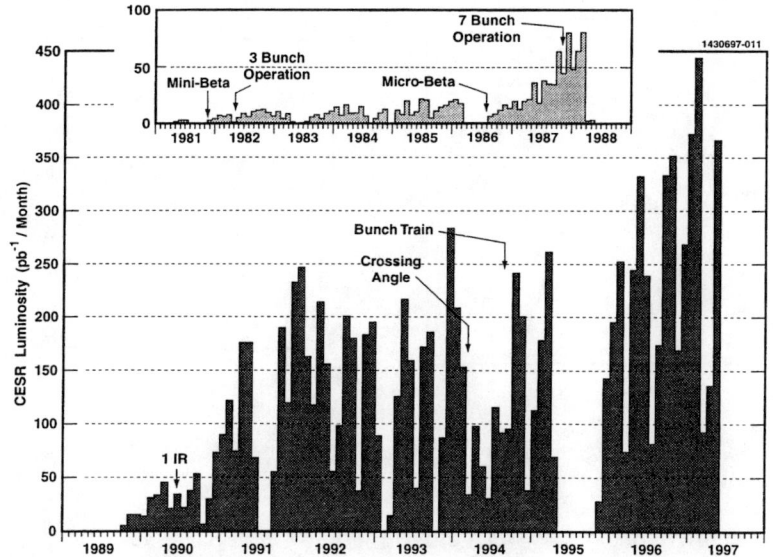
FIGURE 2. CESR integrated luminosity per month from 1980 to summer 1997.

data sample with roughly 5 million $B\bar{B}$ pairs, and even larger numbers of $c\bar{c}$ and $\tau^+\tau^-$ events. Paralleling CESR's evolution there has been steady progress in detector capabilities: tracking upgrades of the original CLEO after a few years of operation; the CLEO II upgrade in the late 1980's, setting a new standard for electromagnetic calorimetry in a general-purpose detector; the recent upgrade of CLEO II to CLEO II.5, providing us with silicon vertexing for the first time; and coming in 1999, CLEO III, with major improvements in tracking, particle ID, data acquisition, and triggering.

In the following sections I trace two of the most important threads that have run through the CESR/CLEO physics program. CLEO studies of semileptonic and rare B decays have been crucial in testing the Standard Model, in determining its parameters, and in setting the stage for the even more exciting studies of B decays that will be conducted in the era of the B factories.

SEMILEPTONIC B DECAYS

Semileptonic decays of B mesons have been our most effective probe of the fundamental parameters that govern b decay. The power of semileptonic B decays derives from their experimental accessibility (high efficiency, low backgrounds), and their comparative simplicity. As was mentioned above, the first demonstration that the upsilons were composed of a new flavor came as evidence of semileptonic B decays at the $\Upsilon(4S)$, an unambiguous weak-interaction process [2]. Measurements of inclu-

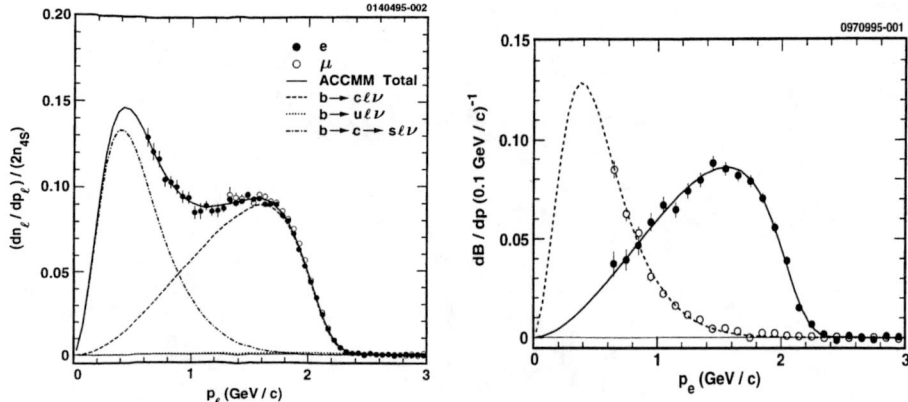

FIGURE 3 Inclusive single-lepton momentum spectrum (left) and lepton-tagged spectrum (right) measured with 2 fb^{-1} of CLEO II data.

sive and exclusive decays $b \to c\ell\nu$ determine the CKM parameter $|V_{cb}|$, and provide incisive tests of our understanding of hadronic interaction effects through measurements of the branching fractions, form factors, and other details. Studies of inclusive and exclusive charmless decays $b \to u\ell\nu$ provide quantitative information about $|V_{ub}|$, an essential ingredient for rigorous testing of the Standard Model.

There have been difficulties and mysteries, however, in our measurements of semileptonic B-meson decays. While the inclusive lepton signal is large and clear, its interpretation is in many ways subtle. Figure 3 shows the semileptonic momentum spectra measured with 2 fb^{-1} of CLEO II data by two separate techniques [3,4]. The fully inclusive single-lepton measurement must be interpreted by fitting to theoretical predictions of both primary ($b \to c\ell\nu$) and secondary ($b \to c \to s\ell\nu$) decays, resulting in severe model dependence. The lepton-tagged analysis, inspired by an elegant ARGUS analysis of a much smaller data sample [5], used charge and angular correlations to separate the primary and secondary components without reliance on models. Both techniques give branching fractions in the vicinity of 10.5%, significantly and uncomfortably smaller than the theoretical expectation. More has been presented on this topic at b20 by Klaus Honscheid.

The main objective of studies of $b \to c\ell\nu$ is the determination of $|V_{cb}|$. The most straightforward approach is to use the inclusive semileptonic branching fraction measurement along with B lifetime measurements from LEP and Fermilab. This procedure is statistically quite precise, but theoretically somewhat uncertain, although there is disagreement among theorists about the size of this uncertainty. The result is $|V_{cb}| = 0.040 \pm 0.001 \pm (0.002 - 0.004)$, where the first error is the total experimental uncertainty (statistical and systematic), and the second gives the range of opinion about the thoretical uncertainty. Further clarification is in the hands of the theorists.

Heavy Quark Effective Theory (HQET) and exclusive measurements provide

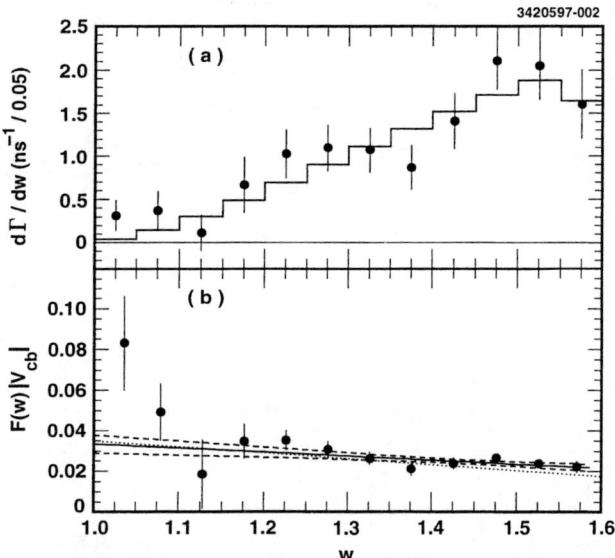

FIGURE 4. (a) $d\Gamma/dw$ distribution for data (points) and fit (solid line). (b) Measured values of $\mathcal{F}_D(w)|V_{cb}|$ (points), the results of a fit (solid line) and statistical errors (dashed line), along with the corresponding function $\mathcal{F}_{D^*}(w)|V_{cb}|$ from the CLEO $\bar{B} \to D^*\ell\bar{\nu}$ analysis (dotted line).

what has been the preferred route to $|V_{cb}|$ for the past several years. Fits to the differential decay rate $d\Gamma/dw$ for $B \to D^{(*)}\ell\nu$ (where $w = (M_B^2 + M_{D^{(*)}}^2 - q^2)/(2M_B M_{D^{(*)}})$) are used to extrapolate to the zero-recoil point, where the normalization is $|V_{cb}|$ times a theoretical factor. Our confidence in the use of HQET in this procedure has been enhanced by measurements of details of semileptonic B decays, especially the form factors in $B \to D^*\ell\nu$ [6]. The mode $B \to D^*\ell\nu$ [7] is somewhat more powerful for this measurement, and CLEO's 1995 result for this mode is $|V_{cb}| = 0.0386 \pm 0.0021 \pm 0.0024$, consistent with similar measurements from the LEP experiments [8]. CLEO has also recently published a measurement using $B \to D\ell\nu$ (Figure 4) which gives $|V_{cb}| = 0.040 \pm 0.005 \pm 0.004 \pm 0.003$ [9]. Overall there is excellent agreement among the inclusive and exclusive measurements, considering the significant and largely independent systematic uncertainties.

Among the parameters of the Standard Model V_{ub} has been of particular interest [10]. The viability of the Standard Model's description of CP violation depends on the magnitudes of all elements of the CKM matrix being nonzero, and for V_{ub} this remained in serious doubt until the first experimental observations of charmless semileptonic B decays in mid-1989. The discovery of $b \to u\ell\nu$ was reported almost simultaneously by CLEO [11] and ARGUS [12]. In both cases the evidence was lepton production above the kinematic limit for $b \to c\ell\nu$. The best information on inclusive $b \to u\ell\nu$ production was presented by CLEO in 1993 and is shown in Figure 5. This measurement was made with 1 fb^{-1} of CLEO II data [13],

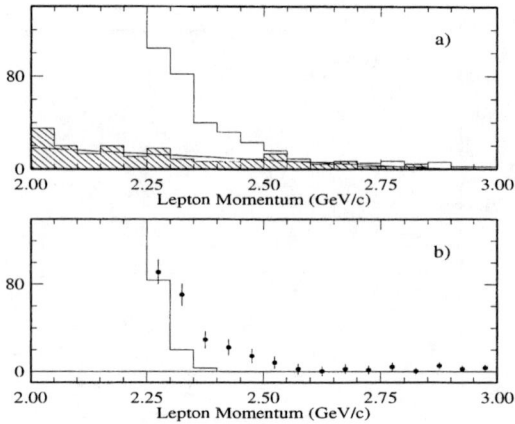

FIGURE 5. End point of the B semileptonic momentum spectrum from 1 fb^{-1} of CLEO II data: (a) $\Upsilon(4S)$ decays and off-resonance continuum (hatched), and (b) continuum-subtracted spectrum with expected $b \to c\ell\nu$ (histogram).

and with input from several inclusive and exclusive models for $b \to u\ell\nu$ leads to $|V_{ub}/V_{cb}| = 0.08 \pm 0.02$, where model dependence is the primary limitation. Using information about B lifetimes that was not available in 1993, one can extract $|V_{ub}|$ directly [14]. Results for several models are given in Table 1. The consistency among ARGUS, CLEO and CLEO II results is somewhat improved, largely because the effect of differences in handling $b \to c\ell\nu$ has been eliminated. Both the mean of the CLEO II results in Table 1 (0.0032) and the spread among the models are consistent with $|V_{ub}/V_{cb}| = 0.08 \pm 0.02$ and $|V_{cb}| = 0.04$.

The measurement of specific charmless final states in semileptonic B decays is the natural next step in the program to refine our knowledge of V_{ub}. The experimental and theoretical systematic uncertainties of an exclusive measurement are

TABLE 1. The $|V_{ub}|$ values obtained by normalizing to the B lifetime ($\tau_B = 1.6$ ps). For ARGUS, a partial branching fraction for $b \to u\ell\nu$ in the momentum range $2.3 - 2.6$ GeV/c of $(1.9 \pm 0.4) \times 10^{-4}$ was used. In averaging the squared error was scaled by χ^2/N.

Model	ARGUS ($\times 10^{-4}$)	CLEO (1990) ($\times 10^{-4}$)	CLEO (1993) ($\times 10^{-4}$)	Average ($\times 10^{-4}$)
KS [16]	32 ± 3	38 ± 6	24.9 ± 2.4	29 ± 3
WSB [17]	43 ± 5	48 ± 8	34.4 ± 3.3	39 ± 4
ACCMM [18]	36 ± 4	37 ± 6	32.7 ± 3.2	34 ± 3
ISGW2 [19]	43 ± 5	45 ± 8	34.3 ± 3.2	38 ± 3

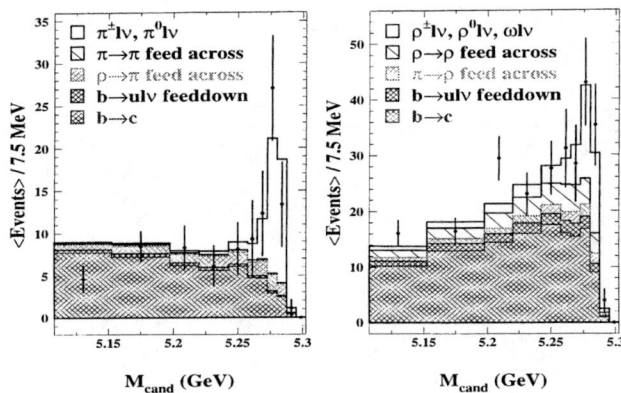

FIGURE 6. Invariant mass distribution for events in the ΔE signal region for $B \to \pi\ell\nu$ and $B \to \rho\ell\nu$. Data points are shown as solid circles and fitted signals are unshaded histograms. Other histograms show background components in fits: $b \to c\ell\nu$ (large hatch), higher-mass $b \to u\ell\nu$ (left hatch), $\pi\ell\nu$ feed-across (small hatch), and $\rho\ell\nu$ feed-across (right hatch).

quite different from the inclusive technique. Furthermore, the detailed information provided by an exclusive measurement allows testing and refinement of the theoretical models on which the determination of $|V_{ub}|$ relies.

CLEO has recently completed and published the exclusive $b \to u\ell\nu$ analysis that was first presented in 1995 [15]. The technique is full reconstruction of five modes ($B^0 \to \pi^\mp\ell^\pm\nu$, $B^0 \to \rho^\mp\ell^\pm\nu$, $B^\pm \to \pi^0\ell^\pm\nu$, $B^\pm \to \rho^0\ell^\pm\nu$, and $B^\pm \to \omega\ell^\pm\nu$), where the neutrino is "detected" by careful measurement of missing momentum and energy. This analysis relies on the usual tools for B reconstruction at the $\Upsilon(4S)$, with B mesons approximately at rest in the lab and the constraints $\Delta E = E_{m\ell\nu} - E_{beam} \simeq 0$ and $M_{m\ell\nu} = \sqrt{E_{beam}^2 - |\vec{p}_{m\ell\nu}|^2} \simeq M_B$. Its great challenge is the reliable determination of the kinematic properties of the neutrino. This is accomplished with selection criteria designed to detect real particles and photons with high efficiency while rejecting spurious ones, and with cuts that eliminate events in which some particle or photon was missed, as indicated by more than one lepton (and hence neutrino), nonzero net charge, or missing mass inconsistent with a neutrino. Under ideal conditions the achieved resolution in the neutrino momentum and direction are 110 MeV (about 5%) and 6°, respectively. The enormous $b \to c\ell\nu$ background is suppressed with momentum cuts, as in the inclusive measurement, but these cuts ($p_\ell > 1.5$ GeV/c for $B \to \pi\ell\nu$ and $p_\ell > 2.0$ GeV/c for $B \to \rho/\omega\ell\nu$) admit most of the signal spectra and do not introduce serious model dependence.

Distributions of the candidate mass $M_{m\ell\nu}$ are shown in Figure 6 for combinations which fall in the ΔE signal region, $-0.15 \leq \Delta E < 0.25$ GeV. Clear signals are apparent above background for both modes. To extract the signals the data are fitted simultaneously in $M_{m\ell\nu}$, ΔE and, for $\rho/\omega\ell\nu$, in $m_{\pi\pi}$ and $m_{3\pi}$. Isospin and quark-symmetry constraints are imposed. From the fit we obtain the yields

TABLE 2. Yields and backgrounds for the five modes, efficiencies found with the ISGW2 model, and fit results. Errors for $\pi^\pm$ and π^0, and for $\rho^\pm$, ρ^0 and ω, are completely correlated.

Title	$\pi^\pm$	π^0	$\rho^\pm$	ρ^0	ω
$\Upsilon(4S)$ Yield	46	19	47	73	7
Bkg.	9.8 ± 2.1	1.5 ± 0.5	9.5 ± 2.1	5.8 ± 1.2	0.3 ± 0.8
Efficiency	0.023	0.015	0.015	0.024	0.006
Signal yield	26.1 ± 6.1	8.6 ± 2.0	19.5 ± 3.3	15.1 ± 2.5	3.5 ± 0.6
$b \to c$	7.0 ± 1.2	2.9 ± 0.8	15.2 ± 1.8	21.5 ± 2.2	4.6 ± 1.1
$b \to u$ BG	0.5 ± 0.1	0.2 ± 0.1	2.7 ± 0.2	2.9 ± 0.2	0.5 ± 0.1
crossfeed	4.1 ± 0.8	1.5 ± 0.3	4.9 ± 0.9	13.4 ± 2.5	0.8 ± 0.2

$N_{\pi\ell\nu}$ and $N_{\rho/\omega\ell\nu}$, and the normalizations for background contributions. The data distributions are very well fitted, with a typical value of χ^2 of 145 for 169 d.o.f. The results are summarized in Table 2.

The $\pi\pi$ and 3π mass distributions are well fitted without nonresonant contributions, which can account for no more than 20% (5%) of the $\rho\ell\nu$ ($\pi\ell\nu$) signal. The yield of $B \to \omega\ell\nu$ is consistent both with no signal and with a signal of the size expected from the $B \to \rho\ell\nu$ result.

The opportunity to study details of exclusive charmless semileptonic decays allows real testing of theoretical models, as well as consistency checks for the interpretation of the signals. The behavior of the $B \to \pi/\rho/\omega\ell\nu$ signal is consistent with theoretical expectations for such distributions as lepton momentum and q^2. The direct measurement of the pseudoscalar/vector ratio can be used to confront theoretical models. Those that are incompatible can be excluded, narrowing the range of theoretical uncertainty in the ultimate value of $|V_{ub}|$.

Averaging over a variety of models that are consistent with the data, we find $\mathcal{B}(B^0 \to \pi^-\ell^+\nu) = (1.8 \pm 0.4 \pm 0.3 \pm 0.2) \times 10^{-4}$, $\mathcal{B}(B^0 \to \rho^-\ell^+\nu) = (2.5 \pm 0.4^{+0.5}_{-0.7} \pm 0.5) \times 10^{-4}$, and $|V_{ub}| = (0.0033 \pm 0.0002^{+0.0003}_{-0.0004} \pm 0.0007)$, where the errors are statistical, experimental systematic, and model-dependent systematic, respectively. For the value of $|V_{ub}|$ there is strikingly good agreement with the inclusive result.

RARE B DECAYS

The essence of the search for new physics is that the *unexpected* is most likely to emerge where the rate for the *expected* is small. The small size of V_{ub} and V_{cb} make B decays an ideal place to pursue this strategy. Rare B decays include exotic processes that give final states that are inaccessible in the Standard Model, processes that vanish at the tree level in the Standard Model but that occur via penguins and box diagrams, and processes that are strongly CKM suppressed.

The search for rare B decays has been a primary objective for CLEO since the beginning, with a number of early papers setting limits on flavor-changing neutral

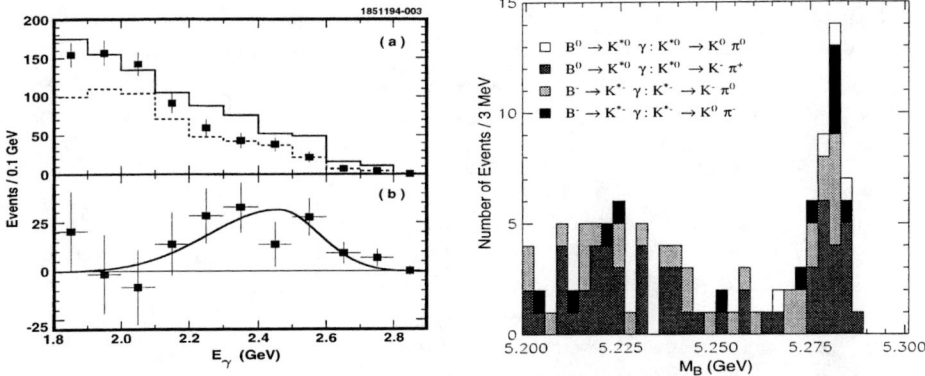

FIGURE 7 Left: CLEO photon energy distribution from the 1995 CLEO II inclusive $b \to s\gamma$ measurement. Right: mass distribution for $B \to K^*\gamma$ candidate events in CLEO II data.

currents, exotic decays and other processes that are still very much of interest. It is only recently that we have begun to find rare B decays, however. In the past few years the focus has been very much on penguins, both the electromagnetic variety exemplified by $b \to s\gamma$, and the hadronic penguins that are the main source of two-body decays such as $B \to K\pi$.

Figure 7 shows one of two inclusive signals for $b \to s\gamma$ from CLEO's 1995 PRL [21], which reported a branching fraction of $\mathcal{B}(b \to s\gamma) = (2.32 \pm 0.57 \pm 0.35) \times 10^{-4}$. Also shown in Figure 7 is the signal for the exclusive decay $B \to K^*\gamma$ [20], updated as of February, 1997. Both of these results have been very influential in constraining possible extensions to the Standard Model. While the inclusive result is somewhat easier to interpret because of the relative absence of hadronic effects, the exclusive approach holds out the hope of an independent measurement of $|V_{td}/V_{ts}|$. Further improvements of both measurements are expected in the near future.

The greatest recent excitement has been in the area of rare hadronic B decays, where results are being finalized using the full 3.3-million $B\bar{B}$ CLEO II data sample collected before the CLEO II.5 upgrade. Clear signals have now been obtained for a number of modes. The ability to explore high mass scales through virtual loops in penguin graphs is a powerful tool in searching for physics beyond the Standard Model. In addition, specific information about branching ratios and indications of the magnitude of so-called "penguin pollution" have obvious ramifications for future studies of CP violation.

In 1993 CLEO reported a statistically significant signal in the combined decay mode $B \to K\pi/\pi\pi$ with a branching fraction consistent with Standard Model expectations [22]. Neither mode showed a significant signal separately, however. That ambiguity has now been resolved with a new set of CLEO II measurements based on 30% more data and enhanced analysis procedures [23]. Like most B-

reconstruction analyses at the $\Upsilon(4S)$ this measurement relies on the beam-energy constraint, giving a mass resolution of ~ 3 MeV/c^2, and uses an energy conservation ΔE cut to reject background and discriminate between $B \to \pi\pi$ and $B \to K\pi$ final states. Further π/K separation is achieved with dE/dx information from the tracking system. There is considerable continuum background for two-body decay modes. This is suppressed with a variety of event-shape inputs, combined for optimal separation into a Fisher discriminant. A likelihood analysis is used to obtain the most probable signal yields for individual modes.

Figure 8 shows the likelihood function results for both neutral and charged B decays into $\pi\pi$ and $K\pi$ final states. Branching fractions are in units of 10^{-5}. As is summarized in Table 3, there are clear signals for several modes, including $B^0 \to K^\pm \pi^\mp$ and $B^\pm \to h^\pm \pi^0$, where h signifies either a π or a K. The absence of significant signals and consequently small branching fractions for the $\pi\pi$ modes are not good news for B-factory measurements of $\sin 2\alpha$.

A number of other rare B modes that have recently been updated are also listed in Table 3. Among these are statistically significant signals for final states with ω and η'. Of particular note is the decay mode $B \to \eta' K$, for which the branching fraction of $7.8^{+2.7}_{-2.2} \pm 1.0 \times 10^{-5}$ is significantly larger than theoretical expectations. Accompanying this is a strong inclusive signal for $B \to \eta' X_s$, where X_s consists of a kaon and up to four pions.

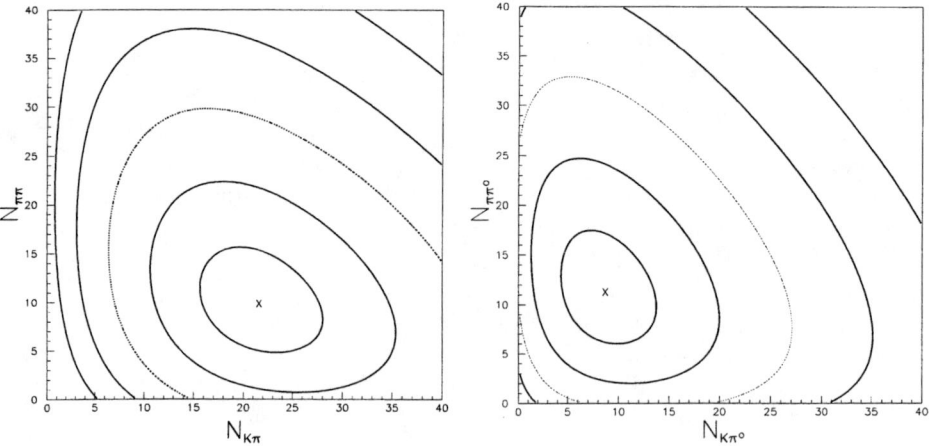

FIGURE 8 Left: contour plots of the likelihood functions for $B^0 \to \pi^\pm \pi^\mp$ vs. $B^0 \to K^\pm \pi^\mp$ and right: contour plots for $B^\pm \to \pi^\pm \pi^0$ vs. $B^\pm \to K^\pm \pi^0$.

TABLE 3 Summary of signal yields, significance and measured branching fraction or upper limit for rare two-body decay modes analyzed with CLEO II data.

Mode	Amplitude	Eff	Yield	Signif	BR (10^{-5})	UL
$K^\pm\pi^\mp$	$-(T'+P')$	44	$21.7^{+6.8}_{-6.0}$	5.6σ	$1.5^{+0.5+0.1}_{-0.4-0.1}\pm 0.1$	
$K^\pm\pi^0$	$-(T'+C'+P')/\sqrt{2}$	37	$8.7^{+5.3}_{-4.2}$	2.7σ		1.6
$K^0\pi^\pm$	P'	12	$9.2^{+4.3}_{-3.8}$	3.2σ	$2.3^{+1.1+0.2}_{-1.0-0.2}\pm 0.2$	4.4
$K^0\pi^0$	$-(C'-P')/\sqrt{2}$	7	$2.3^{+2.2}_{-1.5}$			4.0
$\pi^\pm\pi^\mp$	$-(T+P)$	44	$10.0^{+6.8}_{-6.0}$	2.2σ		1.5
$\pi^\pm\pi^0$	$-(T+C)/\sqrt{2}$	37	$11.3^{+6.3}_{-5.2}$	2.8σ		2.0
$\pi^0\pi^0$	$-(C-P)/\sqrt{2}$	26	$1.2^{+1.7}_{-0.9}$			0.9
$K^\pm K^\mp$	E	44	$0.0^{+1.3}_{-0.0}$	0.0σ		0.4
$K^\pm K^0$	P	12	$0.6^{+3.8}_{-0.6}$	0.2σ		2.1
$K^0\bar{K}^0$	P	2.2	0			1.7
$h^\pm\pi^\mp$		44	$31.7^{+8.4}_{-7.3}$	7.8σ	$2.2^{+0.6}_{-0.5}$	
$h^\pm\pi^0$		37	$20.0^{+6.8}_{-5.9}$	5.5σ	$1.6^{+0.5+0.2}_{-0.5-0.2}\pm 0.1$	
$h^\pm K^0$		12	$9.8^{+4.5}_{-4.0}$	4.4σ	$2.4^{+1.1+0.2}_{-1.0-0.2}\pm 0.2$	
$\omega\pi^\pm$		23	$9.5^{+5.3}_{-4.2}$	2.9σ	$1.2^{+0.7}_{-0.5}\pm 0.2$	
$\omega K^\pm$		20	$8.6^{+4.9}_{-3.9}$	3.3σ	$1.2^{+0.7}_{-0.5}\pm 0.2$	
$\eta' K^\pm$		4.7	$12.0^{+4.1}_{-3.4}$	5.5σ	$7.8^{+2.7}_{-2.2}\pm 1.0$	
$\eta'\pi^\pm$		4.9	$1.4^{+2.0}_{-1.1}$			4.4

Conclusions and Acknowledgments

Someday we may conclude after a talk like this that the cracks in the Standard Model have finally become evident, and that the route to a more profound understanding lay clear before us. Alas, not yet. We have begun to probe the edges of the Standard Model, especially in rare B decays, and the continuing search for processes beyond the Standard Model and precise measurement of parameters through processes like $b \to u\ell\nu$ and CP violation, may lead to the breakthrough we all seek. Perhaps $b30$ will be an even more rewarding occasion than $b20$ has been.

I would like to thank Dan Kaplan and the other organizers of the $b20$ symposium for a wonderfully nostalgic celebration. I would also like to express my gratitude to the CESR accelerator staff for making everything CLEO has done possible, and to the hundreds of current and former CLEO colleagues who have shared this adventure, especially Ed Thorndike and Al Silverman.

REFERENCES

1. R. A. Poling, Ph.D. thesis, Univ. of Rochester, 1981 (unpublished).
2. C. Bebek et al. (CLEO Collaboration), Phys. Rev. Lett. **46**, 84 (1981); K. Chadwick et al. (CLEO Collaboration), Phys. Rev. Lett. **46**, 88 (1981).
3. B. Barish et al. (CLEO Collaboration), Phys. Rev. Lett. **76**, 1570 (1996).
4. R. Wang, Ph.D. thesis, Univ. of Minnesota, 1994 (unpublished).

5. H. Albrecht *et al.* (ARGUS Collaboration), Phys. Lett. **B318**, 397 (1993).
6. A. Anastassov *et al.* (CLEO Collaboration), preprint CLEO CONF 96-8 (1996).
7. B. Barish *et al.* (CLEO Collaboration), Phys. Rev. D **51**, 1014 (1995).
8. R.W. Forty, to appear in *Proceedings of the 2nd International Conference on B Physics and CP Violation (BCONF 97)*, Honolulu, HI, 24-28 Mar 1997.
9. M. Athanas *et al.* (CLEO Collaboration), Phys. Rev. Lett. **79**, 2208 (1997).
10. R.A. Poling, to appear in *Proceedings of the 2nd International Conference on B Physics and CP Violation (BCONF 97)*, Honolulu, HI, 24-28 Mar 1997.
11. R. Fulton *et al.* (CLEO collaboration), Phys. Rev. Lett. **64**, 16 (1990).
12. H. Albrecht *et al.* (ARGUS collaboration), Phys. Lett. **B234**, 409 (1990); H. Albrecht *et al.* (ARGUS collaboration), Phys. Lett. **B255**, 297 (1991).
13. J. Bartelt *et al.* (CLEO collaboration), Phys. Rev. Lett. **71**, 4111 (1993).
14. J. R. Patterson in *Proceedings of the 28th International Conference on High-Energy Physics*, Warsaw, Poland, Z. Ajduk and A.K. Wroblewski, eds., World Scientific (1997).
15. J. P. Alexander *et al.* (CLEO Collaboration), Phys. Rev. Lett. **77**, 5000 (1993).
16. J. Körner and G. Schuler, Z. Phys. **C 38**, 511 (1988).
17. M. Wirbel, B. Stech and M. Bauer, Z. Phys. **C 29**, 637 (1985).
18. G. Altarelli *et al.*, Nucl. Phys. **B 208**, 365 (1982).
19. N. Isgur and D. Scora, Phys. Rev. D **52**, 2783 (1995).
20. R. Ammar *et al.* (CLEO Collaboration), Phys. Rev. Lett. **71**, 674 (1993).
21. M. S. Alam *et al.* (CLEO Collaboration), Phys. Rev. Lett. **74**, 2885 (1995).
22. M. Battle *et al.* (CLEO Collaboration), Phys. Rev. Lett. **71**, 3922 (1993).
23. R. Godang *et al.* (CLEO Collaboration), preprint CLNS-97-1522 (1997).

Beauty and the Beast
Hadronic B Decay and QCD

Klaus Honscheid[1]

Department of Physics
The Ohio State University
Columbus, Ohio 43210

Abstract. We review experimental results on hadronic decays of hadrons containing b quarks. The theoretical implications of these results are also considered.

I INTRODUCTION

B meson decays occur primarily through the CKM favored $b \to c$ transition. In such decays the dominant weak decay diagram is the spectator diagram where the virtual W^- materializes into either a lepton and an anti-neutrino or a $\bar{u}d$ or $\bar{c}s$ quark pair. In hadronic decays, the quark pair forms one of the final state hadrons while the c quark pairs with the spectator anti-quark to form the other hadron. In semileptonic decays, on the other hand, the c quark and spectator antiquark hadronize independently of the leptonic current.

The extraction of Standard Model parameters from experimental results is complicated by the fact that only B hadrons can be studied and not isolated b quarks. The light quarks and the gluons surrounding the b quark in the B meson lead to significant corrections that have to be taken into account. Since leptons do not interact strongly, semileptonic B meson decays are less affected by these QCD corrections and the theoretical calculations are believed to be more reliable.

Turning the argument around, hadronic decays of B mesons provide an ideal environment to test our understanding of perturbative and non-perturbative QCD, of hadronization, and of Final State Interaction effects. After reviewing the experimental situation we present several examples of the intersection of experiment and theory.

---

[1] This work is sponsored in part by the U.S. Department of Energy and the Alfred P. Sloan Foundation.

II THE EXPERIMENTAL STUDY OF B DECAY

Experimental b physics began in 1977 when the CFS collaboration at Fermilab observed a narrow resonance at an energy of about 9.5 GeV in the reaction $p + \text{nucleus} \rightarrow \mu^+\mu^- + X$ [1]. This resonance was named $\Upsilon(9460)$ and was subsequently identified as the 1^3S_1 state of the $b\bar{b}$ or bottomonium system. A second resonance at a mass near 10.0 GeV was isolated in the Fermilab data and later identified as a radial excitation of the $b\bar{b}$ state. For almost two decades following the discovery of the Υ resonance, this was the last significant contribution by a hadron machine to B physics. Within a year of its discovery, the Υ resonance had been confirmed by experiments at DORIS and at CESR. Most of our current knowledge of B mesons is based on analyses of data collected at these two e^+e^- storage rings [4]. In recent years, advances in detector technology, in particular the introduction of high resolution silicon vertex detectors, have allowed experiments at high energy colliders (*i.e.* LEP, SLC and the TEVATRON) to reconstruct b quarks efficiently. This has led to many exciting results on $B^0\bar{B}^0$ oscillations, precise lifetime measurements, and the discovery of new b-flavored hadrons.

TABLE 1. Evolution of $\mathcal{B}(B^- \rightarrow D^0\pi^-)$.

Year	Experiment	Data Sample [pb^{-1}]	No. of Events	Branching Ratio
1983	CLEO	40.2	2	$4.2 \pm 4.2\%$
1990	ARGUS	≈ 150	12	$0.2 \pm 0.08 \pm 0.06\%$
1994	CLEO II	≈ 890	290	$0.55 \pm 0.05 \pm 0.06\%$
1997	CLEO II	≈ 3100	≈ 1000	$0.473 \pm 0.025 \pm 0.040\%$

For hadronic B decay and multi-body final states the branching ratio measurements are dominated by results from the CLEO collaboration. After more than 10 years of experimental work, we now have a set of data precise enough to challenge theoretical models. As an example, Table 1 shows how the measurement of $B^- \rightarrow D^0\pi^-$ improved over the last 15 years. Recent CLEO results on exclusive low multiplicity hadronic B decays are shown in Fig. 1.

The current world averages for exclusive B^- and $\bar{B}^0$ decays are listed in Table 2. Adding the individual channels we find that about 19% of the B^- total width and about 12% of the $\bar{B}^0$ total width have been exclusively reconstructed in hadronic decays.

In our brief survey of the experimental situation in B physics it is important to note that for many of the decay channels the error is dominated by uncertainties in the branching fraction of the underlying charm decay.

Experimental techniques have been steadily improved over the past years. This summer CLEO has presented several analyses based on partial reconstruction. In this technique, D^* mesons are not fully reconstructed but rather tagged by the presence of the characteristic slow pion from the $D^* \rightarrow D^0\pi$

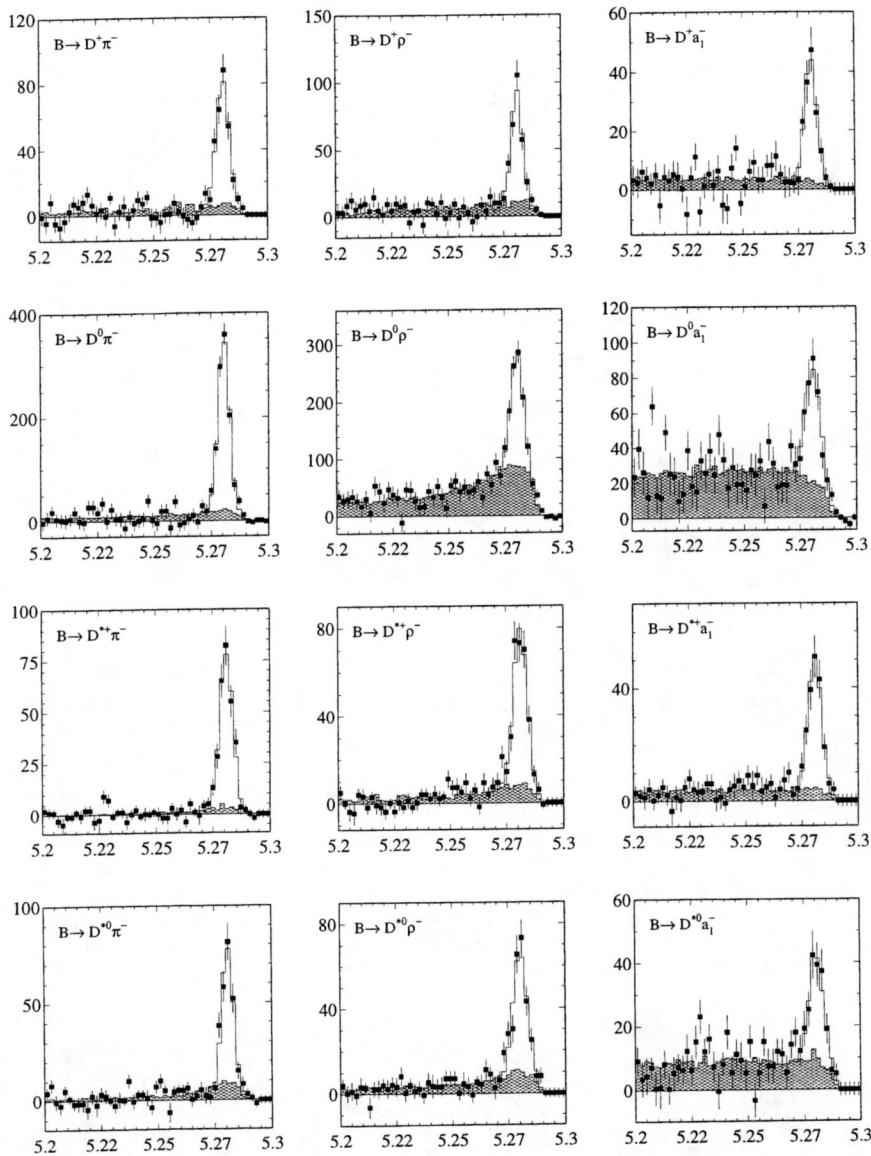

FIGURE 1. Continuum-subtracted beam constrained mass distributions for twelve B modes as seen by the CLEO experiment. The hatched histograms show the background spectrum from other $B\bar{B}$ decays.

TABLE 2. World average B^- and $\bar{B}^0$ branching fractions [%].

B^- Mode	Branching Fraction	$\bar{B}^0$ Mode	Branching Fraction
$B^- \to D^0\pi^-$	$0.46 \pm 0.04 \pm 0.01$	$\bar{B}^0 \to D^+\pi^-$	$0.26 \pm 0.03 \pm 0.02$
$B^- \to D^0\rho^-$	$0.95 \pm 0.12 \pm 0.03$	$\bar{B}^0 \to D^+\rho^-$	$0.79 \pm 0.11 \pm 0.06$
$B^- \to D^0\pi^+\pi^-\pi^-$	$1.25 \pm 0.35 \pm 0.04$	$\bar{B}^0 \to D^+\pi^-\pi^-\pi^+$	$0.81 \pm 0.23 \pm 0.06$
$B^- \to D^{*0}\pi^-$	$0.43 \pm 0.04 \pm 0.01$	$\bar{B}^0 \to D^{*+}\pi^-$	$0.27 \pm 0.02 \pm 0.01$
$B^- \to D^{*0}\rho^-$	$1.27 \pm 0.18 \pm 0.04$	$\bar{B}^0 \to D^{*+}\rho^-$	$0.74 \pm 0.10 \pm 0.02$
$B^- \to D_J^{(*)0}\pi^-$	$0.13 \pm 0.05 \pm 0.00$		
$B^- \to D^{*+}\pi^-\pi^-\pi^0$	$1.65 \pm 0.73 \pm 0.05$		
$B^- \to D_J^{(*)0}\rho^-$	$0.32 \pm 0.21 \pm 0.01$		
$B^- \to D^{*0}\pi^-\pi^-\pi^+$	$0.93 \pm 0.26 \pm 0.03$	$\bar{B}^0 \to D^{*+}\pi^-\pi^-\pi^+$	$0.78 \pm 0.13 \pm 0.02$
$B^- \to D^0 a_1^-$	$0.90 \pm 0.16 \pm 0.03$	$\bar{B}^0 \to D^+ a_1^-$	$0.83 \pm 0.14 \pm 0.06$
$B^- \to D^{*0} a_1^-$	$1.62 \pm 0.35 \pm 0.05$	$\bar{B}^0 \to D^{*+} a_1^-$	$1.17 \pm 0.20 \pm 0.03$
$B^- \to D^0\bar{D}^0 K^-$	< 0.51 (90% C.L.)&		
$B^- \to D^0\bar{D}^{*0} K^-$	< 0.82 (90% C.L.)		
$B^- \to D^{*+}\bar{D}^{*-} K^-$	< 0.92 (90% C.L.)		
$B^- \to D^{*0}\bar{D}^0 K^-$	0.554 ± 0.322	$\bar{B}^0 \to D^{*+}\bar{D}^0 K^-$	0.462 ± 0.240
$B^- \to D^{*0}\bar{D}^{*0} K^-$	1.488 ± 0.789	$\bar{B}^0 \to D^{*+}\bar{D}^{*0} K^-$	1.334 ± 0.629
$B^- \to D^+\pi^-\pi^-$	< 0.14 (90% C.L.)	$\bar{B}^0 \to D^0\pi^+\pi^-$	< 0.16 (90% C.L.)
$B^- \to D^{*+}\pi^-\pi^-$	$0.20 \pm 0.07 \pm 0.01$		
$B^- \to D^{**0}(2420)\pi^-$	$0.15 \pm 0.05 \pm 0.00$		
$B^- \to D^{**0}(2420)\rho^-$	< 0.14 (90% C.L.)		
$B^- \to D^{**0}(2460)\pi^-$	< 0.13 (90% C.L.)	$\bar{B}^0 \to D^{**+}(2460)\pi^-$	< 0.21 (90% C.L.)
$B^- \to D^{**0}(2460)\rho^-$	< 0.46 (90% C.L.)	$\bar{B}^0 \to D^{**+}(2460)\rho^-$	< 0.48 (90% C.L.)
$B^- \to D^0 D_s^-$	$1.32 \pm 0.28 \pm 0.32$	$\bar{B}^0 \to D^+ D_s^-$	$0.72 \pm 0.21 \pm 0.18$
$B^- \to D^0 D_s^{*-}$	$0.92 \pm 0.31 \pm 0.22$	$\bar{B}^0 \to D^+ D_s^{*-}$	$1.11 \pm 0.41 \pm 0.27$
$B^- \to D^{*0} D_s^-$	$1.15 \pm 0.35 \pm 0.28$	$\bar{B}^0 \to D^{*+} D_s^-$	$0.92 \pm 0.23 \pm 0.22$
$B^- \to D^{*0} D_s^{*-}$	$2.63 \pm 0.79 \pm 0.64$	$\bar{B}^0 \to D^{*+} D_s^{*-}$	$1.95 \pm 0.52 \pm 0.48$
$B^- \to \psi K^-$	0.100 ± 0.010	$\bar{B}^0 \to \psi K^0$	0.084 ± 0.013
$B^- \to \psi' K^-$	0.070 ± 0.024	$\bar{B}^0 \to \psi' K^0$	< 0.08 (90% C.L.)
$B^- \to \psi K^{*-}$	0.144 ± 0.031	$\bar{B}^0 \to \psi \bar{K}^{*0}$	0.131 ± 0.021
$B^- \to \psi' K^{*-}$	< 0.30 (90% C.L.)	$\bar{B}^0 \to \psi' \bar{K}^{*0}$	0.151 ± 0.091
		$\bar{B}^0 \to \psi K^-\pi^+$	0.117 ± 0.058
$B^- \to \psi K^-\pi^+\pi^-$	0.140 ± 0.077		
		$\bar{B}^0 \to \psi' K^-\pi^+$	< 0.11 (90% C.L.)
$B^- \to \psi' K^-\pi^+\pi^-$	0.207 ± 0.127		
$B^- \to \chi_{c1} K^-$	0.104 ± 0.040	$\bar{B}^0 \to \chi_{c1} K^0$	< 0.27 (90% C.L.)
$B^- \to \chi_{c1} K^{*-}$	< 0.21 (90% C.L.)	$\bar{B}^0 \to \chi_{c1} \bar{K}^{*0}$	< 0.21 (90% C.L.)
$B^- \to \psi\pi^-$	0.0057 ± 0.0026	$\bar{B}^0 \to \psi\pi^0$	< 0.006 (90% C.L.)
$B^- \to \psi\rho^-$	< 0.077 (90% C.L.)	$\bar{B}^0 \to \psi\rho^0$	< 0.025 (90% C.L.)
		$\bar{B}^0 \to \psi\omega^0$	< 0.027 (90% C.L.)
$B^- \to \psi a_1^-$	< 0.120 (90% C.L.)		

transition. CLEO finds

$$\mathcal{B}(\bar{B}^0 \to D^{*+}\pi^-) = (2.81 \pm 0.11 \pm 0.21 \pm 0.05) \times 10^{-3}$$

$$\mathcal{B}(B^- \to D^{*0}\pi^-) = (4.81 \pm 0.42 \pm 0.40 \pm 0.21) \times 10^{-3}$$

$$\mathcal{B}(B^- \to D_1(2420)\pi^-) = (1.17 \pm 0.24 \pm 0.16 \pm 0.03) \times 10^{-3}$$

$$\mathcal{B}(B^- \to D_2^*(2460)\pi^-) = (2.1 \pm 0.8 \pm 0.3 \pm 0.05) \times 10^{-3}$$

The results are still preliminary. Equipped with this comprehensive set of experimental results we can now confront theoretical models of B meson decay.

III   HADRONIC B DECAY

In hadronic decays of B mesons the underlying weak decay of the b quark is overshadowed by strong interaction effects caused by the surrounding cloud of light quarks and gluons. Since leptons don't interact strongly, these QCD effects are reduced in semileptonic decays. Therefore by comparing suitable hadronic and semileptonic B decays, we can attempt to isolate the hadronic corrections and learn something about the strong interaction. In hadronic decays, the simple (external) spectator diagram is modified by hard gluon exchanges between the initial and final quark lines leading to "color-suppressed" decays which have a different set of quark pairings in the final state. Observation of $B \to \psi X_s$ decays, where X_s is a strange meson, gives experimental evidence for the presence of this diagram. Further information on the size of the color-suppressed contribution can be obtained from $\bar{B}^0 \to D^0$ (or D^{*0})X^0 transitions, where X^0 is a neutral meson. In B^- decays, both internal and external spectator diagrams are present and the amplitudes interfere. By comparing the rates for B^- and $\bar{B}^0$ decays, the size and the relative sign of the color-suppressed amplitude can be determined.

A   Factorization

It was suggested that in analogy to semileptonic decays, two body hadronic decays of B mesons can be expressed as the product of two independent hadronic currents, one describing the formation of a charm meson and the other the hadronization of the remaining $\bar{u}d$ (or $\bar{c}s$) system from the virtual W^-. Qualitatively, for a B decay with a large energy release, the $\bar{u}d$ pair, which is produced as a color singlet, travels fast enough to leave the interaction region without influencing the second hadron formed from the c quark and the spectator anti-quark. The assumption that the amplitude can be expressed as the product of two hadronic currents is called "factorization" in this paper. It is expected that the simple approximation of the strong interaction effects by the factorization hypothesis will be more reliable in B meson decays than in the equivalent D meson decays due to the larger characteristic energy transfers and the consequent suppression of final state interactions.

There are several phenomenological models of the nonleptonic two-body decays of heavy flavors [2]. The model of Bauer, Stech, and Wirbel (BSW) is widely used [3]. In addition to factorization, the BSW model uses hadronic currents instead of quark currents and allows the coefficients a_1, a_2 of the products of currents to be free parameters determined by experimental data. The effective Hamiltonian becomes

$$H = \frac{G_F}{\sqrt{2}} V_{cs}^* V_{ud} [a_1 (\bar{u}d)_H (\bar{s}c)_H + a_2 (\bar{s}d)_H (\bar{u}c)_H] \tag{1}$$

The relation between a_1, a_2 and the QCD coefficients c_1, c_2 is:

$$a_1 = c_1 + \xi c_2 \tag{2}$$
$$a_2 = c_2 + \xi c_1$$

where the factor $\xi (= 1/N_c)$ is the color matching factor.

1 Factorization Tests

The factorization hypothesis can be tested by comparing hadronic exclusive decays to the corresponding semileptonic modes. These tests can be performed for exclusive hadronic decays of B mesons.

As an example, we consider the specific case of $\bar{B}^0 \to D^{*+}\pi^-$. The amplitude for this reaction is

$$A = \frac{G_F}{\sqrt{2}} V_{cb} V_{ud}^* a_1 \langle \pi^- | (\bar{d}u)|0\rangle \langle D^{*+}|(\bar{c}b)|\bar{B}^0\rangle. \tag{3}$$

The CKM factor $|V_{ud}|$ arises from the $W^- \to \bar{u}d$ vertex. The first hadron current that creates the π^- from the vacuum is related to the pion decay constant, f_π, by

$$\langle \pi^-(p)|(\bar{d}u)|0\rangle = -if_\pi p_\mu. \tag{4}$$

The other hadron current can be determined from the semileptonic decay $\bar{B}^0 \to D^{*+}\ell^-\bar{\nu}_\ell$. Here the amplitude is the product of a lepton current and the hadron current that we seek to insert in Equation (3).

Factorization can be tested experimentally by verifying that the relation

$$\frac{\Gamma\left(\bar{B}^0 \to D^{*+}\pi^-\right)}{\frac{d\Gamma}{dq^2}\left(\bar{B}^0 \to D^{*+}\ell^-\bar{\nu}_\ell\right)\Big|_{q^2=m_\pi^2}} = 6\pi^2 a_1^2 f_\pi^2 |V_{ud}|^2, \tag{5}$$

TABLE 3. Ingredients for Factorization Tests.

f_π	131.74 ± 0.15 MeV
f_ρ	215 ± 4 MeV
f_{a_1}	205 ± 16 MeV
V_{ud}	0.9744 ± 0.0010
$\|a_1\|$	1.12 ± 0.1
$\frac{dB}{dq^2}(B \to D^+ \ell \, \nu)\|_{q^2=m_\pi^2}$	$(0.35 \pm 0.06) \times 10^{-2}$ GeV^{-2}
$\frac{dB}{dq^2}(B \to D^+ \ell \, \nu)\|_{q^2=m_\rho^2}$	$(0.33 \pm 0.06) \times 10^{-2}$ GeV^{-2}
$\frac{dB}{dq^2}(B \to D^* \ell \, \nu)\|_{q^2=m_\pi^2}$	$(0.237 \pm 0.026) \times 10^{-2}$ GeV^{-2}
$\frac{dB}{dq^2}(B \to D^* \ell \, \nu)\|_{q^2=m_\rho^2}$	$(0.250 \pm 0.030) \times 10^{-2}$ GeV^{-2}
$\frac{dB}{dq^2}(B \to D^* \ell \, \nu)\|_{q^2=m_{a_1}^2}$	$(0.335 \pm 0.033) \times 10^{-2}$ GeV^{-2}
$\frac{dB}{dq^2}(B \to D^* \ell \, \nu)\|_{q^2=m_{D_s}^2}$	$(0.483 \pm 0.033) \times 10^{-2}$ GeV^{-2}
$\frac{dB}{dq^2}(B \to D^* \ell \, \nu)\|_{q^2=m_{D_s^*}^2}$	$(0.507 \pm 0.035) \times 10^{-2}$ GeV^{-2}

TABLE 4. Test of factorization by comparing hadronic and semileptonic decay rates.

	$R_{\text{Exp}}(B \to D^*)$ [GeV^2]	$R_{\text{Exp}}(B \to D)$ [GeV^2]	R_{Theo} [GeV^2]
$\bar{B}^0 \to D^{(*)+}\pi^-$	1.18 ± 0.21	0.94 ± 0.30	1.22 ± 0.15
$\bar{B}^0 \to D^{(*)+}\rho^-$	2.92 ± 0.70	2.63 ± 0.88	3.26 ± 0.42
$\bar{B}^0 \to D^{(*)+}a_1^-$	3.8 ± 1.0		3.0 ± 0.5

is satisfied. Here q^2 is the four-momentum transfer from the B meson to the D^* meson. Since q^2 is also the mass of the lepton-neutrino system, by setting $q^2 = m_\pi^2 = 0.019 \, GeV^2$, we are requiring that the lepton-neutrino system has the same kinematic properties as does the pion in the hadronic decay. There is some uncertainty on which value to use for a_1. For our tests of the factorization hypothesis we will assume $a_1 = 1.12 \pm 0.1$ which corresponds to the value of the Wilson coefficient c_1 evaluated at a scale $\mu = m_b$ [5]. The error in a_1 reflects the uncertainty in the mass scale at which the coefficient a_1 should be evaluated. In the following discussion we will denote the left hand side of Equation (5) by R_{Exp} and the right hand side by R_{Theo}.

The factorization tests can be applied to several B decays using the modes $\bar{B}^0 \to D^{*+}X^-$ or $\bar{B}^0 \to D^+ X^-$ decays, with $X = \pi^-$, ρ^- or a_1^-.

To obtain numerical predictions for R_{Theo}, we must interpolate the observed differential q^2 distribution [7] for $B \to D^* \ell \, \nu$ to $q^2 = m_\pi^2$, m_ρ^2, and $m_{a_1}^2$, respectively. Until this distribution is measured more precisely we have to use theoretical models to perform this interpolation. The differences between the extrapolations using models for $B \to D^* \ell \, \nu$ are small, on the order of 10-20%. The measurement of this differential distribution published by CLEO II can be combined with the earlier results from the ARGUS and CLEO 1.5 experiments [6,8]. The values of $d\Gamma/dq^2(B \to D^*\ell\nu)$ used for the factorization test are given in Table 3. The results of this test are listed in Table 4.

More subtle tests of the factorization hypothesis can be performed by examining the polarization in B (or D) meson decays into two vector mesons [9]. Again, the underlying principle is to compare the hadronic decays to the appropriate semileptonic decays evaluated at a fixed value in q^2. For instance, the ratio of longitudinal to transverse polarization (Γ_L/Γ_T) in $\bar{B}^0 \to D^{*+}\rho^-$ should be equal to the corresponding ratio for $B \to D^*\ell\nu$ evaluated at $q^2 = m_\rho^2 = 0.6$ GeV^2,

$$\frac{\Gamma_L}{\Gamma_T}(\bar{B}^0 \to D^{*+}\rho^-) = \frac{\Gamma_L}{\Gamma_T}(B \to D^*\ell\nu)|_{q^2=m_\rho^2} \tag{6}$$

The advantage of this method is that it is not affected by QCD corrections [10].

For $B \to D^*\ell\nu$ decay, longitudinal polarization dominates at low q^2, whereas near $q^2 = q^2_{max}$ transverse polarization dominates. There is a simple physical argument for the behavior of the form factors near these two kinematic limits. Near $q^2 = q^2_{max}$, the D^* is almost at rest and its small velocity is uncorrelated with the D^* spin, so all three D^* helicities are equally likely and we expect $\Gamma_T/\Gamma_L = 2$. At $q^2 = 0$, the D^* has the maximum possible momentum, while the lepton and neutrino are collinear and travel in the direction opposite to the D^*. The lepton and neutrino helicities are aligned to give $S_z = 0$, so near $q^2 = 0$ longitudinal polarization is dominant.

For $\bar{B}^0 \to D^{*+}\rho^-$, we expect 88% longitudinal polarization from the argument described above [15]. Similar results have been obtained by Neubert [16], Rieckert [17], and Kramer et al. [21]. Using the measured q^2 distribution for $B \to D^*\ell\nu$ Neubert [16] calculates the transverse and longitudinal polarization in $B \to D^*\ell\nu$ decays. Using his result, we find Γ_L/Γ to be 85% at $q^2 = m_\rho^2 = 0.6$. The agreement between these predictions and the experimental result

$$\Gamma_L/\Gamma(\bar{B}^0 \to D^{*+}\rho^-) = 93 \pm 5 \pm 5\% \tag{7}$$

supports the factorization hypothesis in hadronic B meson decay for q^2 values up to m_ρ^2.

2 Factorization and Internal Spectator Decays

At the present level of precision, there is good agreement between the experimental results and the expectation from factorization for hadronic B decays in the q^2 range $0 < q^2 < m_{a_1}^2$. Note that it is possible that factorization will be a poorer approximation for decays with smaller energy release or larger q^2. For internal spectator decays the validity of the factorization hypothesis is also questionable and requires experimental verification. The naive color transparency argument used in the previous sections is not applicable to decays such as $B \to \psi K$, and there is no corresponding semileptonic decay to compare to. For internal spectator decays one can only compare experimental observables to quantities predicted by models based on factorization. Two such quantities are the production ratio

$$\mathcal{R} = \frac{\mathcal{B}(B \to \psi K^*)}{\mathcal{B}(B \to \psi K)}$$

and the amount of longitudinal polarization Γ_L/Γ in $B \to \psi K^*$ decays. Previous experimental results, $\mathcal{R} = 1.68 \pm 0.33$ and $\Gamma_L/\Gamma = 0.78 \pm 0.04$, were inconsistent with model predictions. The theory had difficulties in accommodating a large longitudinal polarization and a large vector-to-pseudoscalar production ratio. Non-factorizable contributions that reduce the transverse amplitude were proposed to remedy the situation. New experimental results, however, make this apparent breakdown of the factorization hypothesis less likely. The CLEO collaboration published new data on $B \to$ charmonium transitions. Their values,

$$\mathcal{R} = 1.45 \pm 0.20 \pm 0.17 \;,\; \Gamma_L/\Gamma = 0.52 \pm 0.07 \pm 0.04 \;,$$

are perfectly consistent with factorization-based models.

B Aside: CP Engineering

Only very few of the B decay modes that are expected to be useful for CP violation studies at future B factories have actually been observed experimentally. The most prominent mode, of course, is $B^0 \to \psi K_s$. Using the entire run II data sample of 110 pb^{-1}, CDF has 239±22 events in this decay channel. For the first time, CLEO observed a signal of 17.2 ± 5.4 events in the corresponding K_L mode, $B^0 \to \psi K_L$ (Figure 2).

The decay $B^0 \to J/\psi K^{*0}, K^{*0} \to K_S^0 \pi^0$ is a CP eigenstate if the orbital angular momenta present in the B decay are only even or only odd. The CP-even component has a structure similar to that of the CP eigenstate $J/\psi K_S^0$. A purely CP-even decay is thus suitable for measuring $\sin 2\beta$, where β is one of the angles in the unitarity triangle; the same holds for purely CP-odd decays. Since S, P, and D waves may be present, the final state could be an admixture of CP-even and CP-odd eigenstates, which would greatly reduce the sensitivity of the measurement. It is expected that the CP-odd component is small. To answer this question the polarization in $B \to \psi K^*$ decays has to be measured and CLEO has presented the first full angular analysis of this decay mode [12] [11] [13].

Instead of using the traditional helicity basis to describe the angular distribution it turned out to be advantageous to switch to a different basis that is more convenient for the extraction of the parity-odd and parity-even components [14]. In this basis, called the transversity basis, the angular distribution for $B \to \psi K^*$ decays takes the following form:

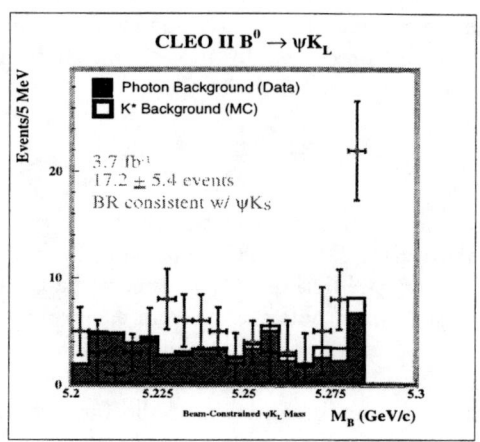

FIGURE 2. First observation of $B^0 \to \psi K_L$ decays by CLEO.

$$\frac{1}{\Gamma} \frac{d^3 \Gamma}{d\cos\theta_{tr}\, d\cos\theta_{K^*}\, d\phi_{tr}} = \frac{9}{32\pi} \{ 2\cos^2\theta_{K^*}(1 - \sin^2\theta_{tr}\cos^2\phi_{tr}) |A_0|^2$$
$$+ \sin^2\theta_{K^*}(1 - \sin^2\theta_{tr}\sin^2\phi_{tr}) |A_\||^2$$
$$+ \sin^2\theta_{K^*}\sin^2\theta_{tr}\sin^2\phi_{tr} |A_\perp|^2$$
$$- \sin^2\theta_{K^*}\sin 2\theta_{tr}\sin\phi_{tr}\, \mathrm{Im}(A_\|^* A_\perp)$$
$$+ \frac{1}{\sqrt{2}}\sin 2\theta_{K^*}\sin^2\theta_{tr}\sin 2\phi_{tr}\, \mathrm{Re}(A_0^* A_\|)$$
$$+ \frac{1}{\sqrt{2}}\sin 2\theta_{K^*}\sin 2\theta_{tr}\cos\phi_{tr}\, \mathrm{Im}(A_0^* A_\perp) \}. \quad (8)$$

The decay amplitudes $A_{0,\|,\perp}$ are related to the helicity amplitudes H_λ by

$$H_\pm = \frac{1}{\sqrt{2}}(A_\| \pm A_\perp), \qquad H_0 = -A_0. \quad (9)$$

Thus, for the $\bar{B}$ decays, $A_\perp$ has to change sign. This can be accomplished by replacing ϕ_{tr} by $\phi_{tr} + \pi$. The relations between the angles in the transversity basis and the helicity basis are

$$\cos\theta_\psi = -\sin\theta_{tr}\cos\phi_{tr},$$
$$\sin\theta_\psi \sin\chi = \cos\theta_{tr},$$
$$\sin\theta_\psi \cos\chi = -\sin\theta_{tr}\sin\phi_{tr}. \quad (10)$$

The transversity basis has some nice properties. $|A_0|^2 + |A_\||^2$ describes the parity-even (S-wave plus D-wave) and $|A_\perp|^2$ the parity-odd (P-wave) components. Integrating over ϕ_{tr} and $\cos\theta_{K^*}$ yields

$$\frac{1}{\Gamma}\frac{d\Gamma}{d\cos\theta_{tr}} = \frac{3}{8}(1-|A_\perp|^2)(1+\cos^2\theta_{tr}) + \frac{3}{4}|A_\perp|^2\sin^2\theta_{tr}. \qquad (11)$$

The parity-odd component can then be extracted from a one parameter fit.

CLEO performs an unbinned maximum likelihood fit to this expression and extracts the parameters of the angular distribution as well as the branching fractions. Averaged over all ψK^* modes, they find for the fractions of longitudinal polarization and parity-odd transverse polarization

$$\Gamma_L/\Gamma = |A_0|^2 = 0.52 \pm 0.07 \pm 0.04 \qquad (12)$$
$$|P|^2 = |A_\perp|^2 = 0.16 \pm 0.08 \pm 0.04. \qquad (13)$$

Setting $\phi(A_0) \equiv 0$, the phases of the other decay amplitudes are

$$\phi(A_\perp) = -0.11 \pm 0.46 \pm 0.03 \qquad (14)$$
$$\phi(A_\parallel) = 3.00 \pm 0.37 \pm 0.04. \qquad (15)$$

The small value for the parity-odd component indicates that the decay $B^0 \to \psi K^{*0}$ followed by $K^{*0} \to K^0_S \pi^0$ will be useful for CP violation studies at asymmetric B factories and for obtaining a measurement of the unitarity angle β. The phases of the decay amplitudes are measured to be close to zero or π, giving no evidence for strong final state interactions.

The CLEO collaboration has searched for other B decays to CP eigenstates and presented the following preliminary upper limits(90% C.L.) for decays to two D mesons.

$$\mathcal{B}(B^0 \to D^{*+}D^{*-}) < 2.2 \times 10^{-3}$$
$$\mathcal{B}(B^0 \to D^{*+}D^-) < 1.8 \times 10^{-3}$$
$$\mathcal{B}(B^0 \to D^+D^-) < 1.2 \times 10^{-3}$$

C Color-Suppression in B Decay

In the decays of charm mesons, the effect of color suppression is obscured by the effects of FSI or reduced by nonfactorizable effects. Because of the larger mass of the b quark, a more consistent pattern of color-suppression is expected in the B system, and current experimental results seem to support that color-suppression is operative in hadronic decays of B mesons. Besides $B \to$ charmonium transitions no other color-suppressed decay has been observed experimentally. Using a data sample of 3.1 fb^{-1} and an improved analysis with better background handling CLEO has placed stringent upper limits on B^0 decay modes with $D^{0(*)}$ and neutral mesons in the final state. The preliminary results shown in Table 5 demonstrate that color-suppression is at work in B decay.

TABLE 5. Preliminary CLEO II results on color-suppressed B decays.

Decay Mode	Upper Limit (90% C.L.) [%]	BSW Model [%]
$\mathcal{B}(\bar{B}^0 \to D^0\pi^0)$	< 0.012	0.012
$\mathcal{B}(\bar{B}^0 \to D^0\rho^0)$	< 0.040	0.008
$\mathcal{B}(\bar{B}^0 \to D^0\eta)$	< 0.068	0.006
$\mathcal{B}(\bar{B}^0 \to D^0\eta')$	< 0.086	0.002
$\mathcal{B}(\bar{B}^0 \to D^0\omega)$	< 0.056	0.008
$\mathcal{B}(\bar{B}^0 \to D^{*0}\pi^0)$	< 0.044	0.012
$\mathcal{B}(\bar{B}^0 \to D^{*0}\rho^0)$	< 0.060	0.013
$\mathcal{B}(\bar{B}^0 \to D^{*0}\eta)$	< 0.069	0.007
$\mathcal{B}(\bar{B}^0 \to D^{*0}\eta')$	< 0.027	0.002
$\mathcal{B}(\bar{B}^0 \to D^{*0}\omega)$	< 0.080	0.013

D The Ratio of Internal and External Spectator Decay Amplitudes

In the BSW model [18], the branching fractions of the $\bar{B}^0$ normalization modes are proportional to a_1^2 while the branching fractions of the $B \to \psi$ decay modes depend only on a_2^2. In decays of charged B mesons to open charm

TABLE 6. Ratios of branching fractions of charged and neutral B mesons.

Mode	Measured Ratio
$B^- \to D^0\pi^- / \bar{B}^0 \to D^+\pi^-$	1.756 ± 0.275
$B^- \to D^0\rho^- / \bar{B}^0 \to D^+\rho^-$	1.206 ± 0.251
$B^- \to D^0 a_1^- / \bar{B}^0 \to D^+ a_1^-$	1.082 ± 0.284
$B^- \to D^{*0}\pi^- / \bar{B}^0 \to D^{*+}\pi^-$	1.586 ± 0.179
$B^- \to D^{*0}\rho^- / \bar{B}^0 \to D^{*+}\rho^-$	1.707 ± 0.331
$B^- \to D^{*0} a_1^- / \bar{B}^0 \to D^{*+} a_1^-$	1.383 ± 0.382

(*i.e.* not to charmonium states) both amplitudes are present and interfere. By comparing measured branching fractions for charged and neutral B mesons, the relative sign of the two amplitudes can be extracted. A fit to the six ratios listed in Table 6 and using the model of Neubert *et al.* yields

$$|a_1| = 1.03 \pm 0.027 \pm 0.06 , \qquad (16)$$

and a fit to the modes with ψ mesons in the final state gives

$$|a_2| = 0.23 \pm 0.01 \pm 0.01 , \qquad (17)$$

The first error on $|a_1|$ and $|a_2|$ includes the uncertainties from the charm or charmonium branching ratios, the experimental systematics associated with detection efficiencies and background subtractions as well as the statistical errors from the branching ratios. The second error quoted is the uncertainty due to the B meson production fractions and lifetimes. We have assumed that the ratio of B^+B^- and $B^0\bar{B}^0$ production at the $\Upsilon(4S)$ is one [19], and have assigned an uncertainty of 10% to it.

The magnitude of the amplitude for external spectator processes, $|a_1|$ can also be determined from $B \to D^{(*)}D_s^{(*)}$ decays. Since these transitions are not subject to interference with the internal spectator amplitude, we can combine B^- and $\bar{B}^0$ decays to reduce the statistical error. We obtain

$$|a_1|_{DD_s} = 0.98 \pm 0.06 \pm 0.04 \tag{18}$$

It is interesting to note that this value of $|a_1|$ agrees with the result of the fit to the $B \to D^{(*)}\pi$ and $B \to D^{(*)}\rho$ modes (see Equation 16). In general, $|a_1|$ could be different for exclusive $b \to c\bar{u}d$ and $b \to c\bar{c}s$ processes.

By comparing branching ratios of B^- and $\bar{B}^0$ decay modes it is possible to determine the sign of a_2 relative to a_1. The experimental results listed in Table 6 are all greater than 1 indicating that for all the decay channels used in this study the interference is constructive.

A least squares fit using the results from Table 6 and the model by Neubert et al. gives

$$a_2/a_1 = 0.22 \pm 0.04 \pm 0.06 \,, \tag{19}$$

where we have ignored uncertainties in the theoretical predictions. The second error is due to the uncertainty in the B meson production fractions(f_+, f_0) and lifetimes (τ_+, τ_0) that enter into the determination of a_1/a_2 in the combination $(f_+\tau_+/f_0\tau_0)$. As this ratio increases, the value of a_2/a_1 decreases. The allowed range of $(f_+\tau_+/f_0\tau_0)$ excludes a negative value of a_2/a_1. Other uncertainties in the magnitude [20] of f_D, f_{D^*} and in the hadronic form factors can change the magnitude of a_2/a_1 but not its sign.

The magnitude of a_2 determined from this fit to the ratio of B^- and B^0 modes is consistent with the value of a_2 determined from the fit to the $B \to \psi$ decay modes.

The observation that the coefficients a_1 and a_2 have the same relative sign in B^+ decay came as a surprise, since destructive interference was observed in hadronic charm decay. The sign of a_2 disagrees with the theoretical extrapolation from the fit to charm meson decays using the BSW model [22]. It also disagrees with the expectation from the $1/N_c$ rule [23], [24]. The result may be consistent with the expectation of perturbative QCD [25]. A possible explanation for the changing sign has been presented by Berthold Stech at this conference [35].

Although constructive interference has been observed in all the B^+ modes studied so far, these comprise only a small fraction of the total hadronic rate. If the constructive interference which is observed in B^+ decay is present at the same level in the remainder of hadronic B^+ decays, then we would expect a lifetime ratio $\tau_{B^+}/\tau_{B^0} \sim 0.83$ unless there is a large compensating contribution from W-exchange to B^0 decay [26]. It is also possible that there is no interference in the higher multiplicity B decays that have not yet been reconstructed. It is therefore important to measure a_1 and a_2 for a large variety of decay modes.

It is intriguing that a_1 determined from $B \to D^{(*)}\pi$, $D^{(*)}\rho$ modes agrees well with the value of a_1 extracted from $B \to DD_s$ decays. The observation of color-suppressed decays such as $\bar{B}^0 \to D^0\pi^0$ would give another measure of $|a_2|$ complementary to that obtained from $B \to$ charmonium decays.

In summary, experimental results on exclusive B decay match very nicely with theoretical expectations. Unlike charm the b quark appears to be heavy enough so that corrections due to the strong interaction are small. Factorization and color-suppression are at work. The intriguing pattern of constructive interference in charge B decays, however, came as a surprise.

IV THE SEMILEPTONIC BRANCHING RATIO, CHARM COUNTING, $B \to C\bar{C}S$ PROBLEM(S)

Over the last years inclusive B decays have become an area of intensive studies, experimentally as well as theoretically. Since the hadronization process to specific final state mesons is not involved in inclusive calculations the theoretical results and predictions are generally believed to be more reliable. Before we confront theory with experiment we briefly review the current experimental situation.

A Results on Inclusive B Decay

1 Semileptonic Decays

The experimentally measured semileptonic branching ratio is determined to be $(10.35 \pm 0.17 \pm 0.35)\%$ in the model independent dilepton analysis [4]. Comparable results are also obtained from the analysis of the single lepton spectrum. Measurements by the LEP collaborations give a slightly higher value of $11.16 \pm 0.20\%$ for the inclusive semileptonic branching fraction. However, these analyses need still to be corrected for the recently observed production of charmed mesons during the W fragmentation (upper vertex). This will reduce the semileptonic branching fraction.

2 Inclusive Hadronic Decays

The CLEO collaboration has presented a new measurement of inclusive $B \to D^0$ decays

$$\mathcal{B}(B \to D^0 X) = 0.636 \pm 0.014 \pm 0.019 \pm 0.018 ,$$

where the last error is due to the uncertainties in $\mathcal{B}(D^0 \to K^-\pi^+)$. New results in the $B \to$ baryons sector include an improved measurement of Ξ_c production (CLEO II, preliminary)

$$\mathcal{B}(\bar{B} \to \Xi_c^0 X)\mathcal{B}(\Xi_c^0 \to \Xi^-\pi^+) = (0.144 \pm 0.048 \pm 0.021) \times 10^{-3} ,$$

$$\mathcal{B}(\bar{B} \to \Xi_c^+ X)\mathcal{B}(\Xi_c^0 \to \Xi^-\pi^+) = (0.453 \pm 0.096^{+0.085}_{-0.065}) \times 10^{-3} .$$

The results on inclusive $b \to c$ transitions can be used to extract the number of charm quarks, n_c, produced per b decay. Naively we expect $n_c = 1.15$ with the additional 15% coming from the fragmentation of the W boson to $\bar{c}s$. Using CLEO II results we can perform the calculation shown in the Table IV A 2. Modes with 2 charm quarks in the final state are counted twice. For the unobserved $B \to \eta_c X$ decay we take the experimental upper limit. ALEPH

TABLE 7. Charm yield per B decay.

		Channel	Branching Fraction [%]
		$B \to D^0 X$	63.6 ± 3.0
+		$B \to D^+ X$	23.5 ± 2.7
+		$B \to D_s^+ X$	12.1 ± 1.7
+		$B \to \Lambda_c^+ X$	2.9 ± 2.0
+		$B \to \Xi_c^{+,0} X$	2.0 ± 1.0
+	2×	$B \to \psi_{direct} X$	1.6 ± 0.16
+	2×	$B \to \psi'_{direct} X$	0.7 ± 0.1
+	2×	$B \to \chi_{c1} X$	0.74 ± 0.14
+	2×	$B \to \chi_{c2} X$	0.5 ± 0.2
+	2×	$B \to \eta_c X$	< 1.8 (90%C.L.)
		n_c	110 ± 5.2

and OPAL find $n_c = 120 \pm 7\%$ and $n_c = 113 \pm 6\%$, respectively. The increase in charm yield can be accounted for by the larger yield in D_s mesons caused by B_s meson decays that are not present at the $\Upsilon(4S)$.

3 Inclusive $b \to c\bar{c}s$ Transitions

Using the same data we can also extract the number of $b \to c\bar{c}s$ transitions, $n_{c\bar{c}}$. The contribution from $B \to \Xi_c^0 X$ is reduced by 1/3 to take into account the fraction that is not produced by the $b \to c\bar{c}s$ subprocess but by $b \to c\bar{u}d + s\bar{s}$ quark popping. This yields

$$n_{cc} = 16 \pm 2\% .$$

B   The Problem

The "Baffling Semileptonic Branching Fraction" problem – a term coined by I. Bigi – first arose when QCD calculations within the parton model [27] were performed. It was then realized that theory cannot accommodate a semileptonic branching fraction below 12.5%. This is significantly above the measured value. Since the semileptonic and hadronic widths are connected via

$$1/\tau = \Gamma = \Gamma_{\text{Semileptonic}} + \Gamma_{\text{Hadronic}}$$

we must look for an enhancement of the hadronic rate to resolve this discrepancy. The hadronic width can be expressed as

$$\Gamma_{\text{Hadronic}} = \Gamma(b \to c\bar{c}s) + \Gamma(b \to c\bar{u}d) + \Gamma(b \to sg) .$$

Several explanations of this $n_c/\mathcal{B}_{\text{sl}}$ discrepancy have been proposed:

1. enhancement of $b \to c\bar{c}s$ due to large QCD corrections or a breakdown of local duality;

2. enhancement of $b \to c\bar{u}d$ due to non-perturbative effects;

3. enhancement of $b \to sg$ and/or $b \to dg$ due to New Physics;

4. systematic problem in the experimental results;

or the problem could be caused by some combination of the above. There has been a fairly large amount of theoretical and experimental work on this topic of the the past 2-3 years, and in the following section we review the current situation by addressing each of the possible solutions separately.

C   Enhancement of $\Gamma(b \to c\bar{c}s)$

Large QCD corrections or a breakdown of local duality could lead to an enhancement of $\Gamma(b \to c\bar{c}s)$ [27], [29], [28] [30]. This explanation affects all three of our observables ($\mathcal{B}_{\text{sl}}, n_c, n_{cc}$). While the calculated semileptonic branching fraction would be closer to the measurement, both the charm yield and n_{cc} would increase, to values around 1.30 and 30% respectively – significantly more than the current world averages listed in the previous section. Several mechanisms have been proposed to search for the missing $b \to c\bar{c}s$ transitions experimentally. Palmer and Stech [29] suggested that $b \to c\bar{c}s$ followed by

$c\bar{c} \to$ gluons, which in turn hadronize into a final state with no charm, has a large branching ratio. The charm content for this mechanism would not be properly taken into account. Recent experimental data from CLEO rule out a suggestion by Dunietz et al. [28] that $b \to c\bar{c}s$ is a major contributor to $B \to$ baryon decays, but instead a significant rate for $B \to D\bar{D}X$ events has been observed by CLEO, ALEPH and OPAL.

1 Charm Meson - Lepton Correlations

It was previously assumed that the conventional $b \to c\bar{u}d$ mechanism accounts for all D meson production in B decay. Buchalla et al. [33] suggested that a significant fraction of D mesons could also arise from $b \to c\bar{c}s$ transitions with light quark popping at the upper vertex. The two mechanisms can be distinguished experimentally by the different final states they produce. In the first case the final state includes only D mesons whereas in the second case two D mesons are produced, one of which is a $\bar{D}$.

Both routes have been exploited experimentally. Two D mesons in the final state are required by CLEO's search for exclusive $B \to D\bar{D}K$ decays. The observed signal is shown in Figure 3 and the preliminary results are listed in Table IV C 1. While the observation of these decays proves the existence

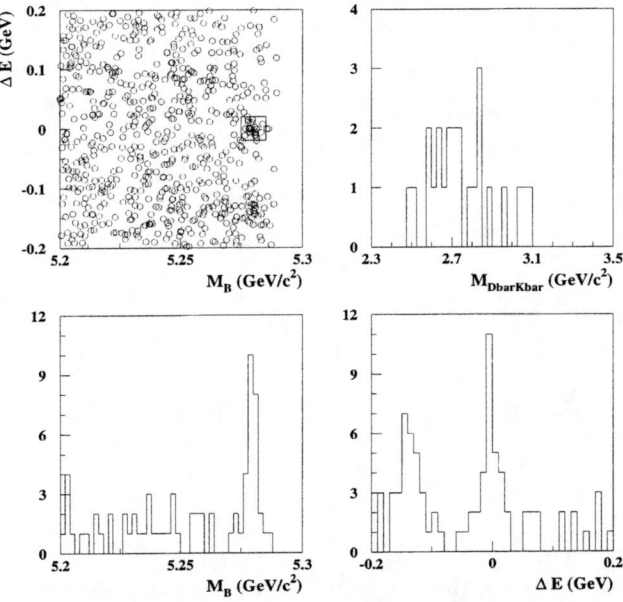

FIGURE 3. Sum of all seven $B \to D\bar{D}XK^-$ modes in CLEO II data. The box in the M_B vs. ΔE plot shows the signal region used for the M_{DK} distribution.

TABLE 8. Preliminary CLEO results on $B \to DDK$ decays.

$\mathcal{B}(\bar{B} \to D^{*+}\bar{D}^0 K^-)$	=	$0.45^{+0.25}_{-0.19} \pm 0.08\%$
$\mathcal{B}(B^- \to D^{*0}\bar{D}^0 K^-)$	=	$0.54^{+0.33}_{-0.24} \pm 0.12\%$
$\mathcal{B}(\bar{B} \to D^{*+}\bar{D}^{*0} K^-)$	=	$1.30^{+0.61}_{-0.47} \pm 0.27\%$
$\mathcal{B}(B^- \to D^{*0}\bar{D}^{*0} K^-)$	=	$1.45^{+0.78}_{-0.58} \pm 0.36\%$

of $\bar{D}$ meson production at the upper vertex, a more inclusive measurement is needed to estimate the overall magnitude of this effect. A recent CLEO analysis exploits the fact that the flavor of the final state D meson tags the decay mechanism. High momentum leptons ($p_\ell > 1.4$ GeV/c) are used to classify the flavor of the decaying B meson. $b \to c\bar{u}d$ transitions lead to $D\ell^+$ combinations while the observation of $\bar{D}\ell^+$ identifies the new $b \to c\bar{c}s$ mechanism. Angular correlations are used to remove combinations with both particles coming from the same B meson. CLEO finds (preliminary)

$$\frac{\Gamma(\bar{B} \to \bar{D}X)}{\Gamma(\bar{B} \to DX)} = 0.100 \pm 0.026 \pm 0.016 ,$$

which implies a new contribution to n_{cc} of

$$\mathcal{B}(\bar{B} \to \bar{D}X) = 0.081 \pm 0.026 .$$

This increases n_{cc} to 0.241 ± 0.032, which is much closer the the theoretical expectations, *but* the charm yield n_c is unchanged. It is interesting to note that now n_{cc} is greater than $n_c - 1 \approx 0.1$ which could indicate a problem with $\Gamma(c\bar{u}d)$ or a very large $\mathcal{B}(b \to sg)$. Supporting evidence for the CLEO results come from ALEPH,

$$\mathcal{B}(\bar{B} \to \bar{D}DX) = 0.138 \pm 0.038 ,$$

as well as DELPHI,

$$\mathcal{B}(\bar{B} \to D^{*+}D^{*-}X) = 0.01 \pm 0.002 \pm 0.003 .$$

D Enhancement of $\Gamma(b \to c\bar{u}d)$

Non-perturbative effects such as constructive interference in B^- decays would reduce the theoretical expectation for the semileptonic branching ratio. A small contribution from W exchange to $\bar{B}^0$ decays would keep the lifetime ratio close to unity and satisfy the experimental constraints on this quantity [26], [34]. This solution would leave n_c and n_{cc} unchanged.

In a next to leading order calculation Bagan et al. [31] find

$$r_{ud} = \frac{\mathcal{B}(b \to c\bar{u}d)}{\mathcal{B}(b \to c\ell\nu)} = 4.0 \pm 0.4 \to \mathcal{B}(b \to c\bar{u}d)_{\text{Theory}} = 42 \pm 4.2\%.$$

Experimentally, we can extract this quantity in the following way:

$$\begin{array}{rl|l}
\mathcal{B}(b \to c\bar{u}d)_{\text{Exp}} = & \mathcal{B}(B \to (D+\bar{D})X) & 87.1 \pm 4.0\% \\
- & \mathcal{B}(B \to \bar{D}X)_{\text{Upper Vertex}} & 8.1 \pm 2.6\% \\
- & \mathcal{B}(B \to D_s X) & 12.1 \pm 1.7\% \\
+ & \mathcal{B}(B \to D_s X)_{\text{Lower Vertex}} & 1.8 \pm 0.9\% \\
+ & \mathcal{B}(B \to \text{baryons} X) & 4.6 \pm 2.1\% \\
- & 2.25 \times \mathcal{B}(b \to c\ell\nu) & 23.6 \pm 1.0\% \\
\hline
 & & 49.7 \pm 5.6\%
\end{array}$$

There is resonable agreement but the errors are still large.

E   Enhancement of $\Gamma(b \to sg)$

A branching fraction $\mathcal{B}(b \to s)$ as large as 10-15% would increase the total hadronic width sufficiently to obtain agreement between the measured and calculated semileptonic branching fractions. Since no charm quarks are involved, n_c and n_{cc} would remain unchanged. Standard Model predictions for $\mathcal{B}(b \to sg)$ are in the few percent region. As discussed by Kagan [32] a gluonic penguin rate of 10% could be interpreted a sign of new physics.

Using $D - -\ell$ correlations as before CLEO obtained a measurement of the ratio $\Gamma(\bar{B} \to DX)_{\text{Lower Vertex}}/\Gamma_{\text{tot}}$. This quantity should be 1 minus corrections for charmonium production, $b \to u$ transitions, $B \to$ baryons, and D_s production at the lower vertex. Most importantly, the $b \to sg$ must also be subtracted. To remove uncertainties due to $\mathcal{B}(D^0 \to K^-\pi^+)$ CLEO normalizes to $\Gamma(\bar{B} \to DX\ell\nu)/\Gamma(\bar{B} \to X\ell\nu)$. Their preliminary result is

$$\frac{\Gamma(\bar{B} \to DX)_{\text{Lower Vertex}}/\Gamma_{\text{tot}}}{\Gamma(\bar{B} \to DX\ell\nu)/\Gamma(\bar{B} \to X\ell\nu)} = 0.901 \pm 0.034 \pm 0.014$$

whereas $0.903 \pm 0.018 - (b \to sg)$ was expected. This corresponds to an upper limit of $\mathcal{B}(b \to sg) < 6.8\%$.

F   Experimental Systematics

Our $n_c, \mathcal{B}_{\text{sl}}$ problem could also be caused by a systematic experimental flaw in the computation of the yield of charm quarks from b decay. Dunietz [36] proposed a lower branching fraction for $D^0 \to K^-\pi^+$ as a possible solution to the charm counting problem, but new, consistent measurements by ALEPH and CLEO, some of which use differing analysis technique, make this a less likely candidate. The absolute branching fraction scales for the D_s meson and Λ_c baryons are also still quite uncertain. Since the inclusive branching ratios to these particles are small, a substantial change to the branching ratio scale would be required to significantly modify the charm yield.

V SUMMARY

Significant progress in the physics of B mesons has been made in the last several years. While we still have to wait for larger data samples to determine the angles of the CKM triangle, the experimentalists at CLEO, LEP, and CDF have tightened the uncertainties on the lengths of the sides of the triangle.

Good progress has been made in the understanding of hadronic B decay. New measurements of $B \to \psi K^{(*)}$ decays and an analysis of the full angular distribution in $B \to \psi K^*$ transition give a more consistent picture of this color-suppressed decay. Experimental tests have verified the factorization hypothesis for B decays with large energy release. It was also confirmed that color-suppression is operative in the B meson system. Somewhat surprisingly, CLEO observed constructive interference in all charged hadronic B decays studied so far. Extrapolating from the charm system the opposite behavior was expected. The improved precision of experimental results has also revealed several discrepancies between theory and experiment in inclusive measurements.

A complete experimental picture of inclusive B decay is now emerging. A new contribution to $b \to c\bar{c}s$ decays has been observed. However, the problem of simultaneously accommodating the low value of n_c and the B semileptonic branching fraction remains. A systematic study of inclusive hadronic B decays to mesons and baryons and an effort to obtain more precise measurements of charm meson absolute branching fractions will be required to fully resolve this problem.

With the end of the Fermilab collider run and the change of the LEP beam energies CLEO and SLD will be the only collider experiments in the next few years to collect data. While this might slow down the current rate of rapid progress in our understanding of heavy flavor physics there are still many answers hidden in the large data samples collected by CDF and the LEP collaborations. This combined with the ever-growing CLEO data sample will provide many new insights into all aspects of B physics.

REFERENCES

1. L. Lederman, contribution to these proceedings.
2. P. Bedaque, A. Das, V.S. Mathur, *Phys. Rev.* D **49**, 269 (1994); B.Y. Blok, M.A. Shifman, *Sov. J. Nucl. Phys.* 45 522 (1987); F. Buccella, M. Lusignoli, G.

Miele, A. Pugliese, and P. Santorelli, *Phys. Rev.* D **51**, 3478 (1995); L.L. Chau, H.Y. Cheng, *Phys. Rev.* D **36**, 137 (1987); M. Gibilisco, G. Preparata, *Phys. Rev.* D **47**, 4949 (1993)

3. M. Bauer, B. Stech, and M. Wirbel, *Z. Phys.* C **29**, 637 (1985); *ibid* **34**, 103 (1987); *ibid* **42**, 671 (1989).
4. T. Browder, K. Honscheid, *Progress in Nuclear and Particle Physics*, Vol. **35**, ed. A. Faessler (1996), and references therein.
5. The error is due to the uncertainty in the scale at which to evaluate the Wilson coefficient.
6. D. Bortoletto and S. Stone, *Phys. Rev. Lett.* **65**, 2951 (1990).
7. Since the form factor for $B \to D^*\ell\nu$ is slowly varying, the width of the ρ^- meson does not significantly modify the result.
8. T.E. Browder, K. Honscheid, and S. Playfer, "Hadronic Decays of B Mesons," 2nd edition, ed. S. Stone, World Scientific (1994).
9. J. Korner and G. Goldstein, *Phys. Lett.* B **89B**, 105 (1979).
10. Peter Lepage (private communication). Also see V. Rieckert, *Phys. Rev.* D **47**, 3053 (1993).
11. M.S. Alam *et al.* (CLEO Collaboration), *Phys. Rev.* D **50** 43 (1994).
12. H. Albrecht *et al.* (ARGUS Collaboration), *Phys. Lett.* B **340** 217 (1994).
13. F. Abe *et al.* (CDF Collaboration), FERMILAB-PUB-96-119-E (1996).
14. A.S. Dighe, I. Dunietz, H.J. Lipkin and J.L. Rosner, *Phys. Lett.* B **369** 144 (1996).
15. J.L. Rosner, *Phys. Rev.* D **42**, 3732 (1990).
16. M. Neubert, *Phys. Lett.* B **264**, 455 (1991).
17. V. Rieckert, *Phys. Rev.* D **47**, 3053 (1993) and V. Rieckert, Thesis, University of Heidelberg (1994), unpublished.
18. M. Neubert, V. Rieckert, Q. P. Xu and B. Stech in *Heavy Flavors*, edited by A. J. Buras and H. Lindner (World Scientific, Singapore, 1992).
19. B. Barish *et al.* (CLEO Collaboration), *Phys. Rev.* D **51**, 1014 (1995).
20. We considered variations of f_D between 120 MeV and 320 MeV. For $f_D = 320$ MeV we find $a_2/a_1 = 0.18$.
21. G. Kramer, T. Mannel and W.F. Palmer, *Z. Phys.* C **55**, 497 (1992).
22. In the fits of Ref. [18] the CLEO 1.5 data favor a positive sign while the ARGUS data prefer a negative sign.
23. B. Blok and M. Shifman, *Nucl. Phys.* B **389**, 534 (1993).
24. I. Halperin *et al.*, *Phys. Lett.* B **349**, 548 (1995).
25. A. J. Buras, *Nucl. Phys.* B **434**, 606 (1995).
26. K. Honscheid, K.R. Schubert and R. Waldi, *Z. Phys.* C **63**, 117 (1994).
27. I. Bigi, B. Blok, M.A. Shifman and A. I. Vainshtein, *Phys. Lett.* B **323**, 408 (1994).
28. I. Dunietz, P.S. Cooper, A.F. Falk and M. Wise, *Phys. Rev. Lett.* **73**, 1075 (1994).
29. W.F. Palmer and B. Stech, *Phys. Rev.* D **48**, 4174 (1993).
30. A.F. Falk, Johns Hopkins University preprint JHU-TIPAC-940016.
31. E. Bagan, P. Ball, V.M. Braun, and P. Gosdzinsky, *Phys. Lett.* B **342**, 362

(1995); Erratum *Phys. Lett.* B **374**, 363 (1996).
32. A.L. Kagan, *Phys. Rev.* D **51**, 6196 (1995).
33. G. Buchalla, I. Dunietz, H. Yamamoto, *Phys. Lett.* B **364**, 188 (1995).
34. M. Neubert and C. T. Sachrajda, CERN-TH/96-19.
35. B. Stech, contribution to these proceedings.
36. I. Dunietz, FERMILAB-PUB-96/104-T.

b Physics at SLD

D. Su

Stanford Linear Accelerator Center, Stanford, California 94309
representing the SLD Collaboration

Abstract. This report summarizes the recent b physics results from SLD. This includes measurements of R_b and A_b as tests of the Standard Model (SM) prediction of b quark coupling to the Z^0, as well as measurements of B lifetimes and mixing. These measurements exploit the special capabilities of SLC/SLD which include a highly polarized electron beam, a CCD pixel vertex detector in conjunction with a small SLC interaction point, and particle identification from the Čerenkov Ring Imaging Detector. The many novel techniques developed for these measurements have led to some unique physics perspectives and competitive results. These developments provide an illustration of the youthful vitality of b physics at its 20th birthday.

THE SLD EXPERIMENT AT SLC

The SLAC Linear Collider (SLC) operating at a center of mass energy on the Z^0 peak is the only e^+e^- linear collider in operation. The SLD experiment residing at the single colliding point of SLC started its physics run in 1992 and accumulated 200,000 Z^0 events by the end of the 1996 run. The general description of the various SLD detector components can be found in [1] [2] [3] [4]. The distinctive aspects of the SLD which have contributed to its competitive b-physics program as well as other measurements are:

Polarized Electron Beam: The development of the strained GaAs cathode for the polarized electron source at SLC was a crucial advance for physics at the Z^0 pole. The 50K Z^0 taken by SLD in 1993 had 63% electron beam longitudinal polarization, while the 150K since 1994 had 77% polarization.

CCD Pixel Vertex Detector: The SLD CCD pixel vertex detector is also a unique feature in high energy physics today. The previous vertex detector (VXD2) with 480 1 cm^2 CCD's used from 1992 to 1995 was replaced in 1996 by an upgrade (VXD3 [2]) with 96 custom made 1.5 cm×8 cm CCDs. VXD3 provides full 3 layer coverage to $|\cos\theta|=0.85$, with layer radial

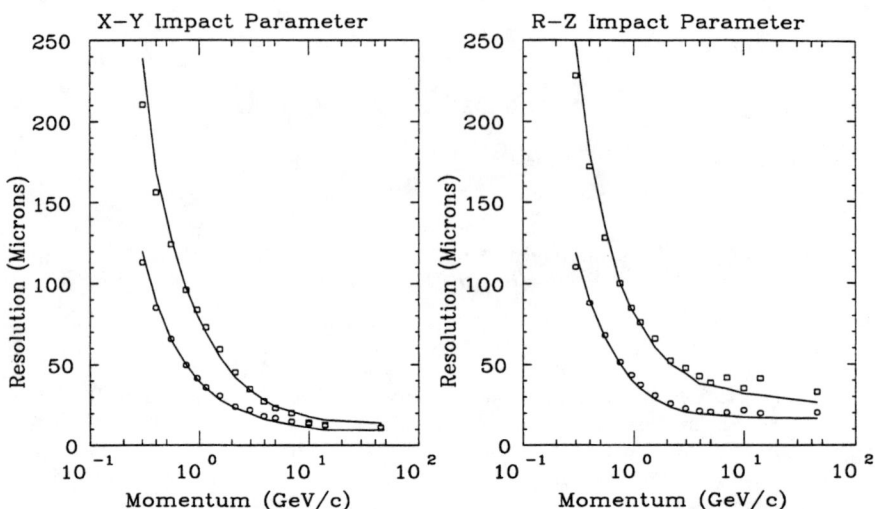

FIGURE 1. Measured impact parameter resolution as a function of momentum for tracks at $\cos\theta = 0$. The squares are VXD2 data, the circles are VXD3 data and the curves are for corresponding MC.

positions from 2.8 cm to 4.8 cm. The spatial resolution achieved with VXD3 is 4.5 μm in both $r\phi$ and z. The track impact parameter resolution at various momenta compared between data and Monte Carlo (MC) for VXD3 and VXD2 is shown in Fig. 1. The 3D spatial information from the CCD pixels ensure clean and efficient tracking without ambiguity, in addition to their high precision.

The Small and Stable SLC Beam Spot: Another important feature at SLC/SLD which further enhances the vertexing-related analyses is the small and stable SLC interaction point (IP). The fact that SLC is colliding micron-sized beams necessitates the position stability of the beams. The average IP xy position is tracked using ~30 sequential hadronic events [1] and is used as the event primary vertex (PV) in xy with a measured resolution of ~7 μm. Besides the benefit of better resolution, the minimal bias from the average IP position especially helps in $b\bar{b}$ events, where the high multiplicities in B decay vertices could sometimes confuse individual event PV finders to mis-assign PV and cause significant systematic effects. Given the known xy beam position, the event-by-event PV z position can be located to 16 μm (32 μm) precision for *udsc* (*b*) events with VXD3.

Particle Identification with CRID: The SLD Čerenkov Ring Imaging Detector (CRID) [3] is one of two such devices in present high energy collider experiments, equipped for hadron identification with high degree of separation. Kaons can be identified with ~60% efficiency in the 1-20 GeV/c momentum range with a pion rejection ratio of 40:1 at low momentum and 10:1 at high momentum.

TESTS OF THE STANDARD MODEL THROUGH THE $Zb\bar{b}$ COUPLING

The measurement of the Z^0 couplings to various fermions is an important part of the precision tests of the SM. Possible new physics beyond the SM may manifest itself through radiative corrections to these couplings. The measurements of Z^0 coupling to the b quark are of especial interest due to the expectation of more pronounced radiative corrections to the $Zb\bar{b}$ vertex both within the SM and from extensions beyond the SM compared to the Z^0 couplings to lighter quarks.

Two observables designed to probe complementary aspects of the $Zb\bar{b}$ coupling are R_b and A_b:

$$R_b = \frac{\Gamma(Z^0 \to b\bar{b})}{\Gamma(Z^0 \to \text{Hadrons})} = \frac{g_L^{b\,2} + g_R^{b\,2}}{\sum_f^{udscb}(g_L^{f\,2} + g_R^{f\,2})},$$

$$A_b = \frac{g_L^{b\,2} - g_R^{b\,2}}{g_L^{b\,2} + g_R^{b\,2}},$$

where g_L^b and g_R^b are the left- and right-handed $Zb\bar{b}$-coupling amplitudes respectively. R_b is a measurement of the $Zb\bar{b}$ coupling strength compared to other flavors, while A_b is a measurement of the extent of parity violation at the $Zb\bar{b}$ vertex. R_b is more sensitive to deviations in the left-handed $Zb\bar{b}$ coupling, while A_b is more sensitive to deviations in the right-handed coupling. The $Zb\bar{b}$ vertex corrections in the SM are particularly large due to the large top quark mass and the fact $|V_{tb}| \sim 1$. These SM vertex corrections act only on the left-handed coupling. Many proposed extensions to the SM also tend to exhibite larger changes in the left-handed $Zb\bar{b}$ coupling than in the right-handed coupling.

R_b Measurement

The formulation of R_b as a ratio of the hadronic cross sections ensures that the oblique (vacuum polarization) electroweak radiative corrections to the Z^0 propagator and the QCD radiative corrections which are common to all flavors largely cancel. This leads to an experimentally viable path to a precision test of the $Zb\bar{b}$ vertex radiative corrections.

The modern R_b measurements generally adopt the double-tag technique to reduce reliance on MC, with R_b and the b-tag efficiency simultaneously extracted from data. Up to winter-96, the ensemble of R_b measurements led to a world average over 3σ higher than the SM [5]. These measurements mostly relied on b-tags exploiting the long B lifetime. However, these measurements, which had typical b-tag purity of $\sim 94\%$, were largely limited by uncertainties in uds, c efficiencies. The exponential nature of the lifetime distribution resulted in a rather difficult task to suppress long lived charm decays from $c\bar{c}$ events using purely geometrical vertex or impact parameter information alone.

A key development in b-tagging to alleviate the remaining charm background tail was to require a large vertex mass in addition to the geometrical vertex information. The charm background has a sharp cut-off in the vertex mass distribution at ~ 1.9 GeV/c. The current SLD measurement of R_b closely follows the vertex mass tag method introduced at Moriond-96 [6] which represented the first utilization of the B mass information in the tag. Following the analysis on 125K Z^0's from 93-95 [7], a new preliminary analysis [8] has been performed on 50K Z^0's from the 96 run with VXD3 based on the same b-tag method.

The b-tag is carried out by first searching for secondary vertices in each hemisphere using a 3D topological vertexing technique [9]. For hemispheres containing secondary vertices, tracks are then further classified as from secondary decay or from PV based on their compatibility with a seed secondary vertex (SV). The raw vertex mass calculated from the secondary tracks already provides a good improvement in b-tag purity, but a significant gain in efficiency can be made to further exploit the decay kinematical information. This is achieved through a 'P_T-corrected' mass by adding a minimal missing P_T into the mass calculation to account for missing neutrals in the B decay. The missing P_T is estimated using the acolinearity between the momentum sum of all secondary tracks and the direction of the B flight path from the PV to the SV. The effectiveness of this correction relies critically on the vertexing resolution as well as the PV resolution. The small and stable IP at SLC is a vital factor in the effective application of this technique. The raw mass and P_T-corrected mass distributions for the 93-95 data and MC are shown in Fig. 2. The hemisphere b-tag efficiency and purity for 93-95 VXD2 data are 35.3% and 98.0% respectively for masses greater than 2 GeV/c, and it improves significantly to 47.9% and 97.6% for the 96 data with VXD3. The b-tag performance is compared with other measurements at LEP in Fig. 3.

The final result obtained from 1993-1995 data and the preliminary result from 1996 data are:

R_b (93-95) $=$ 0.2142 ± 0.0034 (stat) ± 0.0015 (sys) ± 0.0002 (R_c)
R_b (96 prelim.) $=$ 0.2101 ± 0.0034 (stat) ± 0.0022 (sys) ± 0.0003 (R_c)

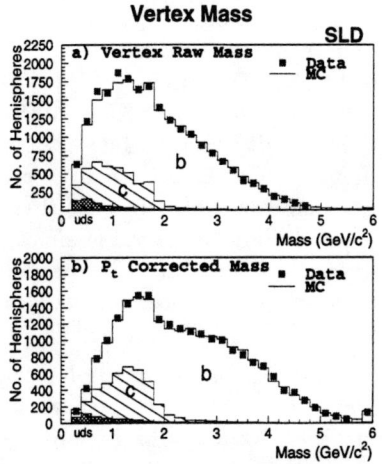
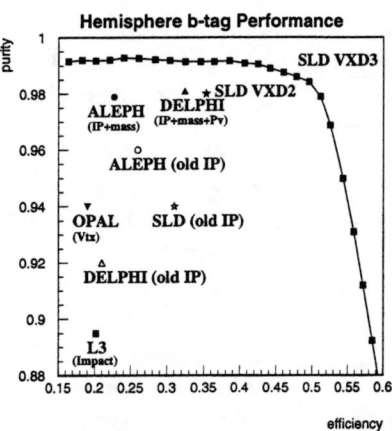

FIGURE 2. The raw vertex mass and P_T-corrected vertex mass distributions compared between data and MC.

FIGURE 3. Comparison of hemisphere b-tag performance for current and previous R_b measurements.

These measurements are in good agreement with the SM prediction of 0.2158 and the most recent measurements from LEP [10], but they disagree at the 2σ level with the old winter-96 world average [5]. It should be noted that the preliminary 96 result with improved b tag using VXD3 has the same statistical precision as the 93-95 result, with only 40% of the 93-95 statistics. The correlated physics systematics related to the uncertainties in uds, c background and b-hemisphere correlations are only ± 0.0006. The dominant systematics at present are detector effects, where the main contribution from detector resolution is conservatively assigned to the full effect of including and excluding resolution correction to the MC. This source of systematics will certainly improve with more data and a better alignment procedure for VXD3 in the future, which should reduce the systematic limit to well below 0.5%.

A_b Measurements

For $Zb\bar{b}$ events with final state b quark at polar angle θ with respect to the electron beam, the Born level expression for the conventional forward-backward b asymmetry is

$$A_{\text{FB}}^b(z) = \frac{\sigma^b(z) - \sigma^b(-z)}{\sigma^b(z) + \sigma^b(-z)} = A_e A_b \frac{2z}{1+z^2},$$

expressed in terms of $z = \cos\theta$. The availability of the electron-beam polarization allows the formation of the left-right forward-backward asymmetry

$$\tilde{A}_{\text{FB}}^b(z) = \frac{[\sigma_L^b(z) - \sigma_L^b(-z)] - [\sigma_R^b(z) - \sigma_R^b(-z)]}{[\sigma_L^b(z) + \sigma_L^b(-z)] + [\sigma_R^b(z) + \sigma_R^b(-z)]} = |P_e| A_b \frac{2z}{1+z^2},$$

where P_e is the electron beam longitudinal polarization. The use of $\tilde{A}^b_{\text{FB}}$ eliminates the dependence on initial state Zee coupling parameter A_e, thus allowing direct measurement of the final state A_b. The high $|P_e|$ of $\sim 77\%$ achieved at SLC/SLD also brings a large gain of $(P_e/A_e)^2 \sim 25$ in statistical power compared to conventional A^b_{FB} in sensitivity to A_b. The effects of detector acceptance and efficiency non-uniformity also cancel to first order for this double asymmetry. For all the A_b measurement results described in this section, $\cos\theta$-dependent QCD corrections including quark mass effects at $\mathcal{O}(\alpha_s)$ level [11] are applied. Individual analyses have also taken into account effects modifying the QCD corrections due to analysis procedures typically suppressing events with hard gluon radiation.

While the direct measurements of A_b at SLD are all based on the left-right forward-backward asymmetry principle, three different techniques are used to tag the $b\bar{b}$ events and determine the b quark direction. The first analysis [12] uses leptons with high P_T with respect to the jet axis to tag b events and to provide $b/\bar{b}$ separation at decay. The actual analysis parameterizes the fractions of various lepton sources in full (P, P_T) space using the MC, and uses a maximum likelihood fit to obtain A_b and A_c simultaneously. The combined preliminary 93-95 result from both electron and muon analyses is $A_b = 0.877 \pm 0.068$ (stat) ± 0.047 (sys). The systematic estimation closely follows the standard procedure recommended by the LEP electroweak working group [13].

The second A_b measurement [14] selects $b\bar{b}$ events using the vertex mass tag as described in the R_b analysis, and assigning the b quark direction using a jet-charge technique. $b\bar{b}$ events are selected with an efficiency of 65% and b purity of 91%. The momentum weighted track charge is calculated for each hemisphere with a typical probability of correct b-charge assignment of 69% per hemisphere. The widths of the two hemisphere charge sum and difference distributions are used to derive the jet-charge analyzing power from the data. Using a maximum likelihood fit which weights events according to the jet-charge value dependent analyzing power, the preliminary result from 93-95 data is $A_b = 0.911 \pm 0.045$ (stat) ± 0.045 (sys). Of the 4.9% total systematic error, the dominant contribution is in fact the data self-calibration statistics of 3.7% which will decrease with more data.

The third A_b measurement [15] also uses the vertex mass tag to select $b\bar{b}$ events in a similar manner as in the jet-charge analysis. The b-quark direction is assigned using the B decay tracks which are identified by the CRID as $K^\pm$. This is the first use of the $b \to c \to s \to K^-$ signature as a $b/\bar{b}$ tag for an asymmetry measurement. Currently using 94-95 data only, this analysis obtains kaon signing of b direction for 39% of the tagged b events with 73% correct signing probability. The kaon signed event thrust axis $\cos\theta$ distributions are shown in Fig. 4. A left-right forward-backward asymmetry fit using the MC as fitting function resulted in a preliminary result of

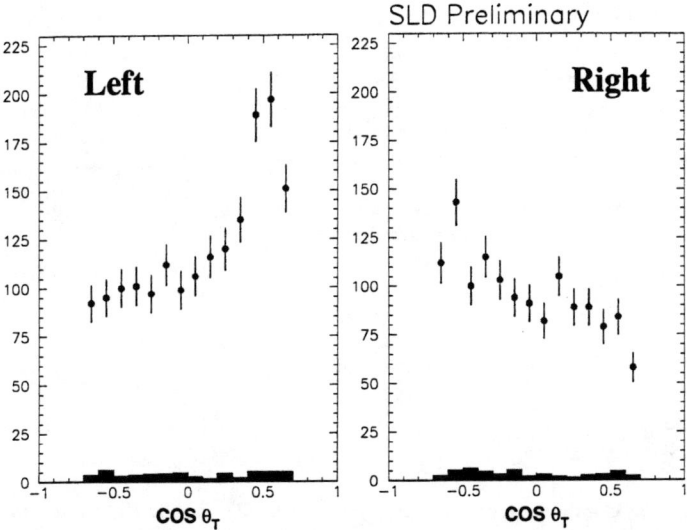

FIGURE 4. Event thrust axis $\cos\theta$ distributions signed by identified kaons in b-tagged events for left-handed and right-handed electron beams. The points are all data events and the shaded histogram is $udsc$ background events estimated from the MC.

$A_b = 0.891 \pm 0.083$ (stat) ± 0.113 (sys). The dominant systematic error by far is the uncertainty in the $B \to K^-/B \to K^+$ production ratio which results in $\delta A_b/A_b = \pm 10.9\%$. The K^-/K^+ production rate uncertainty was estimated based on the ARGUS measurement [16] which gave results in absolute rates for K^+ and K^-. The detailed re-analysis for the production *ratio* uncertainty, where many common systematics cancel, is currently in progress within the ARGUS collaboration.

The above SLD measurements are combined, taking into account the small statistical and systematic correlations, to give an SLD-average direct measurement result of $A_b = 0.898 \pm 0.052$. This can be compared with A_b values derived from A_{FB}^b measurements at LEP [17] as shown in Fig. 5, by dividing out from $A_{FB}^{(b,0)}$ the A_e value of 0.1505 ± 0.0023 obtained by combining all lepton coupling measurements from LEP and SLD.

Takeuchi *et al.* have proposed an interesting scheme for a full analysis of the $Zb\bar{b}$ couplings [18]. The parity violation type variable ζ_b from this analysis is plotted versus $\delta \sin^2 \theta_W^{\text{eff}}$ results as shown in Figure 6. The SM point at (0,0) is defined by m_t=175 GeV, m_H=300 GeV, α_s=0.117 and α_{em}=1/128.96. The thin horizontal band around (0,0) corresponds to the SM m_t, m_H variations indicated in the plot. The 68% and 90% C.L. contours for the best fit to all measurements are also shown. The various measurements involved are not in perfect agreement with the SM and improved measurements in the future will be of great interest. An improved direct measurement of A_b at SLD will have a unique role in helping to disentangle the nature of any persistent discrepancy.

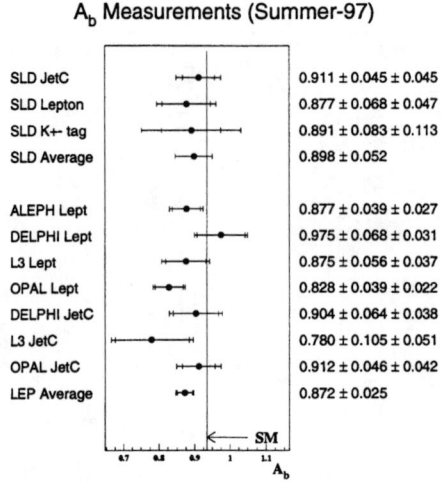
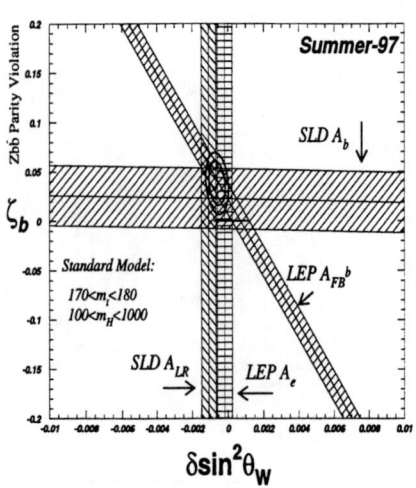

FIGURE 5. Comparison of world A_b measurements.

FIGURE 6. $Zb\bar{b}$ vertex analysis result following Takeuchi, Grant and Rosner.

B LIFETIME MEASUREMENTS

The long B lifetime is a wonderful gift for b physics, and the b physics program at SLD has certainly benefitted greatly from it. The measurement of the B lifetime is not only necessary for the determination of V_{cb}, but also plays a very important role in the understanding of B decay dynamics. The large lifetime difference between D^+ and D^0 prompts a measurement of the B^+/B^0 lifetime ratio in particular. The theoretical expectation of only $\sim 5\%$ deviation [19] from unity for this ratio poses a real experimental challenge.

The fundamental requirement of exclusive B lifetime measurements is the separation of different B species. The most common techniques at present are exclusive reconstruction, e.g. at CDF [20], or semi-exclusive lepton+$D^{(*)}$ correlations used both at LEP [21] and CDF. Since the usable branching ratios are small, it is still difficult to reach 5% precision statistically even starting from very large data samples. The SLD measurements use more inclusive vertex reconstructions to give much higher efficiencies while relying on the very basic difference between B^+ and B^0 in decay vertex charge to achieve the separation.

Two methods are used for B^+, B^0 lifetime measurements at SLD [22]. One analysis starts from a moderately high P_t lepton and searches for other secondary tracks to reconstruct the B semileptonic decay with good charge purity. The second analysis uses an inclusive topological vertexing technique (as described in the R_b measurement section) with an additional secondary track search to improve the B charge reconstruction.

An improved preliminary analysis of 93-96 data has recently been released [23] for the second method which updates the published result in [22] from 93-95 data. The additional improvements in this analysis include: a) using P_t-corrected vertex mass instead of raw vertex mass to increase B selection efficiency; b) weighting events according to the reconstructed mass as charge purity improves at higher mass; c) using large polarized forward-backward b production asymmetry and opposite hemisphere jet charge to purify the charged B sample; d) allowing multiple vertex fits in one B decay to improve decay length reconstruction and reduce B decay mode dependent systematics. The total number of events in the charged and neutral samples are 12841 and 7942 respectively. The sample separation achieved for e.g. the 96 VXD3 data, before the additional B mass and $b/\bar{b}$ tag enhancement weighting gains, is $B^+ : B^0 = 55\% : 33\%$ in the charged sample and $B^+ : B^0 = 24\% : 54\%$ in the neutral sample. The reconstructed B charge distributions compared between data and MC are shown in Fig. 7.

The simultaneous fit to the B^+ and B^0 lifetimes from the two samples gives the following preliminary results:

$$\begin{aligned}
\tau_{B^+} &= 1.698 \pm 0.040 \text{ (stat)} \pm 0.046 \text{ (sys)} \text{ ps,} \\
\tau_{B^0} &= 1.581 \pm 0.043 \text{ (stat)} \pm 0.061 \text{ (sys)} \text{ ps,} \\
\tau_{B^+}/\tau_{B^0} &= 1.072 \pm {}^{0.052}_{0.049} \text{ (stat)} \pm 0.038 \text{ (sys)}.
\end{aligned}$$

These results are among the most precise measurements to date and are consistent with theoretical expectation. The individual B^+ and B^0 lifetime measurements have various significant common B production related systematics

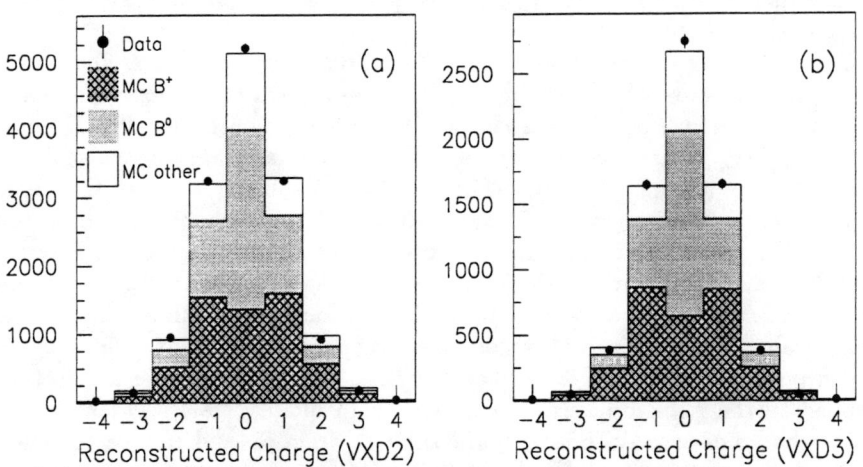

FIGURE 7. Comparison of reconstructed B charge distributions between data and MC for (a) 93-95 VXD2 data and (b) 96 VXD3 data. The MC charge distributions for various types of B hadrons are also shown.

which cancel in the lifetime ratio. This test of the τ_{B^+}/τ_{B^0} ratio will therefore continue to improve with more data.

B MIXING MEASUREMENTS

The measurement of B mixing is currently the best means of accessing $|V_{td}|$. These measurements require tagging the b production initial state and B decay final state flavors in order to measure the mixing mass difference Δm from the B oscillation signal as a function of the decay proper time.

There are 4 preliminary measurements of B_d mixing from SLD using the 93-95 data with different final state tag techniques while sharing the same initial state tag. The highly polarized electron beam provides a unique initial state tag through the large b quark production forward-backward asymmetry. This tag is available to all events and has especially high analyzing power at high $|\cos\theta|$. It is combined with a more commonly used opposite hemisphere jet charge tag to yield a joint initial state tag with an effective correct tag probability of 84%.

The first two analyses [24] are based on semileptonic B decay final state tags. The first analysis reconstructs the B decay vertex from the weighted mean position of the intersection points of other tracks in the hemisphere with a high P_t lepton. This is achieved with essentially no efficiency loss once the high P_t lepton is selected. The second analysis is similar to the semileptonic B lifetime analysis mentioned previously in the B lifetime section, using a lepton and additional tertiary "D" vertex combination. The advantage of this method is the availability of good B charge reconstruction allowing significant enhancement of B_d^0 fraction.

The remaining two analyses [25] are based on more inclusive B decay properties, starting from the b hemisphere tag as described in the R_b measurement section. The first analysis uses charged kaons identified by the CRID among the secondary B decay tracks to tag the decay final state flavor, exploiting again the $b \to c \to s$ signature. This tag is available for $\sim 1/3$ of the tagged B's and correctly signs the B_d final state with 77% probability. The second analysis uses a novel technique to tag final state flavor using the fact that the B decay cascade vertex structure tends to have opposite charge patterns for B and $\overline{B}$ at the B and D vertices. This is measured with a vertex charge dipole calculated from the weighted mean longitudinal position differences between positive and negative B decay tracks. The vertex charge dipole tag is applied to the B candidates formed with high quality tracks which have total B charge of zero, to both enhance the B_d fraction and reduce the effects from missing or misassigned tracks. The $B/\overline{B}$ separation achieved depends on the charge dipole magnitude and reaches 84% correct tag probability at large dipole values.

The Δm_d fits are performed in either decay length or proper time for the four measurements and the results are shown in Fig. 8. Taking into account the statistical and systematic correlations between the measurements, a combined preliminary SLD average of Δm_d is obtained: $\Delta m_d = 0.525 \pm 0.043$ (stat) ± 0.037 (sys) ps^{-1}.

The B_d mixing measurements alone have rather limited precision in determining V_{td} due to the large theoretical uncertainties in the B decay constants. However, a measurement of B_s mixing would determine the ratio $\Delta m_s/\Delta m_d$ where the decay constant uncertainties largely cancel, and therefore provide a clean determination of V_{td}. The measurement of B_s mixing is considerably more demanding than B_d mixing primarily due to the much faster oscillation time scale and also a smaller B_s production fraction. The SLD B_d measurements have already reached comparable precision to existing measurements

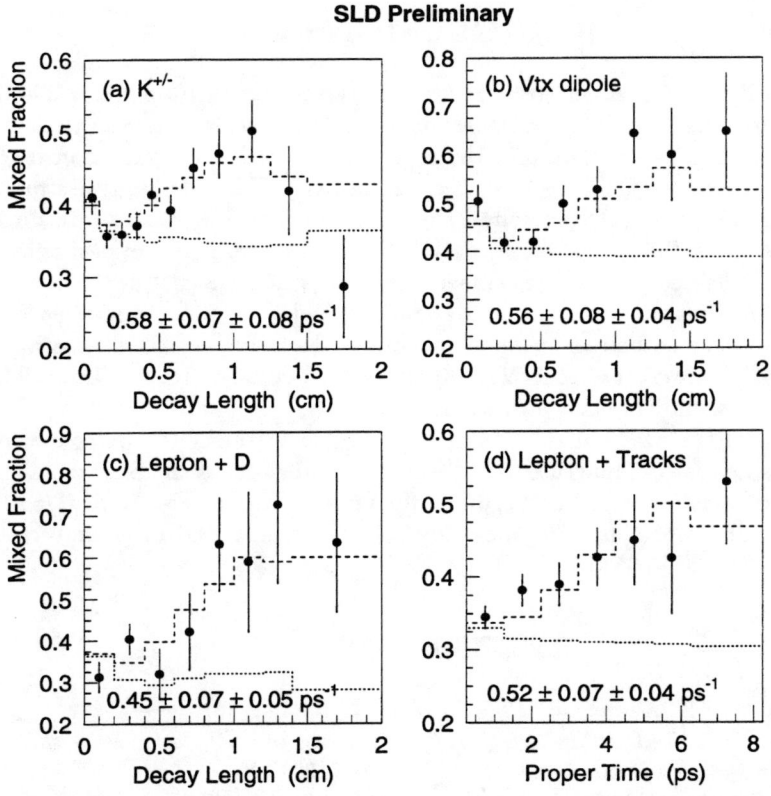

FIGURE 8. The B_d mixing Δm_d fit results for the four different analyses, where points are data, dashed histograms are the best fit MC and dotted histograms represent MC with no B_d mixing.

from LEP [21] and CDF [20], due to the high efficiencies and good analyzing powers in the SLD initial and final state tags, as well as utilization of additional novel techniques. The crucial advantage of SLD in excellent vertexing resolution has not yet exhibited its full potential in the B_d mixing case with its long oscillation time scale, but it will become much more important in measuring B_s mixing. The various SLD measurements described all have core B decay length resolution of $\sim 200\,\mu$m or better for VXD2 data, while improving to $\sim 100\,\mu$m or better for VXD3, despite the inclusive nature of the decay reconstruction. The new technique of vertex charge dipole will not only improve with VXD3 resolution, but it is also more effective for B_s due to the dominant $B_s \rightarrow D_s X$ decays carrying more favorable charge dipole than B_d. Based on the various resolutions achieved from the current analyses and expected gains from known additional tools, the estimated 95% confidence level Δm_s reach for SLD with 500K Z^0's combining all analyses is ~ 16 ps^{-1}.

CONCLUSIONS

This report has described some recent b physics results from SLD. There are still more SLD b physics measurements not covered in this report, such as the search for $B \rightarrow s$ gluon and the search for inclusive CP violation in B decays [26], as well as QCD tests with b quarks [27]. The SLD b-physics measurements have successfully exploited the advantages at SLC/SLD to produce competitive results with modest statistics. There are many new experimental techniques pioneered within these measurements which are significant contributions to the dynamic development of b physics today. A successful 97-98 run holds the promise to deliver more than double the current SLD data statistics, and with the new upgrade VXD3, we can expect significantly improved measurements from SLD in very near future.

On behalf of the SLD collaboration, I would like to thank the personnel of the SLAC accelerator department and the technical staffs of the SLD collaborating institutions for their outstanding effort toward the successful SLC/SLD operations. This work was supported by the US Department of Energy contract DE-AC03-76SF00515.

REFERENCES

1. SLD Collab.: K. Abe *et al.*, *Phys. Rev.* **D53**, 1023 (1996).
2. K. Abe *et al.*, SLAC-PUB-7385 (1997), *submitted to Nucl. Instr. & Meth.*
3. K. Abe *et al.*, *Nucl. Instr. & Meth.* **A343**, 74 (1994).
4. M. D. Hildreth *et al.*, *Nucl. Instr. & Meth.* **A367**, 111 (1995),
   D. Axen *et al.*, *Nucl. Instr. & Meth.* **A328**, 472 (1993),
   A. C. Benvenuti *et al.*, *Nucl. Instr. & Meth.* **A276**, 94 (1989); **A290**, 353 (1990);

M. Woods et al., SLAC-PUB-7320, talk given at the 12th International Symposium on High Energy Spin Physics (SPIN96), Amsterdam, Netherlands, Sep/1996.

5. M. Hildreth, proceedings of XXXI Rencontres de Moriond *"Electroweak Interactions and Unified Theories,"* Mar/96.
6. E. Etzion, SLD Collaboration, proceedings of XXXI Rencontres de Moriond *"Electroweak Interactions and Unified Theories,"* Mar/96.
7. SLD Collab.: K. Abe et al., SLAC-PUB-7481 (1997), *submitted to Phys. Rev. Lett.*
8. SLD Collab.: K. Abe et al., SLAC-PUB-7585, contribution (EPS-118) to the International Euro-Physics Conference on High Energy Physics, Jerusalem, Aug/97.
9. D. Jackson, *Nucl. Inst. & Meth.* **A388**, 247 (1997).
10. ALEPH Collab.: R. Barate et al., *Phys. Lett.* **B401** 150, 163, (1997)
    DELPHI Collab.: G. J. Barker et al., Contribution (EPS-419) to the International Euro-Physics Conference on High Energy Physics, Jerusalem, Aug/97.
11. J. B. Stav and H. A. Olsen, *Phys. Rev.* **D52**, 1359 (1995); *Phys. Rev.* **D50**, 6775 (1994).
12. SLD Collab.: K. Abe et al., SLAC-PUB-7637, Contribution (EPS-124) to the International Euro-Physics Conference on High Energy Physics, Jerusalem, Aug/97.
13. The LEP Electroweak Working Group, "Presentation of the LEP Electroweak Heavy Flavour Results for Summer 1996 Conferences", *LEPHF/96-01*, http://www.cern.ch/LEPEWWG/heavy/lephf9601.ps.gz, July 1996.
14. SLD Collab.: K. Abe et al., SLAC-PUB-7629, Contribution (EPS-122) to the International Euro-Physics Conference on High Energy Physics, Jerusalem, Aug/97.
15. SLD Collab.: K. Abe et al., SLAC-PUB-7630, Contribution (EPS-123) to the International Euro-Physics Conference on High Energy Physics, Jerusalem, Aug/97.
16. ARGUS Collab.: H. Albrecht et al., *Z. Phys.* **C62**, 371 (1994).
17. The LEP Electroweak Working Group & the SLD Heavy Flavor group, *LEPEWWG/97-02, SLD Physics Note 63*, "A Combination of Preliminary Electroweak Measurements and Constraints on the Standard Model".
18. T. Takeuchi, A. Grant, J. Rosner, FermiLab-Conf-94/279-T, talk presented at the DPF'94 Meeting, Albuquerque, NM, Aug/94.
19. I. Bigi, et al., in *B Decays*, editted by S. Stone (World Scientific, New York, 1994), p. 132.
20. A. Barry Wicklund, *talk presented at the b20 Symposium, June/97*, in these proceedings.
21. Vivek Sharma, *talk presented at the b20 Symposium, June/97*, in these proceedings.
22. SLD Collab.: K. Abe et al., *Phys. Rev. Lett.* **79**, 590 (1997).
23. SLD Collab.: K. Abe et al., SLAC-PUB-7635, Contribution (EPS-127) to the International Euro-Physics Conference on High Energy Physics, Jerusalem,

Aug/97.
24. SLD Collab.: K. Abe *et al.*, SLAC-PUB-7228, Contribution (PA08-026A) to the XXVIII International Conference on High Energy Physics, Warsaw, Jul/96; SLD Collab.: K. Abe *et al.*, SLAC-PUB-7229, Contribution (PA08-026B) to the XXVIII International Conference on High Energy Physics, Warsaw, Jul/96.
25. SLD Collab.: K. Abe *et al.*, SLAC-PUB-7230, Contribution (PA08-027/028) to the XXVIII International Conference on High Energy Physics, Warsaw, Jul/96.
26. Mourad Daoudi, representing the SLD Collaboration, talk presented at the International Euro-Physics Conference on High Energy Physics, Jerusalem, Aug/97.
27. SLD Collab.: K. Abe *et al.*, SLAC-PUB-7570/7572/7573, Contributions to the XXVIII International Conference on High Energy Physics, Jerusalem, Aug/97.

Status of KEKB and BELLE

Daniel R. Marlow

Joseph Henry Laboratories, Princeton University, Princeton NJ 08544

Abstract.
The status of the KEKB accelerator and the BELLE detector is summarized.

INTRODUCTION

The KEKB accelerator [1] and the BELLE detector [2] comprise one of two major efforts aimed at the experimental study of CP violation in the neutral B meson system using asymmetric e^+e^- collisions at the $\Upsilon(4S)$ resonance. Although BELLE and KEKB are in many ways similar to the BaBar detector [3,4] and the PEP-II accelerator [5], there are some significant differences, which I will attempt to elucidate in this paper.

The principal physics goal of BELLE is the observation of indirect CP violation and the extraction of the internal angles of the unitarity triangle, ϕ_1, ϕ_2, and ϕ_3.[1] Table 1 lists the processes thought to be most promising for this purpose along with the sensitivity BELLE expects to achieve upon the accumulation of an integrated luminosity of 100 fb^{-1}. The anticipated physics program also includes measurements of the lengths of the sides of the unitarity triangle, as well as the study of charm decays, τ decays, and 2γ reactions.

THE KEKB ACCELERATOR

The KEKB machine consists of two rings, with up to 1.1 A of 8 GeV electrons circulating in one and 2.6 A of 3.5 GeV positrons circulating in the other. The target luminosity is 10^{34} cm^{-2}s^{-1}. The machine employs novel RF cavity designs and fast feedback to dampen coupled bunch instabilities arising from the large beam currents. Two types of cavities will be used: superconducting cavities (SCCs) and a special type of warm cavity, the ARES, which employs a large storage cavity to reduce the effects of beam loading.

---

[1] These angles are sometimes referred to as β, α, and γ, respectively.

TABLE 1. Estimated event yields, backgrounds, and errors on angles of the unitarity triangle for selected indirect CP violation reactions.

ϕ_i	Mode	final states	BR(B)	N_{obs}	N_{BG}	$\delta \sin 2\phi$
ϕ_1	$B^0 \to J/\psi K_S$	$l^+l^-\pi^+\pi^-$	4.0×10^{-4}	770	3	0.080
	$B^0 \to J/\psi K_S$	$l^+l^-\pi^0\pi^0$	4.0×10^{-4}	138	14	0.198
	$B^0 \to J/\psi K_L$	$l^+l^-K_L$	4.0×10^{-4}	375	69(24[a])	0.137
	$B^0 \to J/\psi K^{*0}$	$l^+l^-\pi^+\pi^-\pi^0$	1.6×10^{-3}	223	60(25[a])	0.191[b]
ϕ_2	$B^0 \to \pi^+\pi^-$	$\pi^+\pi^-$	1.3×10^{-5}	283	40	0.147[c]
	$B^0 \to \rho^\pm\pi^\mp$	$\pi^0\pi^+\pi^-$	6.0×10^{-5}	1000	0[d]	≤ 0.19[e]

Notes: (a) anti-correlated background. (b) 100% polarized K^* is assumed. (c) penguin diagrams are ignored. (d) backgrounds are ignored. (e) penguin diagrams are included.

One significant difference between KEKB and PEP–II is that in PEP–II the beams collide head-on, while in KEKB they collide at an angle of ± 11 mr. This approach allows every RF bucket (the bunch spacing is 60 cm) to be filled without undesirable parasitic collisions, resulting in a larger total beam current for a given number of particles per bunch. It also simplifies the design of the interaction region (IR) by eliminating the need for dipole magnets to separate the beams. This has the side benefit of reducing the severity of synchrotron and spent-beam-particle backgrounds.

This strategy does, however, introduce a certain amount of risk, since having the bunches pass through one another at a finite angle opens up the possibility of coupling between longitudinal and transverse oscillations. Detailed simulation work by the KEKB accelerator group indicates that this should not be a problem [1]. As a backup plan, KEK is developing superconducting "crab" cavities that will rotate the bunches as they enter the IR in such a way that they pass through one another head on. These will be ready about one year after the accelerator begins operation.

Over the past year, the focus of activity has shifted from machine studies in the accumulator ring (AR) test area, where prototypes of various accelerator components (e.g., the RF cavities) and techniques have been tested, to procurement of accelerator components and their installation. At the time of this meeting (June 1997), 70% of the ring magnets have been transported to the rings and are being installed and mass production of the ARES and SCC cavities has commenced. Also, construction of a bypass tunnel for the injection system is proceeding ahead of schedule. Commissioning of the e^+ and e^- rings is scheduled to begin in the autumn of 1998. The physics program is expected to commence in early 1999.

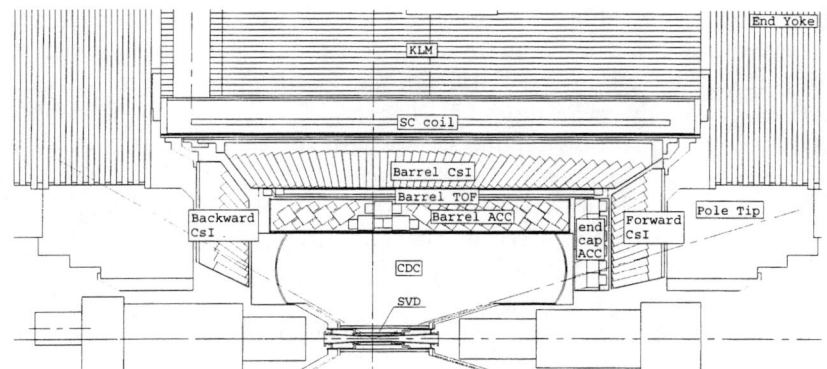

FIGURE 1. Side view of the BELLE detector.

THE BELLE DETECTOR

The BELLE collaboration numbers over 200 physicists from eight countries (Australia, China, India, Japan, Korea, Poland, Russia, Ukraine, and the U.S.). It was formally started in late 1993, although activity in the form of workshops and task forces has been ongoing since 1989.

A side view of one half of the inner part of the detector is shown in Fig. 1. Particles emanating from the IR first encounter a multi-layer silicon vertex detector (SVD), followed by a gas-filled central drift chamber (CDC), an array of aerogel Čerenkov counters (ACC), an array of time-of-flight counters (TOF), a CsI(Tℓ) crystal electromagnetic calorimeter (ECL), a superconducting solenoid, and an iron return yoke (not shown) that is instrumented with resistive plate chambers for the detection of K_L's and muons (KLM). An extreme forward calorimeter (EFC) improves the hermeticity of the detector by extending the coverage to smaller angles. The EFC will also serve as a luminosity monitor. A superconducting solenoid creates a magnetic field of 1.5 T. The general configuration of the detector is typical of e^+e^- detectors, although there is a front/back asymmetry reflecting the difference in beam energies.

The Silicon Vertex Detector

The original version of the BELLE SVD was based on an ambitious design that made use of bump bonding and a newly developed readout IC. Although this design promised important performance advantages, such as on-chip zero-suppression and the provision of trigger signals, it recently became apparent that its implementation involved significant technical and schedule risks. In view of the strong competition from BaBar and other B-physics efforts, the collaboration concluded that it would be best to defer the development of the original system in favor of a design based on existing technologies.

From preliminary efforts along these lines it appears that it will be possible to have a system that is adequate to address the primary physics goals at the beginning of the experiment. Although the details of the design remain to be determined, a likely scenario is that there will be three layers of double-sided strip detectors, providing angular coverage in the range $22° < \theta < 144°$. The most likely candidate for the readout chip is the VA1 chip from the IDE AS company in Oslo, Norway. The VA1 design is closely related to that of the Viking II [6] IC, which is based on the AMPLEX [7] principle.

Central Drift Chamber

The CDC, shown in Fig. 2 employs approximately 8,500 sense wires and 1,500 cathode strips to measure charged tracks [8]. In the region $85 < r < 275$ mm, the endplates have a conical shape, so as to accommodate the space requirements of the most forward accelerator components while maximizing the angular coverage. For $275 < r < 880$ mm the endplates are curved to maximize their strength-to-weight ratio. The endplates are held apart by carbon-fiber re-enforced plastic (CFRP) cylindrical walls.

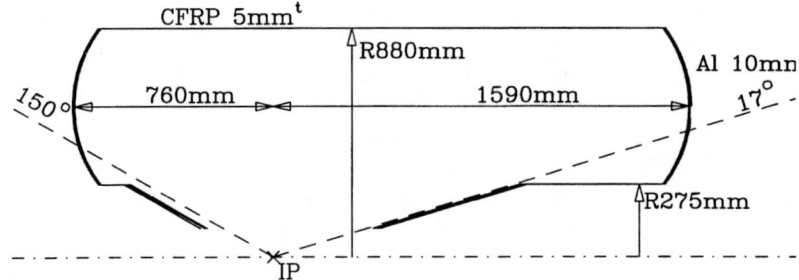

FIGURE 2. Side view of CDC.

To minimize multiple scattering, a He-based gas (50:50 He:C_2H_6) is used and the field wires are 126-μm-diameter bare aluminum. The sense wires are 30-μm-diameter gold-plated tungsten. Aging tests indicate that this combination will survive the anticipated ten-year lifetime of BELLE with a 20% loss of gain and no increase in dark current [9].

The wires are arranged in nearly square cells, with typical maximum drift distances between 8 and 10 mm. There are nine axial layers and five stereo layers plus three cathode-strip layers. The stereo angles of the wires range from 40 to 70 mrad. The cathode-strip layers, which are located at radii of 85, 100, and 101 mm, are used to determine the z-vertex of tracks for the Level 1 trigger. The strips are between 8 and 10 mm wide and run orthogonal to the wire direction.

Beam tests on a small-scale prototype operating in a magnetic field and a full-size prototype tested at $B = 0$ have demonstrated a position resolution of 130 μm and a dE/dx resolution of $\sigma = 5.2\%$. Taking the measured position resolution as input, a Monte Carlo simulation of the CDC predicts a momentum resolution as low as $\sigma_p/p \simeq 0.25\%$ in the momentum region $0.25 < p < 1.0$ GeV/c, gradually increasing to about 0.85% at $p = 5$ GeV/c.

Stringing of the approximately 32,000 wires has started and is proceeding at the rate of about 300 wires per day. That part of the work is expected to be completed in the autumn of this year, with cosmic ray tests expected in the early months of 1998.

Particle Identification

Particle identification is accomplished with a hybrid approach combining time-of-flight (TOF) counters and aerogel Čerenkov counters (ACC). The TOF (along with dE/dx information from the CDC) serves to discriminate between pions and kaons below $p = 1.2$ GeV. This momentum region encompasses most of the kaons used to "tag" the flavor of the opposite-side B^0 meson by tracing back the $b \to c \to s$ cascade. The ACC array provides K/π separation in the region above $p = 1.2$ GeV/c, which is populated by important two-body decays such as $B \to \pi\pi$ and $B \to DK$.

Time of Flight Counters

The TOF system consists of 128 BC408 plastic scintillators with dimensions $l \times w \times t = 2250 \times 60 \times 40$ mm^3. The 40 mm thickness was chosen as a compromise between timing resolution, which improves slightly with thicker scintillator, and the desire to minimize mass in front of the ECL. The light from the scintillators is read using fine-mesh photomultiplier tubes (FMPMT's), which are capable of operation in high magnetic fields. This eliminates the need for light guides and their attendant light loss and mechanical complications.

Each pair of timing counters is arrayed with a thin ($t = 5$ mm) trigger scintillator counter (TSC) made from BC412. The TSC's are read from one end by a single FMPMT. Placing the thin scintillator in coincidence suppresses potential backgrounds from low-energy gamma rays and allows the TOF system to provide a timing strobe for the Level 1 trigger. The resulting coincidence rate from the TOF/TSC array is expected to be of order 10 kHz or less, ensuring proper strobe timing for the vast majority ($\geq 99\%$) of Level 1 triggers.

Beam tests of prototype TOF counters, including the effects of the 1.5 T field, have demonstrated that timing resolutions in the range $65 < \sigma_t < 85$ ps can be obtained along the full length of the timing counters (see Fig. 3).

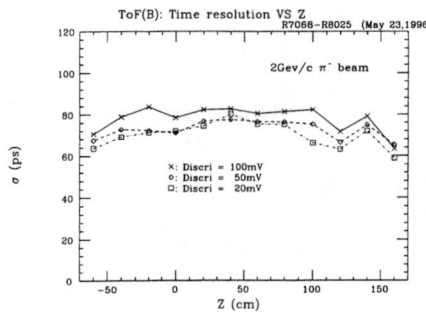
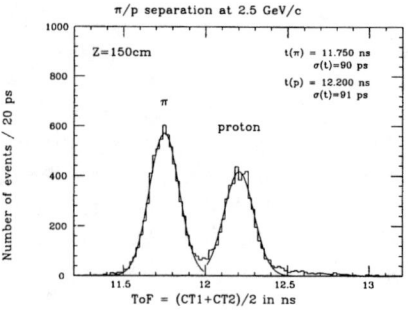

FIGURE 3. Left: Timing resolution versus position of incidence on the TOF counter. Right: A typical test-beam timing spectrum obtained with 2.5 GeV/c pions and protons.

Aerogel Čerenkov Counters

The barrel ACC array comprises 900 blocks of aerogel radiator, divided into 60 azimuthal and 15 polar-angle segments. A typical barrel radiator consists of five aerogel tiles of volume $12 \times 12 \times 2.5$ cm^3 stacked into an approximately cubical block. In the forward endcap, the maximum momentum from two-body decays extends to 4.0 GeV/c, making it difficult to cover the entire momentum range with a single-index aerogel array. Given the difficulties associated with a two-index system and the limited physics gain of endcap coverage, a design using a single $n = 1.03$ index radiator and no TOF counters was adopted. The endcap array is divided into 228 elements arranged in five concentric rings. Each endcap element is read by a single FMPMT.

The boxes that hold the aerogel are made from 200 μm aluminum sheet lined with white reflective (Goretex) paper. The index of the radiators varies from $n = 1.01$ at forward angles, where the momentum from two-body decays is highest, to $n = 1.03$ at backward angles, where the boost is considerably less. Modules with lower index (the light yield is roughly proportional to $n - 1$) are viewed by two 3-inch-diameter FMPMT's, whereas modules in the backward direction are viewed by a single 2.5-inch FMPMT. Once again, the design strikes a balance between maximizing performance (light yield) and minimizing the number of radiation lengths in front of the calorimeter.

The raw photoelectron yield for a typical counter has been measured to be approximately 20 photoelectrons. However, the effective yield $(N_{\text{eff}} = (\mu/\sigma)^2)$ is reduced by about 50% due to the non-ideal single photoelectron response of the FMPMTs. Test beam measurements using a 3.5 GeV/c hadron beam show that for an ADC threshold low enough to yield 98% efficiency for $\beta \simeq 1$ particles, the efficiency for below-threshold particles (protons in this case) was only 2% (see Fig. 4).

The production of the ACC material for the barrel is complete and the as-

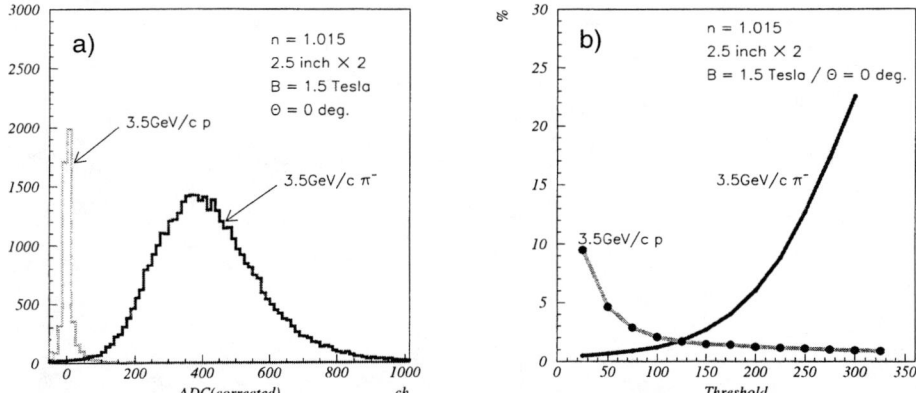

FIGURE 4. (a) Pulse height spectra for 3.5 GeV/c pions and protons. (b) Pion and proton detection efficiency as a function of ADC threshold. The angle θ refers to the angle between the symmetry axis of the FMPMT and the magnetic field.

sembly of counter boxes is well underway. Most of the 1,788 PMT's needed for the barrel and endcap arrays have been delivered. Prototype readout electronics, based on the MQT300A charge-to-time converter IC from LeCroy Corporation, have been tested and work on the production units is well advanced.

The CsI(Tℓ) Calorimeter

The CDC and PID systems are surrounded by a precision electromagnetic calorimeter based on thallium-doped cesium iodide crystals used for the detection photons and π^0's. The barrel array is a roughly cylindrical shell of inner radius 125 cm and thickness 30 cm. The forward and backward endcaps, whose inner surfaces are at $z = 196$ cm and $z = -102$ cm, respectively, extend the coverage to 91% of 4π.

The number of crystals in the barrel and endcap arrays totals 8,816. The typical crystal is 60×60 mm^2 at its midpoint (the crystals are tapered) and 300 mm long. Each crystal is read by a pair of 10×20 mm^2 silicon PIN photodiodes (Hamamatsu S2744-08). J-FET preamplifiers mounted on the crystals result in a noise of about 620 e^-. The complete preamplifier-shaper-ADC chain has an equivalent noise charge of 990 e^-, which corresponds to about 200 keV. Beam test results for a 5×5 crystal array are shown in Fig. 5. The curves quite clearly show that smaller summing clusters work best at low energy, where electronic noise is the dominant contribution to the resolution, whereas larger summing clusters are better at high energy, where shower leakage dominates.

At the time of this symposium approximately 80% of the 6800 barrel crystals

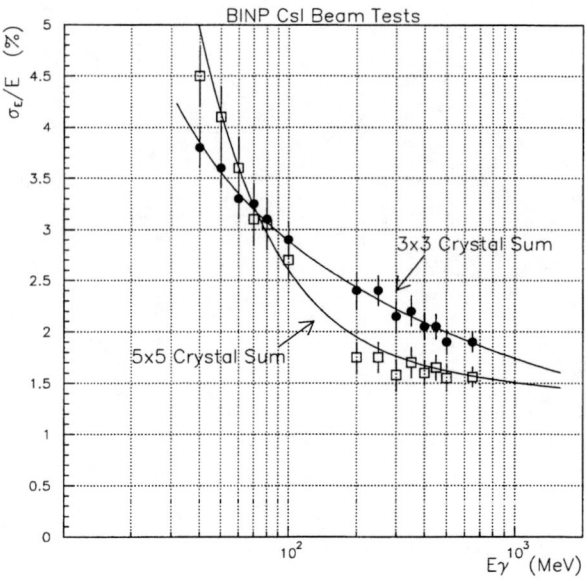

FIGURE 5. Testbeam energy resolution for a prototype 5 × 5 CsI(Tℓ) array.

have been delivered to KEK by vendors in the Ukraine, France, and China. It is expected that all barrel crystals will be on hand by October of this year, at which point delivery of the 2,200 crystals needed for the endcaps will start and continue until June 1998. A total of 14,000 photodiodes has been delivered. Work on the CsI support structure is well advanced, with delivery expected late this summer.

The K_L-μ System

The flux return of BELLE is instrumented with resistive plate chamber (RPC) detectors that are used to track muons as they penetrate the iron and to detect K_L's that interact in the yoke material. The RPC's are placed in 44-mm-wide gaps between the 47-mm-thick iron plates of the flux return. There are 15 radial gaps in the barrel region. In the endcap, there are also 15 gaps, but the iron plates are aligned perpendicular to the beam. Although the array in effect forms a crude hadronic calorimeter, the RPC's do not provide meaningful pulseheight information and are used solely to determine the position of the K_L interaction vertex, from which its direction is inferred.

RPCs are essentially planar spark chambers wherein the avalanche induced by an incident charged particle is quenched when the limited amount of charge

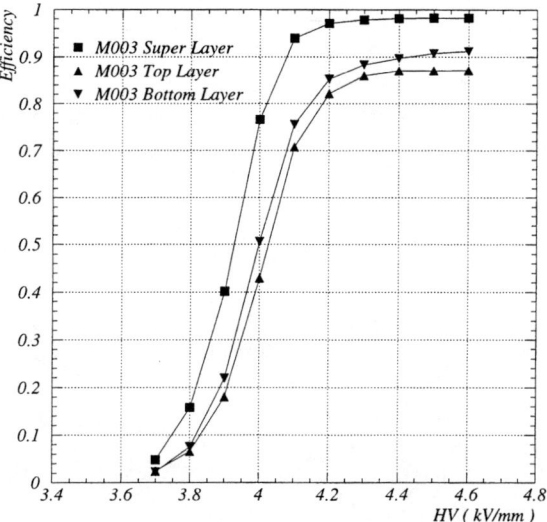

FIGURE 6. Efficiency vs. operating voltage for an endcap RPC module. The triangles show the efficiency that results when HV is applied to only one RPC in a superlayer. The squares show the efficiency when both have HV applied.

on the inner surfaces of highly resistive electrodes is exhausted [10]. The RPC is locally deadened until the inner surfaces can recharge through the resistive material of the plates. Since the highly resistive HV electrodes are effectively transparent to the fast (20-ns-wide) image pulses from the discharges, external electrodes can be used to sense the signals. The induced pulses are typically several tens of mV into 50 Ω, eliminating the need for preamplifiers.

The BELLE RPCs employ glass, which has a bulk resistivity of about 10^{12}-10^{13} Ω-cm. Glass was selected because of its ready availability (standard soda-lime float glass used in household windows is used) and its inherent stability. Since background radiation levels outside of the CsI calorimeter are extremely low, the potential loss of count-rate capability associated with this relatively high resistivity is not an issue.

Each superlayer consists of two RPC structures sandwiched between two sets of pickup electrodes. The external surfaces of the glass are coated with a thin layer of india ink (surface resistivity $\sim 10^6$ $\Omega/\square$), which distributes the high voltage bias. The strips of the external pickup electrodes are approximately 50 mm wide and are oriented at right angles so as to provide two-dimensional position information. Since each RPC is transparent to the image pulses of its partner, either RPC firing is enough to produce a detectable signal on both electrodes (when both RPCs fire the induced signal is effectively doubled).

The typical efficiency for a single RPC is 90%, resulting in a 99% superlayer efficiency. Plateau curves for a typical endcap module are shown in Fig. 6.

To date approximately 80% of the 5,000 m^2 of RPCs required for the endcap and barrel arrays have been fabricated. Most of these have been shipped to KEK and tested. Approximately half of the barrel RPCs have been installed in the BELLE magnet. An order has been placed for the production readout electronics. The balance of the barrel modules and the endcap will be installed in the magnet in the first months of 1998.

REFERENCES

1. KEKB B-Factory Design Report, KEK Report 95-7 (1995).
2. "A Study of CP Violation in B Meson Decays; BELLE Technical Design Report," KEK Report 95-1 (1995).
3. Talk by W. Innes, this Symposium.
4. BaBar Technical Design Report, SLAC-R-95-457 (1995).
5. "An Asymmetric B Factory based on PEP: Conceptual Design Report," edited by M. Zisman, SLAC372 (1991).
6. E. Nygard et al., Nucl. Instr. and Meth. **A301** (1991) 506; P. Aspell et al., Nucl. Instr. and Meth. **A315** (1992) 425.
7. E. Beuville et al., Nucl. Instr. and Meth. **A288** (1990) 157.
8. S. Uno et al., Nucl. Instr. and Meth. **A330** (1993) 55.
9. B Physics Task Force, KEK Report 93-1 (1993) 77.
10. R. Santonico and R. Cardarelli, Nucl. Instr. and Meth. **187** (1981) 377.

THE *b* QUARK AND QCD

Realizing the Potential of Quarkonium

Chris Quigg [1]

Fermi National Accelerator Laboratory [2]
P.O. Box 500, Batavia, Illinois 60510 USA

Abstract. I recall the development of quarkonium quantum mechanics after the discovery of Υ. I emphasize the empirical approach to determining the force between quarks from the properties of $c\bar{c}$ and $b\bar{b}$ bound states. I review the application of scaling laws, semiclassical methods, theorems and near-theorems, and inverse-scattering techniques. I look forward to the next quarkonium spectroscopy in the B_c system.

PROLOGUE

I am very happy to share in this celebration of the twentieth anniversary of the discovery of the b-quark. The upsilon years were a very special time for the development of particle physics. Reviewing the events of two decades ago, I was struck not only by the pace of discovery, but by how easy it was, and how much fun we had. As a young physicist of that time, I am grateful not only for the excellent science, but also for the excellent people that quarkonium quantum mechanics gave me an opportunity to know, work with, and even compete with. They taught me a great deal about physics and life.

THE EMPIRICAL APPROACH

Charmonium quantum mechanics was already well-launched when the Υ came along. Appelquist & Politzer [1] had shown that nonrelativistic quantum mechanics should apply to $Q\bar{Q}$ systems. The Cornell group [2] had shown the predictive power of the nonrelativistic potential-model approach using a "culturally determined" potential,

---

[1] Internet address: quigg@fnal.gov.
[2] Fermilab is operated by Universities Research Association Inc. under Contract No. DE-AC02-76CH03000 with the United States Department of Energy.

$$V(r) = -\frac{\kappa}{r} + \frac{r}{a^2} \; . \qquad (1)$$

Eichten & Gottfried [3] had anticipated the spectroscopy of $b\bar{b}$, predicting

$$M(\Upsilon') - M(\Upsilon) \approx 420 \text{ MeV} \qquad (2)$$
$$\approx \tfrac{2}{3}[M(\psi') - M(\psi)] \; .$$

All this meant that the deductive approach—assuming a form for the interquark potential and calculating the consequences—was in very good hands.

Jon Rosner and I were motivated to take a complementary empirical approach by the way the quarkonium problem came into our consciousness. The facts we had at our disposal in the summer of 1977 were these:

E288	$M(\Upsilon') - M(\Upsilon)$	$M(\Upsilon'') - M(\Upsilon)$
Two-level fit	650 ± 30 MeV	
Three-level fit	610 ± 40 MeV	1000 ± 120 MeV
$M(\psi') - M(\psi)$	≈ 590 MeV	

We were much impressed with the fact that the $\Upsilon' - \Upsilon$ spacing is nearly the same as $\psi' - \psi$. In an off-hand conversation with Bernie Margolis, who was also visiting Fermilab, Jon asked if he knew what kind of potential gave a level spacing independent of the mass of the bound particles. When Bernie said that it was probably some kind of power-law, Jon set off on the path that led to a potential $V(r) = r^\epsilon$ as $\epsilon \to 0$, and eventually to the logarithmic potential. Accurate numerical calculations and a scaling argument showed us that the logarithmic potential, $V(r) = C \log(r/r_0)$ indeed gave a level spacing independent of mass [4].

The logarithmic potential gives a good account of the (then known) ψ and Υ spectra, as shown in Figure 1, but is not unique in doing so. It was an easy matter to produce a modified Coulomb + linear potential that gave equal spacing for the ψ and Υ families, but for no other quark masses. Now, the logarithmic potential is the solution to an idealized statement of the experimental facts. We expect it to be a good representation of the interaction in the region of space that governs the properties of the narrow ψ and Υ states, but we have no reason to attach fundamental significance to it. It is mildly intriguing that a logarithmic confining interaction emerges from the light-front QCD approach when the Hamiltonian is computed to second order [5].

SCALING THE SCHRÖDINGER EQUATION

The Schrödinger equation for the reduced radial wavefunction in a potential $V(r) = \lambda r^\nu$ is

$$\frac{\hbar^2}{2\mu} u'' + \left[E - \lambda r^\nu - \frac{\ell(\ell+1)\hbar^2}{2\mu r^2} \right] u(r) = 0 \; . \qquad (3)$$

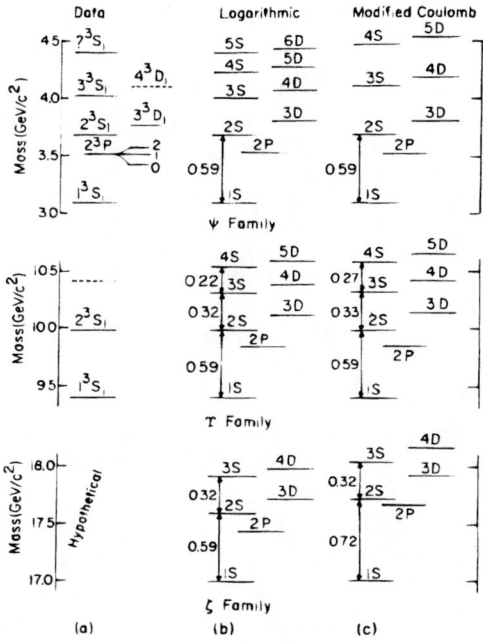

FIGURE 1. Level schemes of the ψ, Υ, and hypothetical ζ families in (a) nature, (b) the logarithmic potential, and (c) a Coulomb + linear potential. (From Quigg & Rosner [4].)

With the substitutions

$$r = \left(\frac{\hbar^2}{2\mu|\lambda|}\right)^{1/(2+\nu)} \rho , \tag{4}$$

$$E = \left(\frac{\hbar^2}{2\mu|\lambda|}\right)^{-2/(2+\nu)} \left(\frac{\hbar^2}{2\mu}\right) \varepsilon , \tag{5}$$

and the identification $w(\rho) \equiv u(r)$, we can bring the Schrödinger equation to dimensionless form [6],

$$w''(\rho) + \left[\varepsilon - \text{sgn}(\lambda)\rho^\nu - \frac{\ell(\ell+1)}{\rho^2}\right] w(\rho) = 0 . \tag{6}$$

The substitution (4) means that quantities with dimension of length scale as $[L] \propto (\mu|\lambda|)^{-1/(2+\nu)}$, whereas (5) tells us that quantities with dimensions of energy scale as $[\Delta E] \propto (\mu)^{-\nu/(2+\nu)}(|\lambda|)^{2/(2+\nu)}$. The scaling behavior in several familiar potentials is shown in Table 1.

175

TABLE 1. How length and energy observables scale with coupling strength $|\lambda|$ and reduced mass μ in power-law and logarithmic potentials.

Potential	$[L]$	$[E]$
Coulomb	$(\mu|\lambda|)^{-1}$	$\mu|\lambda|^2$
Log: $V(r) = C\log r$	$(C\mu)^{-1/2}$	$C\mu^0$
Linear	$(\mu|\lambda|)^{-1/3}$	$\mu^{-1/3}|\lambda|^{2/3}$
SHO	$(\mu|\lambda|)^{-1/4}$	$\mu^{-1/2}|\lambda|^{1/2}$
Square well	$(\mu|\lambda|)^{0}$	μ^{-1}

Quarkonium Decays

The scaling laws have immediate applications to the matrix elements that govern quarkonium decays. Electric and magnetic multipole matrix elements have dimensions

$$\langle n'|Ej|n\rangle \sim L^j, \tag{7}$$

$$\langle n'|Mj|n\rangle \sim L^j/\mu. \tag{8}$$

Radiative widths are given by

$$\Gamma(Ej \text{ or } Mj) \sim p_\gamma^{2j+1}|\langle n'|Ej \text{ or } Mj|n\rangle|^2, \tag{9}$$

so that transition rates scale with mass as

$$\Gamma(E1) \sim \mu^{-(2+3\nu)/(2+\nu)}, \tag{10}$$

$$\Gamma(M1) \sim \mu^{-(4+5\nu)/(2+\nu)}. \tag{11}$$

Probability densities have dimensions L^{-3}. Accordingly, the wave function squared at the origin scales as

$$|\Psi(0)|^2 \sim \mu^{3/(2+\nu)}, \tag{12}$$

so the leptonic width of a vector meson scales as

$$\Gamma(\mathcal{V}^0 \to \ell^+\ell^-) = 16\pi\alpha^2 e_Q^2 |\Psi(0)|^2 / M(\mathcal{V}^0)^2 \tag{13}$$
$$\sim \mu^{-(1+2\nu)/(2+\nu)}.$$

Combining (10) and (11) with (13), we see that

$$\Gamma(E1)/\Gamma(\ell^+\ell^-) \sim \mu^{-(1+\nu)/(2+\nu)}, \tag{14}$$

$$\Gamma(M1)/\Gamma(\ell^+\ell^-) \sim \mu^{-3(1+\nu)/(2+\nu)}. \tag{15}$$

For a potential less singular than a Coulomb potential, radiative decays become relatively less important than leptonic decays, as μ increases.

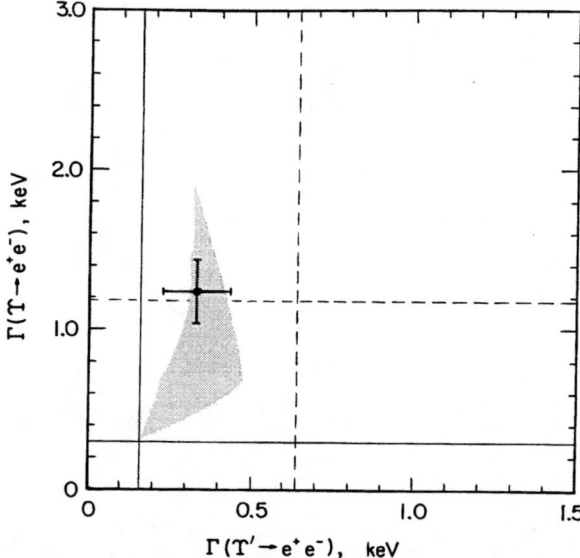

FIGURE 2. Expectations for the leptonic widths of Υ and Υ'. The lower bounds (17) are indicated for $e_b = -\frac{1}{3}$ (solid lines) and $e_b = \frac{2}{3}$ (dashed lines). The shaded region shows the widths predicted for $e_b = -\frac{1}{3}$ on the basis of twenty potentials from [20] that reproduce the ψ and ψ' positions and leptonic widths. The data point represents the 1978 DORIS results.

Measuring the b-quark's charge

For a power-law potential with $\nu \leq 1$, the scaling law (12) implies that

$$|\Psi_b(0)|^2 \geq \frac{m_b}{m_c}|\Psi_c(0)|^2 , \qquad (16)$$

which leads to

$$\Gamma(\Upsilon_n \to \ell^+\ell^-) \geq \frac{e_b^2}{e_c^2} \cdot \frac{m_b}{m_c} \cdot \frac{M(\psi_n)^2}{M(\Upsilon_n)^2}\Gamma(\psi_n \to \ell^+\ell^-) . \qquad (17)$$

The inequality holds for more general potentials than power laws. We can prove it for the ground state for any monotonically increasing potential that is concave downward [7]. For excited states we have given a derivation for the same class of potentials in semiclassical approximation.

The numerical bounds that follow from (17) are indicated in Figure 2. Measurements presented at the 1978 Tokyo Conference by the collaborations working at the DORIS storage ring [8],

$$\begin{aligned}\Gamma(\Upsilon \to \ell^+\ell^-) &= 1.26 \pm 0.21 \text{ keV} , \\ \Gamma(\Upsilon' \to \ell^+\ell^-) &= 0.36 \pm 0.09 \text{ keV} ,\end{aligned}$$

ruled decisively in favor of the $e_b = -\frac{1}{3}$ assignment. The fifth quark was indeed *bottom*.

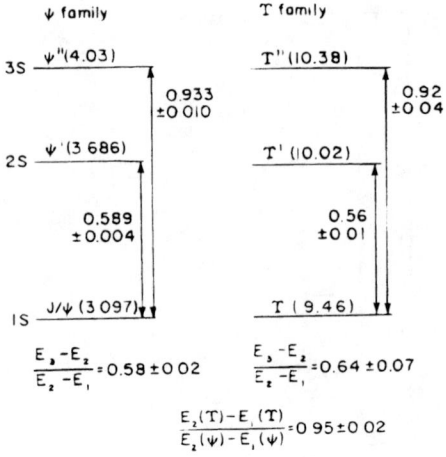

FIGURE 3. Comparison of $^3S_1$ levels of the ψ and Υ families, circa 1979.

The Order of Levels

The discovery in 1975 of the narrow resonances P_c and χ confirmed the theoretical expectation that the 2S and 2P levels are not degenerate in charmonium, as they would be in a pure Coulomb potential. In response to the question of what the 2S–2P splitting says about the interquark potential, André Martin and collaborators (R. Bertlmann, H. Grosse, J.-M. Richard, and others) constructed a series of elegant theorems on the order of levels in potentials [9,10].

After early data on the Upsilons confirmed the similarity of the J/ψ and Υ spectra, as shown in Figure 3, Martin used the scaling laws to deduce a simple power-law potential,

$$V(r) = -8.064 \text{ GeV} + (6.898 \text{ GeV})(r \cdot 1 \text{ GeV})^{0.1} \ . \tag{18}$$

The Martin potential [11] has served as a very useful template for the J/ψ, Υ, and even $\phi(s\bar{s})$ families. In addition, it has led to many informative predictions for the masses of baryons containing charm and beauty.

A Priority Dispute with Isaac Newton

About three years ago, Jon Rosner telephoned me to say that Professor Chandrasekhar had just advised him that we were in a priority dispute with Isaac Newton. "Capitulate at once!" I said. "Newton can be very terrible."

In writing his superb commentary on the *Principia Mathematica* [12], Chandra had found that Newton was the first to explore pairs of dual power-law potentials, and that he had mapped the Kepler problem into the harmonic oscillator. We had shown [13]—three centuries later—that the bound-state spectrum of an infinitely rising power-law potential,

$$V(r) = \lambda r^\nu \quad (\nu > 0) , \tag{19}$$

is connected with that of a singular potential,

$$\bar{V}(r) = \bar{\lambda} r^{\bar{\nu}} \quad -2 < (\bar{\nu} < 0) . \tag{20}$$

The paired Schrödinger equations for the two cases can be written as

$$\frac{\hbar^2}{2\mu} u''(r) + \left[E - \lambda r^\nu - \frac{\ell(\ell+1)\hbar^2}{2\mu r^2} \right] u(r) = 0 , \tag{21}$$

$$\frac{\hbar^2}{2\mu} v''(z) + \left[\bar{E} - \bar{\lambda} z^{\bar{\nu}} - \frac{\bar{\ell}(\bar{\ell}+1)\hbar^2}{2\mu z^2} \right] v(z) = 0 , \tag{22}$$

where $(\nu + 2)(\bar{\nu} + 2) = 4$ and $\bar{E} = \lambda(\bar{\nu}/\nu)^2$, $\bar{\lambda} = -E(\bar{\nu}/\nu)^2$, $(\bar{\ell} + 1/2)^2 \nu^2 = (\ell + 1/2)^2 \bar{\nu}^2$, and $z = r^{1+\nu/2}$.

The familiar quantum-mechanical correspondence between the Coulomb and harmonic oscillator problems emerges as a special case. For circular orbits, Newton cites a relation between ν and $\bar{\nu}$ equivalent to ours. We did capitulate, and so far we have not suffered any indignities at Newton's hands. And because Jon is a scholar, he and Aaron Grant have written an excellent historical review of the classical and quantum-mechanical analyses [14].

COUNTING NARROW LEVELS OF QUARKONIUM

Eichten & Gottfried had argued that a $Q\bar{Q}$ system with $m_Q \gg m_c$ would have at least three narrow $^3S_1$ levels [3]. As Kurt recalled in his talk, this implies a very rich spectroscopy, which figured prominently in the scientific case for CESR. However, the observed $\Upsilon' - \Upsilon$ spacing is much larger than the 420 MeV they predicted. Figure 4 shows that in the logarithmic potential, we would predict three or four narrow $^3S_1$ levels of Υ, in agreement with Eichten & Gottfried's expectation. This circumstance led us to ask how general is the expectation, and on what does it depend?

Semiclassical methods (whose power Taiji Yamanouchi had impressed on us) led us to a remarkable general result [15]: The number of narrow $^3S_1$ levels is

$$n \approx 2 \cdot \left(\frac{m_Q}{m_c} \right)^{1/2} . \tag{23}$$

The derivation is short enough to reproduce in full.

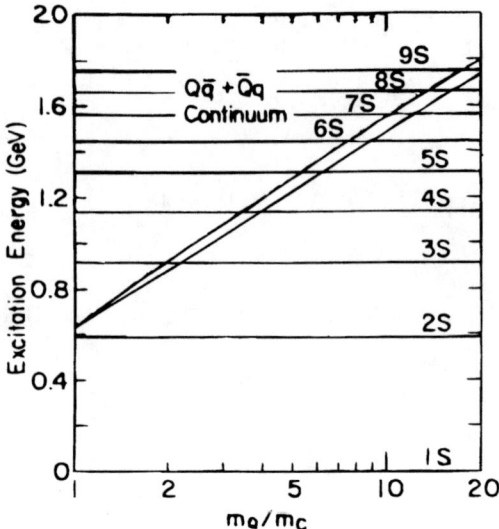

FIGURE 4. Plot of the $Q\bar{q} + \bar{Q}q$ threshold relative to the (ground-state) 1^3S_1 $Q\bar{Q}$ level as a function of the ratio of the heavy-quark mass m_Q to the charmed quark mass m_c for a logarithmic potential. Upper curve: reduced-mass and hyperfine corrections included. Lower curve (straight line): reduced-mass and hyperfine corrections ignored. Horizontal lines denote the 2^3S_1, 3^3S_1, ... $Q\bar{Q}$ levels in this potential. (From Quigg and Rosner [15].)

Set the zero of energy at $2m_Q$. The threshold for the dissociation of quarkonium $(Q\bar{Q}) \to Q\bar{q} + \bar{Q}q$ lies at an excitation energy $\delta(m_Q)$. If $V(r)$ binds $Q\bar{Q}$ states rising at least $\delta(m_Q)$ above $2m_Q$, then the WKB quantization condition is

$$\int_0^{r_0} dr[m_Q(\delta(m_Q) - V(r))]^{1/2} \simeq (n - \tfrac{1}{4})\pi , \qquad (24)$$

where r_0 is the point at which $V(r_0) = \delta(m_Q)$. As Eichten & Gottfried had observed, $\delta(m_Q) \equiv 2m(\text{lowest } Q\bar{q} \text{ state}) - 2m_Q$ approaches a finite limit δ_∞, independent of m_Q, as $m_Q \to \infty$. This means that the only dependence of (24) on the heavy-quark mass is the explicit factor of $\sqrt{m_Q}$ on the left-hand side. We have, by inspection, the general result

$$(n - \tfrac{1}{4}) \propto \sqrt{m_Q} . \qquad (25)$$

This universal behavior is realized in different ways for different potentials, as illustrated in Figure 5. The examples chosen are $V(r) = r$, $V(r) = \ln r$, and $V(r) = -r^{-1/2}$, for which $\Delta E \propto \mu^{-1/3}$, μ^0, and $\mu^{+1/3}$, respectively. All the levels fall deeper into the potential as the reduced mass μ increases, in conformity with the Feynman–Hellmann theorem. For the singular potential with $\nu = -\tfrac{1}{2}$, the levels spread apart as they sink into the well. For the linear

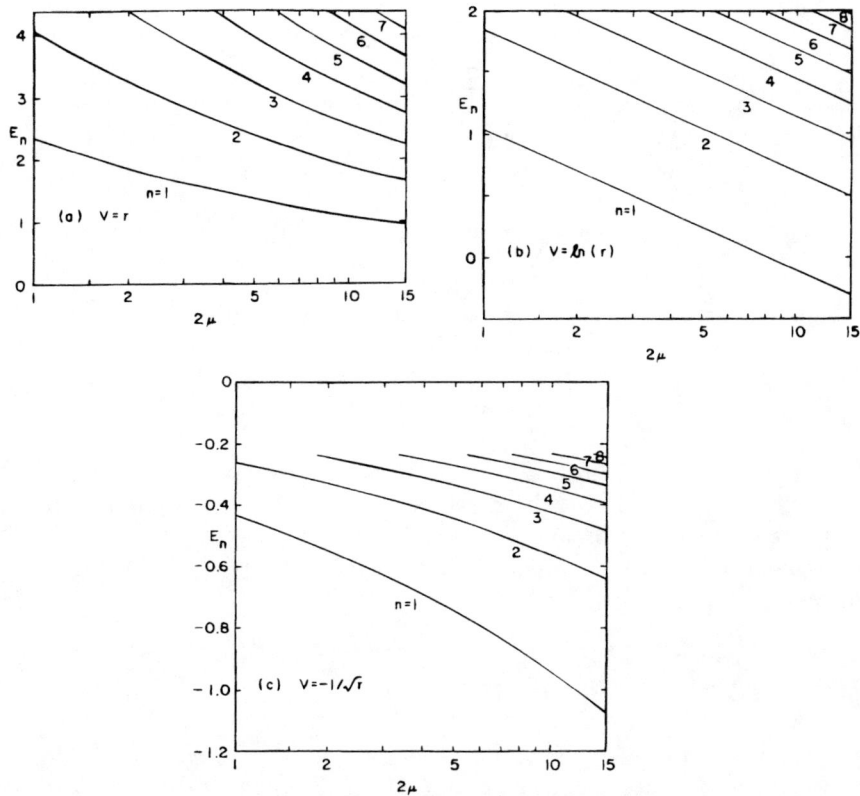

FIGURE 5. Comparison of the reduced-mass dependence of energy levels in three potentials: (a) $V(r) = r$; (b) $V(r) = \ln r$; (c) $V(r) = -r^{-1/2}$.

potential, the levels are packed more densely as μ rises. The logarithmic potential is an intermediate case in which the level spacing is independent of the mass and all levels fall into the well at a common rate given by $E_i(\mu') = E_i(\mu) - \frac{1}{2} \ln(\mu'/\mu)$.

SEMICLASSICAL METHODS AND RESULTS

The power of the WKB approximation for the counting problem encouraged us to explore other applications of semiclassical methods. Evaluating the nonrelativistic connection

$$|\Psi_n(0)|^2 = \frac{\mu}{2\pi} \left\langle \frac{dV}{dr} \right\rangle_n \tag{26}$$

in the semiclassical approximation, we connect the square of the s-wave wave function at the origin to the level density:

$$|\Psi_n(0)|^2 = \frac{(2\mu)^{3/2}}{4\pi^2} E_n^{1/2} \frac{dE_n}{dn} \qquad (27)$$

(for a nonsingular potential) [16]. For example, in a linear potential $V(r) = \lambda r$, $|\Psi_n(0)|^2$ is independent of n, by (26), so $E_n \sim n^{2/3}$.

Using the connection

$$\Gamma_n \equiv \Gamma(\mathcal{V}_n^0 \to \ell^+\ell^-) = 16\pi\alpha^2 e_Q^2 \frac{|\Psi_n(0)|^2}{M(\mathcal{V}_n^0)^2}, \qquad (28)$$

we can derive a variety of semiclassical sum rules, including

$$\sum_{n=\text{narrow}} \frac{\Gamma_n}{M_n^p} \simeq \frac{4\alpha^2 e_Q^2 m_Q^{3/2}}{\pi} \int_0^\delta \frac{dE\, E^{1/2}}{(2m_Q + E)^{2+p}}, \qquad (29)$$

where $\delta = 2M(Q\bar{q}) - 2m_Q$. These are useful in evaluating the heavy-quark mass m_Q, and in calculating the cross section for heavy-quark photoproduction using vector-meson-dominance techniques.

The connection (27) was generalized to higher partial waves by Bell and Pasupathy [17], who found

$$\left|\frac{d^\ell R_{n\ell}(r)}{dr^\ell}\right|^2_{r=0} = \frac{1}{\pi} \left[\frac{\ell!}{(2\ell+1)!!}\right]^2 \left(\frac{2\mu E_{n\ell}}{\hbar^2}\right)^{\ell+1/2} \frac{\partial(2\mu E_{n\ell}/\hbar^2)}{\partial n}, \qquad (30)$$

and generalized to include singular potentials by Moxhay & Rosner [18].

For a monotonically increasing potential, the semiclassical quantization condition

$$\int_0^{r_0} dr [2\mu(E - V(r))]^{1/2} = (n - \tfrac{1}{4})\pi \qquad (31)$$

connects the shape of the potential to the level density:

$$r(V) = \frac{2}{2\mu^{1/2}} \int_0^V dE(V - E)^{1/2} \left[\frac{dE_n}{dn}\right]^{-1}. \qquad (32)$$

Equation (32) is the semiclassical solution to the inverse bound-state problem. Although we never applied it to the quarkonium problem, it stimulated us to think of ways to reconstruct the interquark potential from the properties of the narrow quarkonium levels.

THE INVERSE BOUND-STATE PROBLEM

In one space dimension, binding energies and phase shifts uniquely define a symmetric potential $V(x) = V(-x)$, for which $V(\infty)$ approaches a constant

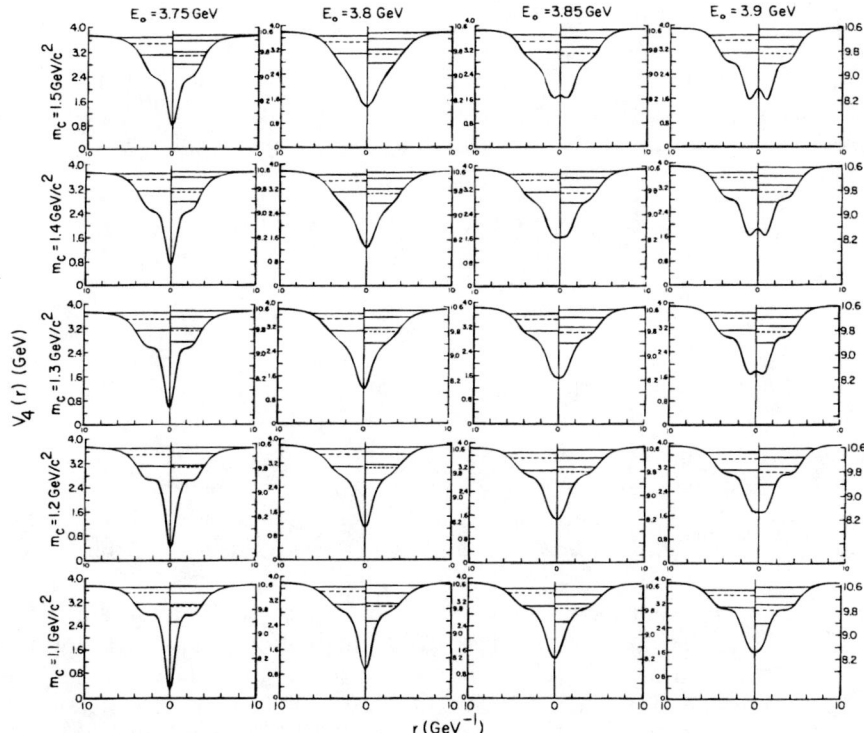

FIGURE 6. Interquark potentials reconstructed from the masses and leptonic widths of $\psi(3.095)$ and $\psi'(3.684)$. The levels of charmonium are indicated on the left-hand side of each graph. Those of the Υ family are shown on the right-hand side of each graph. The solid lines denote $^3S_1$ levels; dashed lines indicate the 2^3P_J levels. (From Thacker, Quigg, and Rosner [20].)

(finite) value. For a "reflectionless" potential (trivial phase shifts), $V(x)$ is an algebraic function of the binding energies [19]. For the s-wave inverse problem *in three dimensions,* the central potential is implied by binding energies and wave functions at the origin.

Thacker, Rosner, and I developed a method of successive approximation to confining potentials in terms of a sequence of reflectionless potentials that support a finite number of bound states [20]. (It is possible to prove interesting statements about the convergence of the procedure [21].) Figure 6 shows our first attempts to reconstruct potentials from what was known about the narrow $c\bar{c}$ levels, and to use those potentials to predict the properties of $b\bar{b}$ states. In time, we were able to determine potentials separately from the ψ and Υ families, and compare them. They agree remarkably well, except at the shortest distances, to which the Υ spectrum has greater sensitivity [22].

The inverse-scattering approach is free from assumptions about the short-distance and long-distance behavior of the potential. It provided additional evidence for flavor independence of the interquark potential, and gave us new information on the shape of the potential and where it is determined—for $0.1 \text{ fm} \lesssim r \lesssim 1 \text{ fm}$. Some important methodological improvements have been achieved using techniques of supersymmetric quantum mechanics [23].

MESONS WITH BEAUTY AND CHARM

The next hurrah for quarkonium physics will be the experimental investigation of the B_c family of $b\bar{c}$ bound states. The B_c family is interesting as a heavy-heavy system that occupies the region of space between the J/ψ and the Υ. Since we know the heavy-quark potential in that region, we should be able to calculate the properties of the $b\bar{c}$ states reliably. Unlike the excited $c\bar{c}$ and $b\bar{b}$ states, all of the $b\bar{c}$ levels below BD threshold are stable against strong decays to light flavors ($b\bar{c} \not\to$ gluons). They cascade by photonic or hadronic transitions to the B_c ground state. The interest in these states is not merely academic. They will soon be discovered and studied at the Tevatron Collider through the decays $B_c \to \psi\pi, \psi\ell\nu, \ldots$

Estia Eichten and I have computed the spectrum of $b\bar{c}$ bound states in a number of interquark potentials [24]. We find that the mass of the ground state should lie in the range $M_{B_c} = 6.258 \pm 0.020 \text{ GeV}/c^2$. The low-lying levels in the Buchmüller–Tye potential [25] are shown in Figure 7. It is noteworthy that approximately 15 s, p, d-wave states lie below BD threshold.

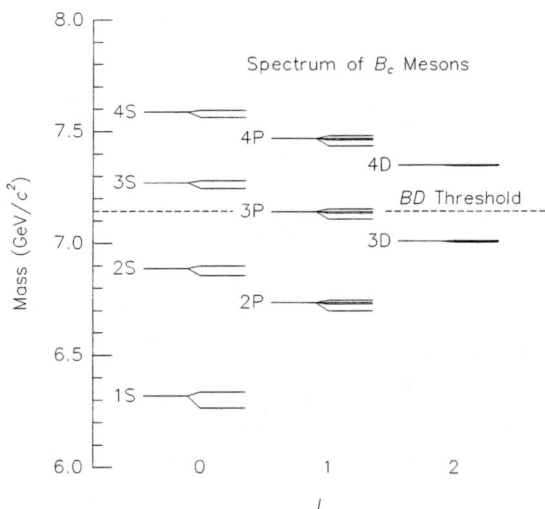

FIGURE 7. The spectrum of $b\bar{c}$ states in the Buchmüller–Tye potential (after Eichten & Quigg [24]).

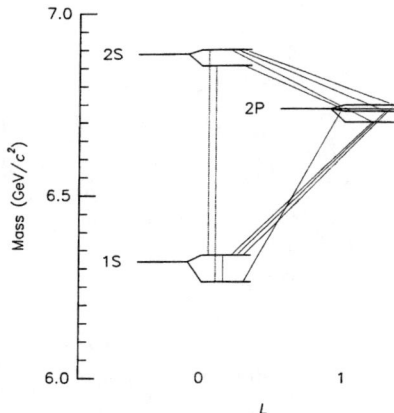

FIGURE 8. Prominent transitions in the B_c spectrum (after Eichten & Quigg [24]).

We have also computed the E1, M1, and hadronic transitions between $b\bar{c}$ levels. The transitions involving $n=2$ and $n=1$ states are shown in Figure 8. The narrow widths of these excited states are gathered in Table 2. We find that the deep binding of the heavy quarks has a profound influence on decay rates of the B_c. For example, we estimate that $f_{B_c} \approx 420$ MeV $\gtrsim 3f_\pi$, which implies that the purely leptonic decay $B_c \to \tau\nu_\tau$ will be unusually prominent.

In view of the CDF Collaboration's success in reconstructing the χ_c and χ_b states, I believe that a reasonable—though challenging—experimental goal will be to map the eight lowest-lying $b\bar{c}$ states (1S, 2S, 2P) through the transitions $2^3S_1 \to 1^3S_1 + \pi\pi$, $2^1S_0 \to 1^1S_0 + \pi\pi$, $B_c + 455$-MeV γs, and $(B_c^* \to B_c\gamma(72 \text{ MeV})) + 353$-, 382-, 397-MeV γs.

Phenomenological issues raised by the B_c family include the systematics of spin splittings for the unequal-mass $b\bar{c}$ system and the importance of relativistic $\mathcal{O}(\beta^2)$ corrections. As a third quarkonium system, B_c should provide a splendid test of *a priori* calculations from lattice QCD.

TABLE 2. Total widths of low-lying excited states of B_c.

$b\bar{c}$ State	Total Width [keV]
1^3S_1	0.135
2^1S_0	55
2^3S_1	90
2^3P_0	79
$2(1^+)$	100
$2(1^{+\prime})$	56
2^3P_2	113

ENVOI

What have we learned from two decades of quarkonium spectroscopy? The first lesson, which underlies all the others, is that nonrelativistic quantum mechanics is an appropriate tool for interpreting the quarkonium spectra. Using this tool, we have been able to demonstrate by comparing the $c\bar{c}$ and $b\bar{b}$ systems that the force between quarks is flavor independent, as we expect from QCD. Moreover, we have been able to map the interaction between heavy quarks in the range $0.1 \text{ fm} \lesssim r \lesssim 1 \text{ fm}$.

The potential-model approach allows a predictive spectroscopy, including calculations of spin splittings, E1 transition rates, the characteristics of 2S → 1S hadronic transitions, and the properties of wave functions at the origin, which are crucial inputs for calculations of quarkonium production in hadronic interactions.

It goes without saying that we have also learned a lot about quantum mechanics!

What can we hope to learn in the years to come? Among experimental goals, we should endeavor to complete the charmonium spectrum: refine our knowledge of the η_c and the ^3P_J states, confirm the $^1P_1$ (h_c) level, find the η_c', and search for narrow D-states. It is also desirable to expand our knowledge of the Υ spectrum: locate the η_b and η_b' and the 1^1P_1 (h_b) level, and search for 1D and 2D states. I am optimistic that we shall soon find the B_c and begin to explore the $b\bar{c}$ spectrum.

On the theoretical side, we should be able to refine our understanding of relativistic effects and spin splittings. (Information from the B_c spectrum should help, in time.) It may be profitable to revisit the coupled-channel effects that influence the spectrum near and above the flavor threshold. There may be lessons of value for B-factory experiments here. And finally, we theorists are determined to solve QCD and predict the interaction between heavy quarks.

ACKNOWLEDGEMENTS

It is a pleasure to thank Dan Kaplan and our other IIT hosts for the invitation to speak and for four pleasant and stimulating days in Chicago. I am grateful to Stew Smith and the Princeton University Physics Department for generous hospitality during the spring semester of 1997. Kyoko Kunori provided valuable calligraphic assistance [).]. I am indebted to Ken Lane, Fermilab Visual Media Services, and the CERN Public Information Office for historical photographs shown in my talk.

I am happy to have this opportunity to thank my quarkonium friends and collaborators, Hank Thacker, Waikwok Kwong, Jonathan Schonfeld, Peter Moxhay, Taiji Yamanouchi, Leon Lederman, André Martin, and especially

Jon Rosner and Estia Eichten. I also want to thank absent friends Ben Lee, John Bell, and S. Chandrasekhar for many important lessons.

REFERENCES

1. T. Appelquist and H. D. Politzer, *Phys. Rev. Lett.* **34**, 43 (1975).
2. See Kurt Gottfried's excellent summary of the work of the Cornell group, *These Proceedings.*
3. E. Eichten and K. Gottfried, *Phys. Lett.* **66B**, 286 (1977).
4. C. Quigg and J. L. Rosner, "Quarkonium Level Spacings," *Phys. Lett.* **71B**, 153 (1977).
5. R. J. Perry, "A Simple Confinement Mechanism for Light-Front Quantum Chromodynamics," (electronic archive: hep-th/9411037); M. M. Brisudova, R. J. Perry, K. G. Wilson, *Phys. Rev. Lett.* **78**, 1227 (1997).
6. C. Quigg and J. L. Rosner, "Scaling the Schrödinger Equation," *Comments Nucl. Part. Phys.* **8A,** 11 (1978).
7. J. L. Rosner, C. Quigg, H. B. Thacker, "Measuring the Fifth Quark's Charge: The Role of Υ Leptonic Widths," *Phys. Lett.* **74B**, 350 (1978).
8. See the summary of Υ and Υ' properties reported in Session B1, on p. 265 of the *Proceedings of the 19th International Conference on High Energy Physics,* Tokyo, 1978, edited by S. Homma, M. Kawaguchi, and H. Miyazawa (Physical Society of Japan, Tokyo, 1979).
9. Their work is surveyed in H. Grosse and A. Martin, "Exact Results on Potential Models for Quarkonium Systems," *Phys. Rep.* **60**, 341 (1980).
10. J. L. Rosner, "Charm and Beauty in Particle Physics," an appreciation on the occasion of André Martin's retirement, (electronic archive: hep-ph/9501291).
11. A. Martin, Phys. Lett. B **93,** 338 (1980); in *Heavy Flavours and High Energy Collisions in the 1-100 TeV Range,* edited by A. Ali and L. Cifarelli (Plenum Press, New York, 1989), p. 141.
12. S. Chandrasekhar, *Newton's Principia for the Common Reader* (Clarendon Press, Oxford, 1995).
13. C. Quigg and J. L. Rosner, "Quantum Mechanics with Applications to Quarkonium," *Phys. Rep.* **56**, 167 (1979).
14. A. Grant and J. L. Rosner, "Classical Orbits in Power-Law Potentials," *Am. J. Phys.* **62,** 310 (1994).
15. C. Quigg and J. L. Rosner, "Counting Narrow Levels of Quarkonium," *Phys. Lett.* **72B**, 462 (1978).
16. C. Quigg and J. L. Rosner, "Semiclassical Sum Rules," *Phys. Rev.* **D17**, 2364 (1978).
17. J. S. Bell and J. Pasupathy, *Phys. Lett.* **83B**, 389 (1979); *Z. Phys. C* **2**, 183 (1979).
18. P. Moxhay and J. L. Rosner, *J. Math. Phys.* **21**, 183 (1980).
19. For the case of confining potentials, see H. B. Thacker, C. Quigg, and

J. L. Rosner, "Inverse Scattering Approach to Quarkonium Potentials. I: One-Dimensional Formalism and Methodology," *Phys. Rev.* **D18**, 274 (1978).

20. H. B. Thacker, C. Quigg, and J. L. Rosner, "Inverse Scattering Approach to Quarkonium Potentials. II: Applications to ψ and Υ Families," *Phys. Rev.* **D18**, 287 (1978).
21. J. F. Schonfeld, *et al.*, "On the Convergence of Reflectionless Approximations to Confining Potentials," *Ann. Phys. (NY)* **128**, 1 (1980).
22. C. Quigg, H. B. Thacker, and J. L. Rosner, "Constructive Evidence for Flavor Independence of the Quark-Antiquark Potential," *Phys. Rev.* **D21**, 234 (1980); C. Quigg and J. L. Rosner, "Further Evidence for Flavor-Independence of the Quark-Antiquark Potential," *Phys. Rev.* **D23**, 2625 (1981).
23. W. Kwong and J. L. Rosner, in the Nambu Festschrift volume, *Prog. Theor. Phys. (Suppl.)* **86**, 366 (1986); W. Kwong, H. Riggs, J. L. Rosner, and H. B. Thacker, *Phys. Rev.* **D39**, 1242 (1989).
24. E. J. Eichten and C. Quigg, *Phys. Rev.* **D49**, 5845 (1994).
25. W. Buchmüller and S.-H. H. Tye, *Phys. Rev.* **D24**, 132 (1981).

Hadronic Production of Heavy Quarks

Matteo Cacciari

Deutsches Elektronen-Synchrotron DESY
D-22603 Hamburg, Germany

Abstract. We review the status of theoretical evaluations of heavy quark and heavy quarkonium hadroproduction cross sections and their comparisons with experimental data.

INTRODUCTION

When, more than twenty years ago, charm was discovered [1] and subsequently interpreted [2] as the first heavy quark ever observed, it came as a big surprise. The discovery of bottom [3] a few years later produced perhaps (I wasn't around at that time...) less excitement, but was surely equally important in extending our knowledge of what is now called the heavy quark sector.

This sector is today well known, and probably recently completed by the discovery of the top quark [4]. Theoretical physics of heavy quarks has therefore shifted gear, and moved from "discovery mode" to "precision physics."

The new name of the game is now testing Quantum Chromodynamics, by checking its predictions against experimental results. The latter have now grown quite accurate, and therefore demand equally precise theoretical calculations. In this talk I shall describe the state of the art of such calculations for heavy quark hadroproduction.

I shall first review the fixed-order next-to-leading order (NLO) QCD calculation, now available for total cross sections and one- and two-particles distributions. It is a consolidated result and provides a benchmark for future developments.

Large logarithms appear in this calculation and potentially make it less reliable in some regimes: $\log(S/m^2)$ and $\log(p_T^2/m^2)$ become large when the center of mass energy $\sqrt{S}$ or the transverse momentum p_T of the observed quark is much larger than its mass. Large $\log(1 - 4m^2/\hat{s})$ appear when the heavy quarks are produced close to the partonic threshold. I shall briefly describe the resummation of $\log(1 - 4m^2/\hat{s})$ and $\log(p_T^2/m^2)$ terms, and the inclusion of non-perturbative fragmentation effects.

I shall finally also briefly comment on the subject of heavy quarkonium production, where our understanding seems to have been greatly increased by a lot of recent theoretical activity.

I NLO CALCULATION

The road to the NLO evaluation of heavy quark hadroproduction cross sections was paved by Collins, Soper and Sterman [5], who argued that the following factorization formula holds:

$$d\sigma(H_1 H_2 \to Q\overline{Q}; m) = \sum_{ij} \int f_{i/H_1} f_{j/H_2} d\hat{\sigma}(ij \to Q\overline{Q}; m) + \mathcal{O}\left(\frac{\Lambda_{QCD}}{m}\right), \quad (1)$$

where the summation over partons i and j runs only over gluons and light quarks, and the heavy quarks are generated only at the perturbative level, by gluon splitting. The cross section explicitly depends on the heavy quark mass m and on all other scales entering the problem (total energy $\sqrt{S}$, transverse momentum p_T, etc.).

Along these lines, explicit calculations were performed by two groups, Nason, Dawson and Ellis [6] on one side and Beenakker, Kuijf, Meng, van Neerven, Schuler and Smith [7] on the other. More recently Mangano, Nason, and Ridolfi (MNR) [8] have presented a Montecarlo integrator, based on the first of the two calculations, which can provide fully exclusive cross sections, thereby allowing detailed comparisons with experimental data. A very extensive collection of such comparisons is presented in a recent review [9], from which we select some plots to be shown here.

The ratios of next-to-leading over leading order predictions for total cross sections depend on $m/\sqrt{S}$ and are about 1.3 for top production at Tevatron energy ($\sqrt{S} = 1800$ GeV) and of order 2 or larger for charm and bottom already at fixed target energy. Large uncertainties, due to monotonic renormalization/factorization scale dependence, are present for charm and bottom, while the prediction is fairly reliable for top ($\pm 10\%$), as shown in figure 1.

One-particle inclusive differential cross sections, usually p_T distributions, can also be considered. Fig. 2 shows on the left a comparison of E769 pion-nucleon fixed target data with NLO QCD plus two non-perturbative contributions: an intrinsic initial transverse momentum k_T of the colliding partons, with $\langle k_T \rangle = 1$ GeV, and fragmentation effects of the charm into the observed charmed hadrons, described with the aid of a Peterson [10] fragmentation function (FF) with $\epsilon = 0.06$. The overall result (the dot-dashed line) can be seen to be slightly softer than the data. Uncertainties are however large: a larger mass, a larger $\langle k_T \rangle$ or a harder FF could help improve the agreement. We should mention that, while the inclusion of the non-perturbative contributions

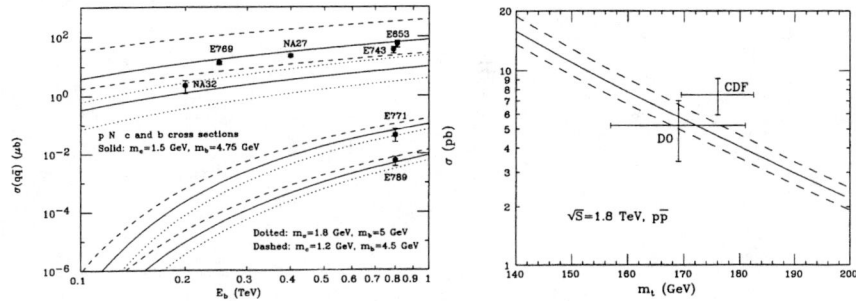

FIGURE 1. Charm and bottom production at fixed target experiments vs. NLO theoretical predictions (left), and top production at the Tevatron (right), from ref. [9].

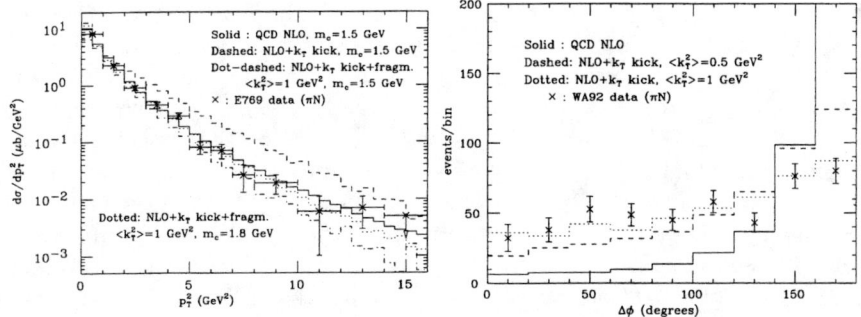

FIGURE 2. Charm p_T inclusive distribution in πN collisions vs. theoretical predictions [NLO calculation + initial k_T ($\langle k_T^2 \rangle = 1$ GeV^2) + non-pert. fragmentation (Peterson, $\epsilon = 0.06$)] on the left, and azimuthal correlation of the $c\bar{c}$ pair on the right. From ref. [9].

might appear unnecessary here, it looks however mandatory when considering two-particle distributions or even p_T distributions in photon-hadron collisions, pure NLO QCD being there clearly unable to describe the data by itself. An example is given in fig. 2, on the right, where WA92 pion-nucleon data for the azimuthal distance $\Delta\phi$ of the c and the $\bar{c}$ are compared to theoretical predictions: the inclusion of the k_T kick brings the curve into agreement with the data. More such comparisons can be found in [9]. The overall picture suggests some consistency between the non-perturbative inputs needed to describe the one- and the two-particle distributions in both photo- and hadroproduction.

More comparisons of p_T distributions can be done with the data for bottom production taken at the Tevatron by the CDF and D0 experiments. These distributions initially caused some concern, as they were markedly higher than the NLO QCD predictions. The data have now come down a bit, thanks to a better understanding of some decay chains used in the analysis (like

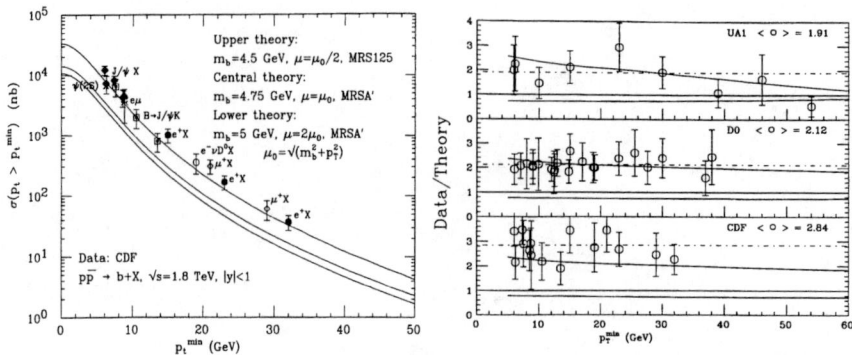

FIGURE 3. Bottom p_T distributions data versus theoretical predictions, from ref. [9].

the $B \to J/\psi \to \mu^+\mu^-$ one) and to the use of microvertex silicon detectors which allow a much better identification and reconstruction of the heavy quark events. But, still, theory and data are not in perfect agreement. Fig. 3 shows on the left the data from CDF and on the right the data/theory ratios from CDF, D0 and UA1. Data can be seen to overshoot the central theoretical predictions by factors between two and three, and to be in fair agreement with the upper edge of the theoretical uncertainty band. One can thus conclude that no serious disagreement is present, but that the situation certainly deserves further investigations, also in the light of data on forward bottom production by the D0 Collaboration, which appear to be about a factor of four higher than the theoretical prediction.

II RESUMMATION OF THRESHOLD EFFECTS

As we mentioned in the Introduction, potentially large logarithms appear in the NLO fixed-order calculation and make it less reliable in some regimes. We deal now with $\log(1 - 4m^2/\hat{s})$ terms, which appear in the partonic cross section in the form

$$\hat{\sigma}(\hat{s}) = \sigma_0(\hat{s}) \left[1 + C\alpha_s \log^2(1 - \frac{4m^2}{\hat{s}}) + \cdots \right]. \qquad (2)$$

These logarithms become large when the cms energy the colliding partons have is close to the invariant mass of the two produced heavy quarks, that is to say, when the quarks are produced close to the threshold. Resummation of such terms can turn out to be phenomenologically important when the quark is very heavy with respect to the hadronic collision energy: top production at the Tevatron, with the top weighing 175 GeV and the machine delivering 1800 GeV, could be such a case.

In order to better investigate this issue three groups have in recent years attempted such a resummation. We are not going to go into any details here, to be found in the respective papers, but simply summarize their conclusions and point to similarities and differences of approaches and results.

Laenen, Smith and van Neerven [11] performed the resummation in x-space, by directly exponentiating the leading log (LL) $\alpha_s \log^2(1-x)$ term, where $x = 4m^2/\hat{s}$. Their formulae necessitate the introduction of an artificial infrared cutoff which serves a double purpose: it avoids hitting the Landau pole when the now z-dependent argument of the running coupling α_s tends to zero as $z \to 1$, and it regulates an otherwise divergent integral of the form $\int_0^1 dz \exp[|a| \log^2(1-z)]$. They predict a 10% increase over the NLO fixed order calculation for top production at the Tevatron, but with a large uncertainty due to the dependence on the unphysical cutoff.

Berger and Contopanagos [12] instead perform the resummation in Mellin moments space, and then invert back to x-space for producing phenomenological predictions. When performing this inversion they apply a so-called principal value prescription to deform the integration contour and avoid hitting the Landau pole. They then discard all the non-leading terms generated by the inverse transform, arguing they are not universal, and only retain the LL ones. The further need for an infrared cutoff to avoid the $\int_0^1 dz \exp[|a|\log^2(1-z)]$ divergence is met by choosing it in such a way that all integrations are confined to a perturbative domain, i.e. to regions where the discarded non-leading terms are really subdominant. Through this procedure they also find an increase of about 10%, but claim a small uncertainty due to the motivated choice of the cutoff.

A third evaluation of soft-gluon resummation effects has been performed by Catani, Mangano, Nason and Trentadue [13]. They also perform the resummation in Mellin space, and avoid the Landau pole by what they call minimal prescription, i.e. a choice of the integration contour which leaves the non-perturbative branch cut to the right. After doing so, they perform the z integration by retaining both the leading and the next-to-leading (NLL) terms. Their argument is that the NLL contributions are generated by momentum conservation and, while formally subleading, discarding them leads to factorially divergent integrals and hence to the need for a cutoff. The phenomenological outcome of their investigation, at variance with previous findings, is that top hadroproduction at the Tevatron energy is predicted to increase by only 1% with respect to the fixed order result. They also claim little uncertainty on this result due to the absence of an explicit cutoff.

Whatever the approach one considers correct, we are however still far from experimentally probing such fine details (see experimental errors in fig. 1).

III LARGE TRANSVERSE MOMENTUM RESUMMATION

Another kind of large logarithmic terms appearing in the NLO calculation and eventually spoiling its convergence are $\log(p_T^2/m^2)$ terms. They also need therefore to be resummed to all orders to allow for a sensible phenomenological prediction. Such a resummation has been performed along the following lines [14].

One observes that in the large-p_T limit ($p_T \gg m$) the only important mass terms are those appearing in the logs, all the others being power suppressed. This means that an alternative description of heavy quark production can be achieved by considering *massless* quarks and providing at the same time perturbative distribution and fragmentation functions also for the heavy quark, describing the logarithmic mass dependence. The factorization formula becomes

$$d\sigma(p_T) = \sum_{ijk} \int F_{i/H_1}(\mu, [m]) F_{j/H_2}(\mu, [m]) d\hat{\sigma}(ij \to k; p_T, \mu) D_k^Q(\mu, m), \quad (3)$$

with parton indices i,j and k also running on Q, taken massless in $\hat{\sigma}$, now an $\overline{\text{MS}}$ subtracted cross section for light parton production. The dependence on m of the parton distribution functions $F_{i/H}$, shown between square brackets in eq. (3), is only there when i or j happens to be the heavy quark Q.

The key point is that the large mass of the heavy quark allows for the evaluation in perturbative QCD (pQCD) of its distribution and fragmentation functions. Initial state conditions for $F_{Q/H}(\mu_0 = m)$ [15] and $D_k^Q(\mu_0 \simeq m)$ [16] can be calculated in pQCD at NLO level in the $\overline{\text{MS}}$ scheme. These initial state conditions can then be evolved with the Altarelli-Parisi equations up to the large scale set by $\mu \simeq p_T$. This evolution will resum to all orders the large logarithms previously mentioned.

It is important to mention that due to the neglecting of power suppressed mass terms this approach becomes unreliable when $p_T \simeq m$. In this region only a case by case comparison with the full NLO massive calculation – here reliable and to be taken as a benchmark – can tell how accurate the resummed result is.

Phenomenological analyses [14] show that the effect of the resummation becomes sizeable only at very large p_T, say greater than 50 GeV for bottom hadroproduction at the Tevatron, resulting in a milder factorization/renormalization scale dependence of the result and in a slightly softer p_T spectrum.

Non-perturbative effects describing, say, the $b \to B$ meson transition can also be included within this formalism. The $b \to B$ fragmentation function can be fitted to LEP e^+e^- data and then used for predicting B cross sections at the Tevatron. Recent analyses [17] show that this FF, when used in

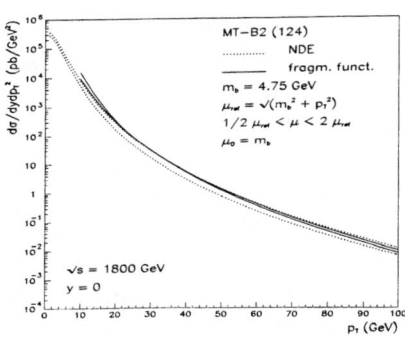 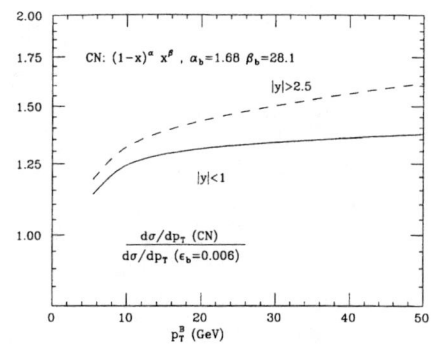

FIGURE 4. Comparison of bottom p_T distributions in the fixed order (NDE) and resummed approaches (left plot, from ref. [14]), and effect of a harder $b \to B$ fragmentation function on bottom production at the Tevatron [18] (right plot).

[14], Reprinted with kind permission of Elsevier Science - NL, Sara Burgerhartstraat 25, 1055 KV Amsterdam, The Netherlands.

connection with a NLO evaluation of bottom production like the MNR one, should probably be taken harder than commonly done in the past. This choice of a harder non-perturbative FF would increase the Tevatron cross sections from twenty to fifty per cent [18] (see fig. 4), helping to reconcile theory and experimental data.

IV HEAVY QUARKONIUM PRODUCTION

The production of heavy quarkonia has been subjected to intense study in the last two or three years, with tens of papers having being produced on the problem of J/ψ, ψ', χ and Υ production in e^+e^-, γp, $p\bar{p}$, pN, πN collisions.

The reason for such a surge in interest was the appearance of a theoretical framework, the so called Factorization Approach (FA) by Bodwin, Braaten and Lepage [19], which seems able to solve the theoretical problems that quarkonium production models faced in the past, and also to reconcile theoretical predictions with experimental data, previously in disagreement up to factors of fifty in some instances.

In this talk I shall not review the Factorization Approach in detail, leaving this theoretical introduction to other sources (see for instance [20]). I shall just recall how the FA writes the quarkonium state H production cross section in the following form:

$$\sigma(ij \to Q\bar{Q} \to H) = \sum_n \hat{\sigma}(ij \to Q\bar{Q}[n])\langle \mathcal{O}^H(n)\rangle. \qquad (4)$$

According to this equation, the cross section for producing the observable quarkonium state H is factorized into two parts. In the short distance part a $Q\overline{Q}$ pair of heavy quarks is produced in the spin/colour/angular momentum state $^{2S+1}L_J^{(c)} \equiv n$ by the scattering of the two light partons i and j. Subsequently this pair hadronizes into the quarkonium H; $\langle \mathcal{O}^H(n) \rangle$ is formally a non-relativistic QCD (NRQCD) matrix element describing this non-perturbative transition.

An important feature of this equation is that also $Q\overline{Q}$ pairs in a colour octet state are allowed to contribute to the production of a colour singlet quarkonium H: their colour is neutralized via a non-perturbative emission of soft gluons. While the corresponding matrix elements are suppressed by the need of such an emission, the short distance coefficients can on the other hand be large, perhaps overcompensating the suppression of the non-perturbative term. This explains why colour octet contributions can play a very important role in predicting the total size of quarkonium production cross sections.

The limited space available doesn't allow any detail about the phenomenological studies which have been performed. Ref. [21] contains a small up-to-date review, with references to the original papers. I shall only mention here that the introduction of the colour octet channels allows for what looks like a "reasonable" description of the data. Colour octet matrix elements have to be fitted to the data themselves, and the uncertainties on these fits are certainly not smaller than a factor of two, due to the many systematics entering their determination: parton distribution functions, charm quark mass, α_s value, higher order QCD corrections, etc. However, the values one gets appear of the correct order of magnitude if compared via NRQCD scaling rules [22] to the colour singlet ones, thereby supporting the underlying picture.

So far all fits have been performed using leading order cross sections. Very recently a next-to-leading order calculation for quarkonium total cross sections, both via singlet and octet channels, has been completed [23]. It will therefore be possible to reduce at least some of the uncertainties previously mentioned and hence obtain more reliable fits.

REFERENCES

1. J.J. Aubert et al., *Phys. Rev. Lett.* **33**, 1404 (1974);
   J.-E. Augustin et al., *Phys. Rev. Lett.* **33**, 1406 (1974).
2. T. Appelquist and H.D. Politzer, *Phys. Rev. Lett.* **34**, 43 (1975);
   A. De Rujula and S.L. Glashow, *Phys. Rev. Lett.* **34**, 46 (1975);
   C.A. Dominguez and M. Greco, *Lett. Nuovo Cim.* **12**, 439 (1975).
3. S.W. Herb et al., *Phys. Rev. Lett.* **39**, 252 (1977).
4. F. Abe et al. (CDF Coll.), *Phys. Rev.* **D50**, 2966 (1994);
   F. Abe et al. (CDF Coll.), *Phys. Rev. Lett.* **74**, 2626 (1995);
   S. Abachi et al. (D0 Coll.), *Phys. Rev. Lett.* **74**, 2632 (1995).
5. J.C. Collins, D. Soper and G. Sterman, *Nucl. Phys.* **B263**, 37 (1986).

6. P. Nason, S. Dawson and R.K. Ellis, *Nucl. Phys.* **B303**, 607 (1988);
   P. Nason, S. Dawson and R.K. Ellis, *Nucl. Phys.* **B327**, 49 (1989), erratum ibid. **B335** (1990) 260.
7. W. Beenakker, H. Kuijf, W.L. van Neerven and J. Smith, *Phys. Rev.* **D40**, 54 (1989);
   W. Beenakker, W.L. van Neerven R. Meng, G.A. Schuler and J. Smith, *Nucl. Phys.* **B351**, 507 (1991).
8. M.L. Mangano, P. Nason and G. Ridolfi, *Nucl. Phys.* **B373**, 295 (1992).
9. S. Frixione, M.L. Mangano, P. Nason and G. Ridolfi, CERN-TH/97-16 (hep-ph/9702287).
10. C. Peterson, D. Schlatter, I. Schmitt and P.M. Zerwas, *Phys. Rev.* **D27**, 105 (1983).
11. E. Laenen, J. Smith and W.L. van Neerven, *Nucl. Phys.* **B369**, 543 (1992);
    E. Laenen, J. Smith and W.L. van Neerven, *Phys. Lett.* **B321**, 254 (1994).
12. E. Berger and H. Contopanagos, *Phys. Lett.* **B361**, 115 (1995);
    E. Berger and H. Contopanagos, *Phys. Rev.* **D54**, 3085 (1996).
13. S. Catani, M.L. Mangano, P. Nason and L. Trentadue, *Phys. Lett.* **B378**, 329 (1996);
    S. Catani, M.L. Mangano, P. Nason and L. Trentadue, *Nucl. Phys.* **B478**, 273 (1996).
14. M. Cacciari, M. Greco, *Nucl. Phys.* **B421**, 530 (1994).
15. J.C. Collins and W.-K. Tung, *Nucl. Phys.* **B278**, 934 (1986).
16. B. Mele and P. Nason, *Nucl. Phys.* **B361**, 626 (1991).
17. G. Colangelo and P. Nason, *Phys. Lett.* **B285**, 167 (1992);
    M. Cacciari and M. Greco, *Phys. Rev.* **D55**, 7134 (1997).
18. M.L. Mangano, private communication.
19. G.T. Bodwin, E. Braaten and G.P. Lepage, *Phys. Rev.* **D51**, 1125 (1995), erratum ibid. **D55** (1997) 5853.
20. E. Braaten, S. Fleming and T.C. Yuan, Ann. Rev. Nucl. Part. Sci. **46**, 197 (1996);
    E. Braaten, Talk given at 3rd International Workshop on Particle Physics Phenomenology, Taipei, Taiwan, 14-17 Nov 1996, OHSTPY-HEP-T-97-004 (hep-ph/9702225);
    M. Beneke, Lecture at the XXXIV SLAC Summer Institute on Particle Physics (August 1996), CERN-TH/97-55 (hep-ph/9703429).
21. M. Cacciari, Proceedings of the XXXII Rencontres de Moriond "QCD and Hadronic Interactions", Les Arcs, March 1997, DESY 97-091 (hep-ph/9706374).
22. G.P. Lepage, L. Magnea, C. Nakhleh, U. Magnea, and K. Hornbostel, *Phys. Rev.* **D46**, 4052 (1992).
23. A. Petrelli, M. Cacciari, M. Greco, F. Maltoni and M.L. Mangano, CERN-TH/97-142 (hep-ph/9707223).

The Beauty of Quarkonia

Aida X. El-Khadra

Department of Physics, University of Illinois,
1110 West Green Street, Urbana, IL, 61801-3080

Abstract. I review the status of lattice QCD calculations of the bottomonium spectrum. These calculations have been used to determine the strong coupling and the b quark mass from experimental measurements of the spin averaged splittings and the Υ mass.

INTRODUCTION

In recent years there has been a lot progress in calculations of hadronic quantities based on lattice QCD. This progress is due in part to the development of improved lattice actions [1].

The quarkonium systems are especially well suited for the development and testing of improved lattice actions, as systematic errors from the lattice calculation are easy to analyze for nonrelativistic bound states [2]. Beautiful historical reviews of our theoretical understanding of the quarkonium systems were given by Kurt Gottfried [3] and Chris Quigg [4] at this conference.

With quantitative control over systematic lattice errors, the lattice QCD calculations of the quarkonium boundstates can be compared to experimental measurements to yield precise determinations of standard model parameters, such as the strong coupling and the heavy quark mass.

THE QUARKONIUM SPECTRUM

Two different formulations for fermions have been used in lattice calculations of the quarkonia spectra. In the nonrelativistic limit the QCD action can be written as an expansion in powers of v^2, where v is the velocity of the heavy quark inside the boundstate [5]; I shall henceforth refer to this approach as NRQCD. Lepage and collaborators [6] have adapted this formalism to the lattice regulator. Several groups have performed numerical calculations of quarkonia in this approach. In Refs. [7,8] the NRQCD action is used to calculate the $b\bar{b}$ and $c\bar{c}$ spectra, including terms up to $\mathcal{O}(mv^4)$ and $\mathcal{O}(a^2)$.

In addition to calculations in the quenched approximation, this group is also using gauge configurations that include 2 flavors of sea quarks with mass $m_q \sim \frac{1}{2} m_s$ to calculate the $b\bar{b}$ spectrum [9,10]. The leading order NRQCD action is used in Ref. [11] for a calculation of the $b\bar{b}$ spectrum in the quenched approximation.

The Fermilab group [12] developed a generalization of previous approaches, which encompasses the nonrelativistic limit for heavy quarks as well as Wilson's relativistic action for light quarks. Lattice-spacing artifacts are analyzed for quarks with arbitary mass. Ref. [13] uses this approach to calculate the $b\bar{b}$ and $c\bar{c}$ spectra in the quenched approximation. We considered the effect of reducing lattice-spacing errors from $\mathcal{O}(a)$ to $\mathcal{O}(a^2)$. The SCRI collaboration [14] is also using this approach for a calculation of the $b\bar{b}$ spectrum using the same gauge configurations as the NRQCD collaboration with $n_f = 2$ and an improved fermion action (with $\mathcal{O}(a^2)$ errors).

All but one group use gauge configurations generated with the Wilson action, leaving $\mathcal{O}(a^2)$ lattice-spacing errors in the results. The lattice spacings, in this case, are in the range $a \simeq 0.05 - 0.2$ fm. Ref. [15] uses an improved gauge action together with a nonrelativistic quark action improved to the same order (but without spin-dependent terms) on coarse ($a \simeq 0.4 - 0.24$ fm) lattices.

Results with 2 flavors of degenerate sea quarks (using the staggered fermion action) have become available from a number of groups [9,16,17,10], with lattice-spacing and finite-volume errors similar to the quenched calculations, significantly reducing this systematic error. This year, calculations including two flavors of degenerate Wilson quarks in the quark sea were presented by the SESAM collaboration [18]. They calculate the $b\bar{b}$ spectrum using a NRCQD action including terms up to $\mathcal{O}(mv^6)$.

The results for the $b\bar{b}$ spectrum from all groups are summarized in Figure 1. The agreement between the experimentally-observed spectrum and lattice QCD calculations is impressive. As indicated in the preceding paragraphs, the lattice artifacts are different for all groups. Figure 1 therefore emphasizes the level of control over systematic errors.

Several systematic effects associated with the inclusion of sea quarks still need to be studied further. Both the SESAM and NRQCD collaborations have studied the dependence of the quarkonia splittings on the sea quark masses. The inclusion of sea quarks with realistic light-quark masses is very difficult. However, the sea-quark mass dependence of the quarkonia splittings can be analyzed with chiral perturbation theory [19] to guide the extrapolation. Both collaborations find the sea-quark mass dependence to be too small to resolve with their statistical errors. The consistency between the results from the two groups indicates that they are independent of the sea-quark action (staggered vs. Wilson).

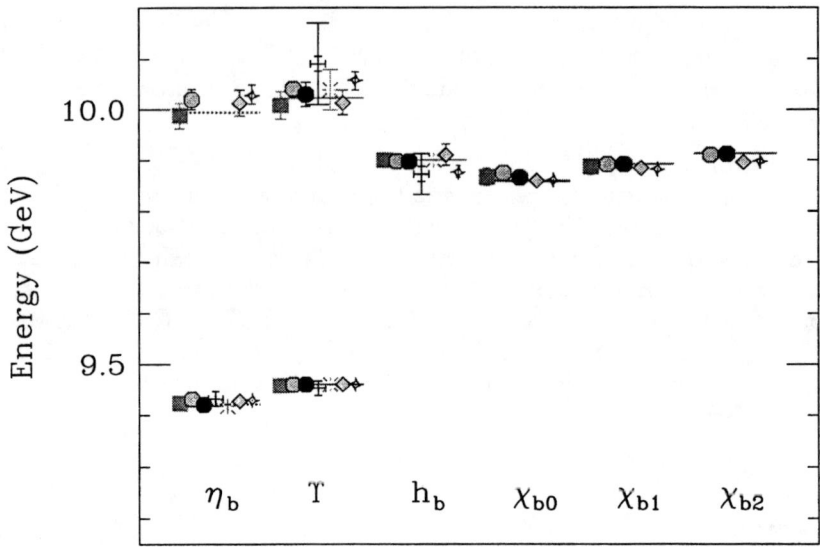

FIGURE 1. A comparison of lattice QCD results for the $b\bar{b}$ spectrum (statistical errors only). -: Experiment; □: FNAL [13]; ○: NRQCD ($n_f = 0$) [7]; •: NRQCD ($n_f = 2$) [9]; +: UK(NR)QCD [11]; ∗: SCRI [14]; ◇: SESAM [18], (shaded) $n_f = 2$, (fancy) $n_f = 0$.

THE STRONG COUPLING

At present, the QCD coupling, α_s, is determined from many different experiments, performed at energies ranging from a few to more than 100 GeV [20]. In most cases perturbation theory is used to extract α_s from the experimental information. Experimental and theoretical progress over the last few years has made these determinations increasingly precise. However, all determinations, including those based on lattice QCD, rely on phenomenologically-estimated corrections and uncertainties from nonperturbative effects. These effects will eventually (or already do) limit the accuracy of the coupling constant determination. When lattice QCD is used the limiting uncertainty comes from the (total or partial) omission of sea quarks in numerical simulations.

Within the framework of lattice QCD the conversion from the bare to a renormalized coupling can, in principle, be made nonperturbatively. In the definition of a renormalized coupling, systematic uncertainties should be controllable, and at short distances, its (perturbative) relation to other conventional definitions calculable. For example, a renormalized coupling can be defined from the nonperturbatively computed heavy-quark potential (α_V) [21].

An elegant approach has been developed in Ref. [22], where a renormalized coupling is defined nonperturbatively through the Schrödinger functional. The authors compute the evolution of the coupling nonperturbatively using a finite size scaling technique, which allows them to vary the momentum scales by an order of magnitude. The same technique has also been applied to the renormalized coupling defined from twisted Polyakov loops [23]. The numerical calculations include only gluons at the moment. However, the inclusion of fermions is possible. Once such simulations become available they should yield very accurate information on α_s and its evolution. The strong coupling can also be computed from the three-gluon vertex, suitably defined on the lattice [24].

An alternative is to define a renormalized coupling through short distance lattice quantities, like small Wilson loops or Creutz ratios which can be calculated perturbatively and by numerical simulation. For example, the coupling defined from the plaquette, $\alpha_P = -3 \ln \langle \text{Tr} \, U_P \rangle / 4\pi$, can be expressed in terms of α_V (or $\alpha_{\overline{MS}}$) by [25]:

$$\alpha_P = \alpha_V(q)[1 - (1.19 + 0.07\, n_f)\alpha_V(q) + \mathcal{O}(\alpha_V^2)] \tag{1}$$

at $q = 3.41/a$, close to the ultraviolet cut-off. α_V is related to the more commonly used $\overline{MS}$ coupling by

$$\alpha_{\overline{MS}}(Q) = \alpha_V(e^{5/6}Q)(1 + \frac{2}{\pi}\alpha_V + \ldots) \, . \tag{2}$$

The size of higher-order corrections associated with the above defined coupling constants can be tested by comparing perturbative predictions for short-distance lattice quantities with nonperturbative results [25]. The comparison of the nonperturbatively calculated coupling of Ref. [22] with the perturbative predictions for this coupling using Eq. (1) is an additional consistency test.

The relation of the plaquette coupling in Eq. (1) to the $\overline{MS}$ coupling has recently been calculated to 2-loops [26,27] in the quenched approximation (no sea quarks, $n_f = 0$). The extension to $n_f \neq 0$ will significantly reduce the uncertainty due to the use of perturbation theory.

Sea Quark Effects

Calculations that properly include all sea-quark effects do not yet exist. If we want to make contact with the "real world", these effects have to be estimated phenomenologically or extrapolated away.

The phenomenological correction necessary to account for the sea-quark effects omitted in calculations of quarkonia that use the quenched approximation gives rise to the dominant systematic error in this calculation [28,29]. By demanding that, say, the spin-averaged 1P-1S splitting calculated on the lattice reproduce the experimentally observed one (which sets the lattice spacing, a^{-1}, in physical units), the effective coupling of the quenched potential is in effect matched to the coupling of the effective 3 flavor potential at the typical momentum scale of the quarkonium states in question. The difference in the evolution of the zero flavor and 3,4 flavor couplings from the effective low-energy scale to the ultraviolet cut-off, where α_s is determined, is the perturbative estimate of the correction.

For comparison with other determinations of α_s, the $\overline{MS}$ coupling can be evolved to the Z mass scale. An average [20] of Refs. [28,29] yields for α_s from calculations in the quenched approximation:

$$\alpha_{\overline{MS}}^{(5)}(m_Z) = 0.110 \pm 0.006 \quad . \tag{3}$$

The phenomenological correction described in the previous paragraph has been tested from first principles in Ref. [16]. The 2-loop evolution of $n_f = 0$ and $n_f = 2$ $\overline{MS}$ couplings – extracted from calculations of the $c\bar{c}$ spectrum using the Wilson action in the quenched approximation and with 2 flavors of sea quarks respectively – to the low-energy scale gives consistent results. After correcting the 2 flavor result to $n_f = 3$ in the same manner as before and evolving $\alpha_{\overline{MS}}$ to the Z mass, they find [16]

$$\alpha_{\overline{MS}}^{(5)}(m_Z) = 0.111 \pm 0.005 \tag{4}$$

in good agreement with the previous result in Eq. (3). The total error is now dominated by the rather large statistical errors and the perturbative uncertainty.

The most accurate result to date comes from the NRQCD collaboration [9,10]. They use results for α_s from the $b\bar{b}$ spectrum with 0 and 2 flavors of (staggered) sea quarks to extrapolate the inverse coupling to the physical 3 flavor case directly at the ultraviolet momentum, $q = 3.41/a$. They obtain a result consistent with the old procedure. Their analysis includes a study of the sea-quark mass dependence. Their result is:

$$\alpha_V^{(3)}(8.2\,\text{GeV}) = 0.195 \pm 0.003 \pm 0.001 \pm 0.004 \quad . \tag{5}$$

The first error is statistics, the second error an estimate of residual cut-off effects and the third (dominant) error is due to the quark mass dependence. The conversion to $\overline{MS}$ (including the 2-loop term of Refs. [26,27]) and evolution to the Z mass then gives:

$$\alpha_{\overline{MS}}^{(5)}(m_Z) = 0.118 \pm 0.003 \quad , \tag{6}$$

where the error now also includes the perturbative uncertainty from Eq. (2). A similar analysis is performed in Ref. [16] on the same gauge configurations but using the Wilson action for a calculation of the $c\bar{c}$ spectrum. The result for the coupling is consistent with Refs. [4,15].

The calculation of the SCRI collaboration [14] ($n_f = 2$) can be combined with the result of Ref. [13]. Using the same analysis as in Ref. [9] gives [10]

$$\alpha_{\overline{MS}}^{(5)}(m_Z) = 0.116 \pm 0.003 \ , \qquad (7)$$

nicely consistent with Eq. (6). Clearly, more work is needed to confirm the results of Eqs. (6) and (7), especially in calculations with heavy quark actions based on Ref. [12]. In particular, the systematic errors associated with the inclusion of sea quarks into the simulation have to be checked, as outlined above.

THE HEAVY QUARK MASS

Because of confinement, the quark masses cannot be measured directly, but have to be inferred from experimental measurements of hadron masses, and depend on the calculational scheme employed.

In lattice QCD quark masses are determined nonperturbatively, by tuning the bare lattice quark mass (m_Q^{lat}) so that, for example, the experimentally observed Υ mass is reproduced by the calculation. Phenomenologically useful quark masses are the perturbatively defined pole and $\overline{MS}$ masses. They can be obtained from the bare lattice mass using (one-loop) perturbation theory:

$$m_Q^{\text{pole}} = Z_m^{\text{pole}} m_Q^{\text{lat}} \ , \qquad m_Q^{\overline{MS}}(m_Q) = Z_m^{\overline{MS}} m_Q^{\text{lat}} \ . \qquad (8)$$

The heavy-quark mass can be determined alternatively from a calculation of the binding energy, E_{bind}. The binding energy can be obtained by subtracting the perturbatively calculable heavy-quark rest energy from the ground-state energy. The pole mass is then:

$$m_Q^{\text{pole}} = \frac{1}{2}(M_{Q\bar{Q}}^{\text{exp}} - E_{\text{bind}}) \qquad (9)$$

This method is insensitive to errors in tuning the bare mass, because the binding energy depends only mildly on the quark mass.

Of course, as always, all systematic errors arising from the lattice QCD calculation need to be under control for a phenomenologically interesting result; in particular, the systematic error introduced by the (partial) omission of sea quarks has to be removed. The short-distance corrections that introduced the dominant uncertainty to the α_s determination from quarkonia are absent for the pole mass determination, because this effective mass does not run for momenta below its mass.

Ref. [30] used both methods described above for a determination of the b quark pole mass from a lattice QCD calculation of the $b\bar{b}$ spectrum. As expected, a comparison of their results with zero and 2 flavors of sea quarks finds compatible results for the pole mass:

$$m_b^{\text{pole}} = (5.0 \pm 0.2) \text{ GeV} \tag{10}$$

For the $\overline{\text{MS}}$ mass, Ref. [30] quotes $m_b^{\overline{\text{MS}}}(m_b) = 4.0(1)$ GeV. The error in both results is dominated by perturbation theory.

A similar analysis is being performed for lattice results obtained with the relativistic action [31].

The B meson system has been used to determine the b-quark $\overline{\text{MS}}$ mass using a discretization of HQET [32]. Their result,

$$m_b^{\overline{\text{MS}}}(m_b) = 4.15(5)(20) \text{ GeV} \tag{11}$$

is consistent with NRQCD result.

CONCLUSIONS

Phenomenological corrections are a necessary evil that enter most coupling constant and quark mass determinations. In contrast, lattice QCD calculations with complete control over systematic errors will yield truly first-principles determinations of α_s and m_Q from the experimentally observed hadron spectrum.

At present, determinations of α_s from the experimentally measured quarkonia spectra using lattice QCD are comparable in reliability and accuracy to other determinations based on perturbative QCD from high energy experiments. They are therefore part of the 1996 world average for α_s [20]. The phenomenological corrections for the most important sources of systematic errors in lattice QCD calculations of quarkonia have already been replaced by first principles calculations. This has led to a significant increase in the accuracy of α_s determinations from quarkonia, and similarly, to a precise determination of the b quark mass.

Still lacking for a first-principles result is the proper inclusion of sea quarks. A difficult problem in this context is the inclusion of sea quarks with physical light quark masses. At present, this can only be achieved by extrapolation (from $m_q \simeq 0.3 - 0.5 m_s$ to $m_{u,d}$). If the light quark mass dependence of the quarkonia spectra is mild, as indicated by the NRQCD and SESAM collaborations, the associated systematic error can be controlled. First-principles calculations of quarkonia could then be performed with currently available computational resources.

ACKNOWLEDGEMENTS

I thank the organizers for an enjoyable conference. This work was supported in part by an Alfred P. Sloan foundation fellowship and by the DOE OJI program under the grant DE-FG02-91ER40677.

REFERENCES

1. For a review, see for example, P. Lepage, *Nucl. Phys.* **B** (Proc. Suppl.) **47** (1996) 3; F. Niedermayr, *Nucl. Phys.* **B** (Proc. Suppl.) **53** (1997) 56. For pedagogical introductions to Lattice Field Theory, see, for example: M. Creutz, *Quarks, Gluons and Lattices* (Cambridge University Press, New York 1985); A. Hasenfratz and P. Hasenfratz, *Annu. Rev. Nucl. Part. Sci.* **35** (1985) 559; A. Kronfeld, in *Perspectives in the Standard Model*, R. Ellis, C. Hill and J. Lykken (eds.) (World Scientific, Singapore 1992), p. 421; P. Lepage, in Schladming 1996, *Perturbative and Nonperturbative Aspects of Quantum Field Theory*, p.1, hep-lat/9607076; For introductory reviews of lattice QCD, see also A. Kronfeld and P. Mackenzie, *Annu. Rev. Nucl. Part. Sci.* **43** (1993) 793; A. El-Khadra, in *Physics in Collision 14*, S. Keller and H. Wahl (eds.) (Editions Frontieres, France 1995), p. 209.
2. P. Lepage, *Nucl. Phys.* **B** (Proc. Suppl.) **26** (1992) 45; B. Thacker and P. Lepage, *Phys. Rev.* **D43** (1991) 196; P. Lepage and B. Thacker, *Nucl. Phys.* **B** (Proc. Suppl.) **4** (1988) 199.
3. K. Gottfried, these proceedings.
4. C. Quigg, these proceedings, Fermilab-CONF-97/266-T, hep-ph/9707493.
5. E. Eichten and F. Feinberg, *Phys. Rev.* **D23** (1981) 2724; W. Caswell and P. Lepage, *Phys. Lett.* **B167** (1986) 437.
6. P. Lepage and B. Thacker, *Phys. Rev.* **D43** (1991) 196; P. Lepage, *et al.*, *Phys. Rev.* **D46** (1992) 4052.
7. C. Davies, *et al.* (NRQCD collaboration), *Phys. Rev.* **D50** (1994) 6963; C. Davies, hep-lat/9705039.
8. C. Davies, *et al.* (NRQCD collaboration), *Phys. Rev.* **D52** (1995) 6519.
9. C. Davies, *et al.* (NRQCD collaboration), *Phys. Lett.* **B345** (1995) 42; *Phys. Rev.* **D56** (1997) 2755; *Nucl. Phys.* **B** (Proc. Suppl.) **47** (1996) 409.
10. J. Shigemitsu, *Nucl. Phys.* **B** (Proc. Suppl.) **53** (191997) 16.
11. S. Catterall, *et al.*, *Phys. Lett.* **B300** (1993) 393; *Phys. Lett.* **B321** (1994) 246.
12. A. El-Khadra, A. Kronfeld and P. Mackenzie, *Phys. Rev.* **D55** (1997) 3933.
13. A. El-Khadra, G. Hockney, A. Kronfeld, P. Mackenzie, T. Onogi and J. Simone, in preparation.
14. S. Collins, R. Edwards, U. Heller, and J. Sloan, hep-lat/9512026.
15. M. Alford, *et al.*, *Nucl. Phys.* **B** (Proc. Suppl.) **42** (1995) 787; *Phys. Lett.* **B361** (1995) 87.
16. S. Aoki, *et al.*, *Phys. Rev. Lett.* **74** (1995) 22.
17. M. Wingate, *et al.*, *Phys. Rev.* **D52** (1995) 307.
18. S. Guesken, Lattice'97 proceedings; A. Spitz, Lattice'97 proceedings; T. Lippert, *et al.*, hep-lat/9707004; N. Eicker, *et al.*, hep-lat/9709002.
19. B. Grinstein and I. Rothstein, *Phys. Lett.* **B385** (1995) 265.
20. R. Barnett, *et al.*, *Phys. Rev.* **D54** (1996) 1, p.77.

21. For a review of α_s from the heavy-quark potential, see K. Schilling and G. Bali, *Nucl. Phys.* B (Proc. Suppl.) **34** (1994) 147; T. Klassen, *Phys. Rev.* **D51** (1995) 5130.
22. M. Lüscher, R. Sommer, P. Weisz, and U. Wolff, *Nucl. Phys.* **B413** (1994) 481.
23. G. de Divitiis, *et al.*, *Nucl. Phys.* **B433** (1995) 390; *Nucl. Phys.* **B437** (1995) 447.
24. C. Bernard, C. Parrinello and A. Soni, *Phys. Rev.* **D49** (1994) 1585; B. Allés, *et al.*, hep-lat/9605033.
25. P. Lepage and P. Mackenzie, *Phys. Rev.* **D48** (1992) 2250.
26. B. Allés, *et al.*, *Phys. Lett.* **B324** (1994) 433; *Nucl. Phys.* **B413** (1994) 553; *Nucl. Phys.* B (Proc. Suppl.) **34** (1994) 501.
27. M. Lüscher and P. Weisz, *Phys. Lett.* **B349** (1995) 165; *Nucl. Phys.* **B452** (1995) 234.
28. A. El-Khadra, G. Hockney, A. Kronfeld and P. Mackenzie, *Phys. Rev. Lett.* **69** (1992) 729; A. El-Khadra, *Nucl. Phys.* B (Proc. Suppl.) **34** (1994) 141
29. The NRQCD Collaboration, *Nucl. Phys.* B (Proc. Suppl.) **34** (1994) 417.
30. C. Davies, *et al.* (NRQCD Collaboration), *Phys. Rev. Lett.* **73** (1994) 2654.
31. A. Kronfeld, Lattice'97 proceedings, and in preparation.
32. V. Gimenez, *et al.*, *Phys. Lett.* **B393** (1997) 124.

Exclusive Hadronic B Decays

Berthold Stech

Institut für Theoretische Physik, Universität Heidelberg, Philosophenweg 16, D-69120 Heidelberg, Germany

Abstract. Exclusive non-leptonic two-body decays are discussed on the basis of a generalized factorization approach which also includes non-factorizeable contributions. Numerous decay processes can be described satisfactorily. The success of the method makes possible the determination of decay constants from non-leptonic decays. In particular, we obtain $f_{D_s} = (234 \pm 25)$ MeV and $f_{D_s^*} = (271 \pm 33)$ MeV. The observed constructive and destructive interference pattern in charged B and D decays, respectively, can be understood in terms of the different α_s-values governing the interaction among the quarks. The running of α_s is also the cause of the observed strong increase of the amplitude of lowest isospin when going to low-energy transitions.

I INTRODUCTION

Since we celebrate today 20 years of beauty physics it may be appropriate to start the discussion of hadronic weak interactions by briefly recalling what was known about this subject in the seventies. In spite of many years of intense research on K and hyperon decays, there was no coherent understanding of non-leptonic decays. For example, the empirically found dominance of $|\Delta \vec{I}| = 1/2$ transitions over $|\Delta \vec{I}| = 3/2$ transitions by a factor 500 was a complete mystery. Moreover, the strongest of all weak decay amplitudes – the $K \to 2\pi$ amplitude – was found to have to vanish in the $SU3$ symmetry limit (Gell-Mann's theorem) and no close relation between K decays and hyperon decays could be seen. In 1974 an important step forward was made: the construction of an effective Hamiltonian which incorporates the effects of hard gluon exchange processes [1]. Still, a factor 20 out of the factor 500 could not be explained, nor could the specific pattern of hyperon decays. The physics at this time, dealing only with u, d and s quarks, was not rich enough. In the corresponding decay processes too few fundamentally different decay channels are open.

The discovery of open charm in 1976 brought hope for enlightenment. Many decay channels could now be studied. But also new puzzles showed up. Unexpectedly, the non-leptonic widths of D^0 and D^+ turned out to differ by a factor 3 and a strong destructive amplitude interference in exclusive decays was found. While D decays occur in a resonance region of the final particles which complicates the analysis, the discovery of beauty precisely 20 years ago gave us particles – the B mesons – which are ideally suited for the study of non-leptonic decays. Again, new interesting effects showed up, in particular and contrary to the case in D decays, a constructive amplitude interference in charged B decays. Recent results [2] of large Penguin-type contributions and sizeable transitions to the η' particle have still to be understood. Moreover, B meson decays give the first realistic possibility to find CP-violating effects outside the K system.

The dramatic effects observed in hadronic weak decays gave rise to many speculations. It was a great challenge to find the correct explanation. Today we know that the strong confining colour forces among the quarks are the decisive factor. These forces are enormously effective in low energy processes and still sizeable even in energetic B decays. Although a strict theoretical treatment of the intricate interplay of weak and strong forces is not yet possible, a semi-quantitative understanding of exclusive two-body decays from K decays to D and B decays has been achieved. The consequences of the QCD-modified weak Hamiltonian can be explored by relating the complicated matrix elements of 4-quark operators to better known objects, to form factors and decay constants.

In the present talk I will describe the generalized factorization method developed recently [3], which also takes non-factorizeable contributions into account and has been quite successful so far. It allows the prediction of many exclusive B decays. I will also show that the interesting and so far puzzling pattern of amplitude interference in B, D and K decays is caused by the different values of α_s acting in these cases.

II THE EFFECTIVE HAMILTONIAN

At the tree level non-leptonic weak decays are mediated by single W-exchange. Hard gluon exchange between the quarks can be accounted for by using the renormalization group technique. One obtains an effective Hamiltonian incorporating gluon exchange processes down to a scale μ of the order of the heavy quark mass. For the case of $b \to c\bar{u}d$ transitions, e.g., the effective Hamiltonian is

$$H_{eff} = \frac{G_F}{\sqrt{2}} V_{cb} V_{ud}^* \left\{ c_1(\mu)(\bar{d}u)(\bar{c}b) + c_2(\mu)(\bar{c}u)(\bar{d}b) \right\} \tag{1}$$

where $(\bar{d}u) = (\bar{d}\gamma^\mu(1-\gamma_5)u)$ etc. are left-handed, colour-singlet quark currents. $c_1(\mu)$ and $c_2(\mu)$ are scale-dependent QCD coefficients known up to next-to-leading order [4]. Depending on the process considered, specific forms of the four-quark operators in the effective Hamiltonian can be adopted. Using Fierz identities one can put together those quark fields which match the constituents of one of the hadrons in the final state of the decay process. Let us consider, as an example, the decays $B \to D\pi$. The corresponding amplitudes are – apart from a common factor –

$$\mathcal{A}_{\bar{B}^0 \to D^+\pi^-} = (c_1 + \frac{c_2}{N_c})\langle D^+\pi^-|(\bar{d}u)(\bar{c}b)|\bar{B}^0\rangle,$$

$$+ c_2 \langle D^+\pi^-|\frac{1}{2}(\bar{d}t^a u)(\bar{c}t^a b)|\bar{B}^0\rangle$$

$$\mathcal{A}_{\bar{B}^0 \to D^0\pi^0} = (c_2 + \frac{c_1}{N_c})\langle D^0\pi^0|(\bar{c}u)(\bar{d}b)|\bar{B}^0\rangle$$

$$+ c_1 \langle D^0\pi^0|\frac{1}{2}(\bar{c}t^a u)(\bar{d}t^a b)|\bar{B}^0\rangle$$

$$\mathcal{A}_{B^- \to D^0\pi^-} = \mathcal{A}_{\bar{B}^0 \to D^+\pi^-} - \sqrt{2}\mathcal{A}_{\bar{B}^0 \to D^0\pi^0} \quad . \tag{2}$$

N_c denotes the number of quark colours and t^a the Gell-Mann colour $SU(3)$ matrices. The last relation in Eq. 2 follows from isospin symmetry of the strong interactions. The three classes of decays illustrated in Eq. 2 are referred to as class I, class II, and class III respectively.

III GENERALIZED FACTORIZATION

How shall we deal with the complicated and scale-dependent four-quark operators? Because the $(\bar{d}u)$ and the $(\bar{c}u)$ currents in (2) can generate the π^- and D^0 mesons, respectively, the above amplitudes contain the scale-independent factorizeable parts

$$\mathcal{F}_{(\bar{B}D)\pi} = \langle \pi^-|(\bar{d}u)|0\rangle\langle D^+|(\bar{c}b)|\bar{B}^0\rangle,$$
$$\mathcal{F}_{(\bar{B}\pi)D} = \langle D^0|(\bar{c}u)|0\rangle\langle \pi^0|(\bar{d}b)|\bar{B}^0\rangle \tag{3}$$

which can be expressed in terms of the decay constants f_π and f_D, and the single current transition form factors $B \to D$ and $B \to \pi$, respectively. For the non-factorizeable contributions we define hadronic parameters $\epsilon_1(\mu)$ and $\epsilon_8(\mu)$ such that the amplitudes (2) take the form [5,3]

$$\mathcal{A}_{\bar{B}^0 \to D^+\pi^-} = a_1 \mathcal{F}_{(\bar{B}D)\pi}$$
$$\mathcal{A}_{\bar{B}^0 \to D^0\pi^0} = a_2 \mathcal{F}_{(\bar{B}\pi)D}$$
$$a_1 = (c_1(\mu) + \frac{c_2(\mu)}{N_c})(1 + \epsilon_1^{(BD)\pi}(\mu)) + c_2(\mu)\epsilon_8^{(BD)\pi}$$
$$a_2 = (c_2(\mu) + \frac{c_1(\mu)}{N_c})(1 + \epsilon_1^{(B\pi)D}(\mu)) + c_1(\mu)\epsilon_8^{(B\pi)D} \quad . \tag{4}$$

The effective coefficients a_1 and a_2 are scale-independent. ϵ_1 and ϵ_8 obey renormalization-group equations and their scale dependence compensates the scale dependence of the QCD coefficients c_1 and c_2 [3]. a_1 and a_2 are process-dependent quantities because of the process dependence of the hadronic parameters ϵ_1 and ϵ_8. So far, then, Eq. (4) provides a parametrization of the amplitudes only and allows no predictions to be made. To get predictions, non-trivial properties of QCD have to be taken into account. We employ at this point the $1/N_c$ expansion of QCD. The large N_c counting rules tell us that $\epsilon_1 = O(1/N_c^2)$ and $\epsilon_8 = O(1/N_c)$. Thus one obtains for a_1 and a_2 in (4)

$$a_1 = c_1(\mu) + c_2(\mu)(\frac{1}{N_c} + \epsilon_8^{(BD)\pi}(\mu)) + O(1/N_c^2)$$
$$a_2 = c_2(\mu) + c_1(\mu)(\frac{1}{N_c} + \epsilon_8^{(B\pi)D}(\mu)) + O(1/N_c^2) \quad . \tag{5}$$

For B decays using $c_1(m_b) = 1 + O(1/N_c^2)$ and $c_2(m_b) = O(1/N_c)$ one finally gets [3]

$$a_1 = c_1(m_b) + O(1/N_c^2)$$
$$a_2 = c_2(m_b) + \zeta^B c_1(m_b) + O(1/N_c^3) \tag{6}$$

with

$$c_1(m_b) \approx 1 \quad \text{and} \quad \zeta^B = \frac{1}{N_c} + \epsilon_8^{(B\pi)D}(m_b) \quad .$$

Now, neglecting $O(1/N_c^2)$ terms, we are left with a single parameter (ζ^B) only. It should be emphasized that putting this parameter equal to $1/N_c$ does not correspond to any consistent limit of QCD. For a_2 the more general expression (6) must be used [6,7].

ζ^B is a dynamical parameter: in general, it will take different values for different decay channels. To deal with this, let us introduce a process-dependent factorization scale μ_f defined by $\epsilon_8(\mu_f) = 0$. The renormalization-group equation then gives [3]

$$\epsilon_8(\mu) = -\frac{4\alpha_s}{3\pi} \ln \frac{\mu}{\mu_f} + O(\alpha_s^2) \quad . \tag{7}$$

For different processes the variation of the factorization scale μ_f is expected to scale with the energy release to the outgoing hadrons in the decay process. With $\mu_f \approx O(m_b)$ one gets from (6), (7)

$$\Delta \zeta^B \approx \frac{4\alpha_s}{3\pi} \frac{\Delta \mu_f}{m_b} \approx \text{few } \% \quad . \tag{8}$$

Thus, the process dependence of ζ^B is expected to be very mild. To a good approximation a single value appears sufficient for the description of two-body B decays. One finds (see section IV) $\zeta^B = 0.45 \pm 0.05$.

A similar discussion also holds for D decays. There one is led to [3]

$$a_1 \approx c_1(m_c) + \zeta'^D c_2(m_c)$$
$$a_2 \approx c_2(m_c) + \zeta^D c_1(m_c)$$
$$\zeta'^D \approx \zeta^D, \qquad (9)$$

and again one expects only a mild process dependence of ζ^D. Indeed, the corresponding description of exclusive D decays brought reasonable success. ζ^D turned out to be very small or zero. There is also theoretical support (using QCD sum rule methods) for a partial or full cancellation of the $1/N_c$ term by non-factorizeable contributions [8]. On the other hand, the corresponding calculation of ζ^B is more involved [9] and has so far not been successful.

IV DETERMINATION OF a_1 AND a_2

The most direct way to determine the effective constant a_1 consists in comparing non-leptonic decay rates with the corresponding differential semi-leptonic rates at momentum transfers equal to the masses of the current generated particles [10]. One gets, for example,

$$\frac{\Gamma(\bar{B}^0 \to D^{(*)+}\rho^-)}{d\Gamma(\bar{B}^0 \to D^{(*)+}\ell^-\bar{\nu})/dq^2|_{q^2=m_\rho^2}} = 6\pi^2 |V_{ud}|^2 f_\rho^2 |a_1|^2 \ . \qquad (10)$$

Because the generated particle is a vector particle like the lepton pair, the form factor combinations occurring in the numerator and denominator cancel precisely. Thus, the ratio (10) is solely determined by $|a_1|$ and the ρ-meson decay constant f_ρ. Taking by convention a_1 real and positive, the measured rates [11] give [3] $a_1 = 1.09 \pm 0.13$ in agreement with the expectation (6). a_1 values obtained from several other processes are in full agreement with the above number. In transitions to pseudoscalar particles the form factor combinations in equations replacing (10) do not cancel. But for $B \to D, D^*$ matrix elements all form factors are well determined using experimental data and the heavy quark effective theory [12]. The latter relates in particular longitudinal form factors to the transverse ones.

Values for $|a_2|$ can be obtained from the analysis of class II transitions. The decays $\bar{B}^0 \to D^{0(*)}h^0$ ($h^0 : \pi^0, \rho^0, a_1^0$) have not yet been observed, but the branching ratios for $\bar{B} \to K^{(*)}J/\psi$ and $\bar{B} \to K^{(*)}\psi(2S)$ are available [11]. The analysis requires model estimates for the heavy-to-light form factors, which enter here. We use the NRSX model [13] which is based on the extrapolation of the BSW form factors [7] at $q^2 = 0$ by appropriate pole and dipole formulae. Where available, more sophisticated calculations agree with these results. (See e.g. Ref. 14). We find [3] $|a_2| = 0.21 \pm 0.01 \pm 0.04$, where the second error accounts for the model dependence.

The relative phase between a_2 and a_1 together with the magnitude of a_2 can be obtained from the decays $B^- \to D^{(*)0}h^-$ where, as seen from (2) and (4), the two amplitudes interfere. The data for the ratios $\Gamma(B^- \to D^{(*)0}h^-)/\Gamma(\bar{B}^0 \to D^{(*)+}h^-)$ give conclusive evidence for constructive interference [11]. Taking a_2 to be a real number (vanishing final state interaction), we find [3] $a_2/a_1 = +0.21 \pm 0.05 \pm 0.04$. Combined with the value for a_1 this gives $a_2 = +0.23 \pm 0.05 \pm 0.04$. The nice agreement between the two determinations of $|a_2|$ shows that the process dependence of this quantity cannot be large. An analysis with an alternative and very simple form factor model gives slightly larger values for a_2 but the results from different processes are again consistent with each other [3].

The positive value for a_2/a_1 in exclusive B decays is remarkable. It is different from the value of the same ratio in exclusive D decays. There a_2/a_1 is negative causing a sizeable destructive amplitude interference. The change of a_2/a_1 inf going from B to D and K decays will be discussed in section VI.

V TESTS AND RESULTS

The B meson, because of its large mass, has many decay channels. We learned from important examples the values of a_1 and a_2 and their near-process-independence in energetic two-body decays. Thus numerous tests and predictions for branching ratios and for the polarizations of the outgoing particles can be made. I will be very brief here and simply refer to Ref. 3 for the compilation of branching ratios in tables, for a detailed discussion and for comparison with the data. Also discussed there is the possible influence of final state interactions. Limits on the relative phases of isospin amplitudes are given. In contrast to D decays final state interactions do not seem to play an important role for the dominant exclusive B decay modes. For the much weaker Penguin-induced transitions, $\bar{B} \to K^{(*)}\pi$ for example, this statement does not hold. Small amplitudes can get an additional contribution from stronger decay channels [7,15]. In the $\bar{B} \to K^{(*)}\pi$ case the decay can proceed via virtual intermediate $D^{(*)}\bar{D}_s^{(*)}$ like channels generated by the $b \to c\bar{c}s$ interaction. The colour-octet $c\bar{c}$ pair, if at low invariant mass, may then turn into a pair of light quarks by gluon exchange. This gives rise to a "long-range Penguin" contribution [15] in addition to the short-distance Penguin amplitude. In future application of our generalized factorization method to rare decays this should be kept in mind. Here, however, I will not discuss this subject further.

Non-leptonic decays to two spin-1 particles also need a separate discussion. Here one has three invariant amplitudes corresponding to outgoing S, P, and D waves. Non-factorizeable contributions to these amplitudes may, in general, have an amplitude composition different from the factorizeable one which cannot be dealt by introducing effective a_1 parameters. Whether or not and to what extent factorization also holds in these more complicated circumstances can be learned from the polarization of the final particles. In class I decays the factorization approximation predicts a polarization identical to the one occurring in the corresponding semi-leptonic decays at the appropriate q^2 value. For $B \to D^* V$ decays the theoretical predictions have very small errors [3]. Another case of particular interest is the polarization of the J/ψ particle in the decay $B \to K^* J/\psi$. Form factor models predict a longitudinal polarization of around 40%. A recent CLEO measurement [16] gives $(52 \pm 7 \pm 4)\%$. It can be shown [17] that small changes of the ratios of form factors obtained in the NRSX model at $q^2 = m^2_{J/\psi}$ are sufficient to get full agreement with the measurements of the longitudinal as well as both transverse polarizations. At present, even with respect to polarization measurements, the generalized factorization approximation is in agreement with the data.

Because of its success, the generalized factorization method, besides allowing many predictions for yet unmeasured decays, can also be used to determine unknown decay constants. A case in point is the determination of the decay constant of the D_s and D_s^* particles. Comparing non-leptonic decays to D_s, D_s^* with those to light mesons, we find [3]

$$f_{D_s} = (234 \pm 25) \text{ MeV}, \quad f_{D_s^*} = (271 \pm 33) \text{ MeV} . \qquad (11)$$

In this determination a_1 cancels and, presumably, also some of the experimental systematic errors. The value for f_{D_s} is in excellent agreement with the value $f_{D_s} = (241 \pm 37)$ MeV obtained from the leptonic decay of the D_s meson [18]. There are several other decay constants which can be measured this way. Of particular interest are the decay constants of P-wave mesons like the a_0, a_1, K_0^*, K_1 particles.

VI   FROM B TO D TO K DECAYS

The process dependence of the coefficients a_1 and a_2 governing exclusive B decays turned out to be very mild. In fact, it is not seen within the errors of the present data. But a_1 and a_2 change strongly in going from B decays to D decays or even down to K decays. In the generalized factorization scheme this is expected because of the different factorization scales and the corresponding $\alpha_s(\mu_f)$ values controlling the strength of the colour forces between the quarks. In Fig. 1 the ratio a_2/a_1 is plotted as a function of $\alpha_s(\mu_f)$. We used for the Wilson coefficients the renormalization-group-invariant definitions of Ref. 19. They appear appropriate for describing the changes of the scale-independent coefficients a_1 and a_2 with the changing particle energy. As seen from the figure the positive value of a_2/a_1 found for exclusive B decays indicates that here small values of α_s govern the colour forces in the first instant of the decay process. This is an impressive manifestation of the colour transparency argument put forward by Bjorken [10]. In D decays the stronger gluon interactions redistribute the quarks: the induced neutral current interaction is already sizeable. We took the corresponding values of a_1 and a_2 from the

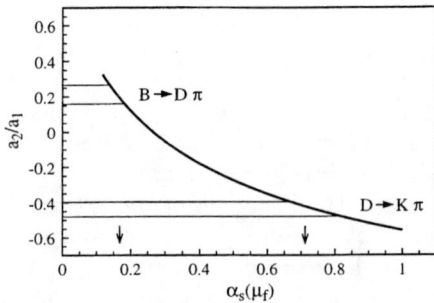

FIGURE 1. The ratio a_2/a_1 as a function of the running coupling constant evaluated at the factorization scale. The bands indicate the phenomenological values of a_2/a_1 extracted from $\bar{B} \to D\pi$ and $D \to K\pi$ decays.

measured isospin amplitudes. They are less affected by final state interactions than the individual amplitudes. The ratio $|A_{1/2}|/|A_{3/2}|$ is already rather large (≈ 4) leading to $a_2/a_1 \approx -0.45$. According to the figure this corresponds to an effective value $\alpha_s \approx 0.7$. The negative value of a_2, and the corresponding destructive amplitude interference in charged D decays, has been known for many years [7,20]. Since the bulk of D decays are two-body or quasi two-body decays, this interference is the main cause for the lifetime difference of D^+ and D^0, in full accord with estimates of the relevant partial inclusive decay rates [21].

Because of the onset of non-perturbative effects one cannot extent Fig. 1 to larger α_s values. However, the trend to smaller and smaller values of the ratio of the Wilson coefficients $c_+(\mu_f)/c_-(\mu_f)$, which is already down to ≈ 0.17 for D decays, is visible. It indicates a strong and, presumably, non-perturbative force in the colour 3^* channel of two quarks, i.e. in the scalar diquark channel [22]. In K decays one is very close to the limiting case $a_2/a_1 = -1$ for which the $|\Delta \vec{I}| = 1/2$ rule would hold strictly.

VII CONCLUSIONS

The matrix elements of non-leptonic exclusive decays are notoriously difficult to calculate. Factorization provides for a connection with better known objects. If combined with the $1/N_c$ expansion method and properly applied and interpreted, it turns out to be very useful, at least for energetic B decays, and has passed many tests. Thus it enables reliable predictions for many decay channels to be made and also permits the determination of decay constants which are difficult to measure otherwise. Factorization does not necessarily hold to the same degree for transitions to two vector particles. These are more sensitive to non-factorizeable contributions and final state interactions.

The constant a_1 is predicted to be one, apart from $1/N_c^2$ corrections, in exclusive B decays and to be practically process-independent. The analysis confirmed these expectations. The particularly interesting parameter a_2, within errors, also does not show a process dependence. The positive value of a_2/a_1 extracted from exclusive B decays is remarkable. The obvious interpretation is that a fast-moving colour-singlet quark pair interacts little with soft gluons. The constructive interference in energetic two-body B^- decays does not imply that the lifetime of the B^- meson should be shorter than the lifetime of the $\bar{B}^0$ meson: the majority of transitions proceed into multi-body final states. For these the relevant scale may be lower than m_b leading to destructive interference. Also, there are many decay channels for which interference cannot occur. The running of a_1 and a_2 with $\alpha_s(\mu_f)$, which in turn depends on the energy release to the final particles, is very interesting. It causes the change from constructive amplitude interference in B^- decays to strong destructive interference in D and K decays. Since exclusive two-body and quasi two-body decays are dominant in D decays this destructive interference is the main cause of the lifetime difference between D^0 and D^+. As the energy decreases the lowest isospin amplitude is seen to become more and more dominant. Strange-particle decays are the most spectacular manifestation of the dramatic changes occuring when the effective α_s gets large. A unified picture of exclusive non-leptonic decays emerges which ranges from very low scales to the large energy scales relevant for B decays.

VIII ACKNOWLEDGEMENT

The work reported here was performed in a fruitful and most enjoyable collaboration with Matthias Neubert which is gratefully acknowledged. The author also would like to thank Dan Kaplan and the other organizers of the b20 symposium for the very pleasent meeting and Matthias Jamin for a useful discussion.

REFERENCES

1. G. Altarelli and L. Maiani, *Phys. Lett.* B **52** (1974) 351;
   M. K. Gaillard and B. W. Lee, *Phys. Rev. Lett.* **33** (1974) 108.
2. F. Würthwein, CLEO Collaboration, hep-ex/9706010;
   R. Poling CLEO Collaboration, these Proceedings.
3. M. Neubert and B. Stech, hep-ph 9705292, to appear in *Heavy Flavours*, Second Edition, ed. A. J. Buras and M. Lindner (World Scientific, Singapore).
4. G. Altarelli, G. Curci, G. Martinelli, and S. Petrarca, *Phys. Lett.* B **99** (1981) 141; *Nucl. Phys.* B **187** (1981) 461;
   A. J. Buras and P. H. Weisz, *Nucl. Phys.* B **333** (1990) 66.

5. H. Y. Cheng, *Phys. Lett. B* **335** (1994) 428;
   J. M. Soares, *Phys. Rev. D* **51** (1995) 3518.
6. N. Deshpande, M. Gronau, and D. Sutherland, *Phys. Lett. B* **90** (1980) 431;
   Nucl. Phys. B **183** (1981) 367.
7. M. Bauer and B. Stech, *Phys. Lett. B* **152** (1985) 380;
   M. Bauer, B. Stech, and M. Wirbel, *Z. Phys. C* **34** (1987) 103.
8. B. Blok and M. Shifman, *Yad. Fiz.* **45** (1987) 221, 478, and 841 [*Sov. J. Nucl. Phys.* **45** (1987) 135, 301, and 522]; **46** (1987) 1310 [**46** (1987) 767].
9. B. Blok and M. Shifman, *Nucl. Phys. B* **389** (1993) 534; **399** (1993) 441 and 459.
10. J. D. Bjorken, *Nucl. Physcis B* (Proc. Suppl.) **11** (1998) 325.
11. T. E. Browder, K. Honscheid, and D. Pedrini, *Ann. Rev. Nucl. Part. Sci.* **46** (1996) 395.
12. For a review see M. Neubert, *Phys. Rep.* **245** (1994) 259; *Int. J. Mod. Phys. A* **11** (1996) 4173.
13. M. Neubert, V. Rieckert, B. Stech, and Q. P. Xu, in *Heavy Flavours*, First Edition, ed. A. J. Buras and M. Lindner (World Scientific, Singapore 1992), p. 286.
14. P. Ball and V. M. Braun, *Phys. Rev. D* **55** (1997) 5561.
15. W. F. Palmer and B. Stech, *Phys. Rev. D* **48** (1993) 4147.
16. J. D. Lewis, Proceedings, B-Physics and CP Violation Conf. Honolulu, HI (1997) to be published.
17. Y. Y. Keum, private communication.
18. J. D. Richman, *Proceedings of the 28th International Conference on High-Energy Physics*, Warsaw, Poland, July 1996 ICHEP 96:143 [hep-ex/9701014].
19. A. J. Buras, M. Jamin, E. Lauterbacher, and P. Weisz, *Nucl. Phys. B* **370** (1992) 104, 501.
20. A. J. Buras, J. M. Gérard, and R. Rückl, *Nucl. Phys. B* **268** (1986) 16.
21. B. Stech, *Nucl. Phys.* (Proc. Suppl.) **B1** (1988) 17;
    B. Stech, in *CP-violation*, ed. C. Jarlskog (World Scientific, Singapore 1989) p. 680.
22. M. Neubert and B. Stech *Phys. Rev. D* **44** (1991) 775;
    B. Stech, *Mod. Phys. Lett. A* **6** (1991) 3113;
    M. Jamin and A. Pich, *Nucl. Phys. B* **425** (1994) 15.

The b-quark and Symmetries of the Strong Interaction

Mark B. Wise

California Institute of Technology, Pasadena, CA 91125 USA [1]

Abstract. Applications of HQET and NRQCD to fragmentation are briefly reviewed. The special role of the b-quark in applications of heavy quark symmetry is discussed. Predictions of HQET for semileptonic B decays to excited charmed mesons are considered.

HQET

The heavy quark effective theory (HQET) is a limit of the theory of the strong interactions appropriate for hadrons containing a single heavy quark Q. In such hadrons the light degrees of freedom typically have momentum of order Λ_{QCD}. Interactions of the heavy quark with the light degrees of freedom cause changes in its four-velocity v of order $\Delta v \sim \Lambda_{QCD}/m_Q$. Consequently for these hadrons it is a reasonable approximation to take the limit of QCD where $m_Q \to \infty$ with the heavy quark's four-velocity fixed.

The part of the QCD Lagrange density involving the heavy quark field is

$$\mathcal{L} = \bar{Q}(i\slashed{D} - m_Q)Q. \qquad (1)$$

The QCD heavy quark field is related to its HQET counterpart by

$$Q = e^{-im_Q v \cdot x}\left[1 + \frac{i\slashed{D}}{2m_Q} + \ldots\right]Q_v, \qquad (2)$$

where

$$\slashed{v}Q_v = Q_v. \qquad (3)$$

Putting eq. (2) into the QCD Lagrange density and using eq. (3) yields

---

[1] Work supported in part by the U.S. Dept. of Energy under Grant No. DE-FG03-92-ER40701.

$$\mathcal{L} = \mathcal{L}_{HQET} + \delta_1\mathcal{L} + \ldots, \tag{4}$$

where the HQET Lagrange density is [1]

$$\mathcal{L}_{HQET} = \bar{Q}_v iv \cdot D Q_v. \tag{5}$$

If there are several heavy flavors a sum over different flavors of heavy quarks is understood. This Lagrange density is independent of the heavy quark mass and spin and has the spin-flavor symmetry [2] of HQET. $\delta_1\mathcal{L}$ contains corrections to the $m_Q \to \infty$ limit suppressed by a single power of the heavy quark mass. Explicitly [3]

$$\delta_1\mathcal{L} = \frac{1}{2m_Q}[O^{(Q)}_{kin,v} + O^{(Q)}_{mag,v}], \tag{6}$$

where the kinetic energy term is

$$O^{(Q)}_{kin,v} = \bar{Q}_v(iD_\perp)^2 Q_v. \tag{7}$$

Here, $D^\mu_\perp = D^\mu - v^\mu(v \cdot D)$ are the components of the covariant derivative perpendicular to the four-velocity. The chromomagnetic energy term is

$$O^{(Q)}_{mag,v} = \bar{Q}_v \frac{g}{2}\sigma_{\alpha\beta} G^{\alpha\beta A} T^A Q_v. \tag{8}$$

Note that the part of $\delta_1\mathcal{L}$ involving $O^{(Q)}_{kin,v}$ breaks the flavor symmetry but not the spin symmetry. $O^{(Q)}_{mag,v}$ breaks both symmetries.

In the limit $m_Q \to \infty$ the angular momentum of the light degrees of freedom,

$$\vec{S}_\ell = \vec{J} - \vec{S}_Q, \tag{9}$$

is conserved [4]. Therefore, in this limit, hadrons occur in doublets with total angular momentum

$$j_\pm = s_\ell \pm 1/2.$$

Here $\vec{J}^2 = j(j+1)$ and $\vec{S}_\ell^2 = s_\ell(s_\ell+1)$. In the case of mesons with $Q\bar{q}$ flavor quantum numbers, the ground state doublet has spin-parity of the light degrees of freedom $s_\ell^{\pi_\ell} = \frac{1}{2}^-$. For $Q = c$ this doublet contains the D and D^* mesons with spin 0 and 1 respectively and for $Q = b$ they are the B and B^* mesons. An excited doublet of mesons with $s_\ell^{\pi_\ell} = \frac{3}{2}^+$ has also been observed. In the $Q = c$ case this doublet contains the $D_1(2420)$ and $D_2^*(2460)$ with spin 1 and spin 2 respectively. The analogous $Q = b$ mesons are called B_1 and B_2^*.

NRQCD

For quarkonia (i.e., $Q\bar{Q}$ hadrons) physical properties are usually predicted using an expansion in v/c where v is the magnitude of the heavy quarks' relative velocity and c is the speed of light [5]. So the appropriate limit of QCD to take in this case is the $c \to \infty$ limit [6]. In eq. (1) the speed of light was set to unity. Making the factors of c explicit it becomes

$$\mathcal{L} = c\bar{Q}(i\not{D} - m_Q c)Q, \qquad (10)$$

where

$$\partial_0 = \frac{1}{c}\frac{\partial}{\partial t}, \qquad (11)$$

and the covariant derivative

$$D_\mu = \partial_\mu + \frac{ig}{c}A_\mu^A T^A. \qquad (12)$$

Note that the strong coupling g has the same units as $\sqrt{c}$. The full QCD heavy quark field Q is related to its NRQCD counterpart by

$$Q = e^{-im_Q c^2 t}\left[1 + \frac{i\not{D}_\perp}{2m_Q c} + \cdots\right]\begin{pmatrix}\psi\\0\end{pmatrix}, \qquad (13)$$

where ψ is a two component Pauli spinor and $D_\perp = (0, \mathbf{D}_\perp)$. Putting eq. (13) into eq. (10) gives

$$\mathcal{L} = \mathcal{L}_{NRQCD} + \cdots, \qquad (14)$$

where

$$\mathcal{L}_{NRQCD} = \psi^\dagger\left(i\left(\frac{\partial}{\partial t} + igA_0^A T^A\right) + \frac{\vec{\nabla}^2}{2m_Q}\right)\psi. \qquad (15)$$

The $c \to \infty$ limit of QCD is called non-relativistic quantum chromodynamics (NRQCD). Since the kinetic energy appears as a leading term in NRQCD this theory does not have a heavy quark flavor symmetry; however, it still has a heavy quark spin symmetry. The gluon field A_0 in eq. (15) is not a propagating field. It gives rise to a Coulomb potential between the heavy quarks. All the interactions of the propagating transverse gluons with the heavy quarks are suppressed by powers of $1/c$. The leading interaction of the propagating transverse gluons with the heavy quarks is also invariant under heavy quark spin symmetry.

SPECIAL ROLE OF THE BOTTOM QUARK

The c, b and t quarks can be considered heavy. Unfortunately the top is so heavy that it decays before forming a hadron. Heavy quark symmetry is not a useful concept for the t-quark. The charm quark mass is not large enough for one to be confident that predictions based on heavy quark symmetry will work well. For charmonium $v^2/c^2 \sim 1/3$ and $\Lambda_{QCD}/m_c \sim 1/7$. However, for the b-quark, corrections to predictions based on heavy quark symmetry should be small. This "special role" of the b-quark is illustrated nicely by comparing with experiment the predictions of heavy quark symmetry for fragmentation.

Heavy quark symmetry implies that the probability $P^{(H)}_{h_Q \to h_s}$ for heavy quark Q with spin along the fragmentation axis (i.e., helicity) h_Q to fragment to a hadron H with spin of the light degrees s_l, total spin s and helicity h_s is [7]

$$P^{(H)}_{h_Q \to h_s} = P_{Q \to s_l} p_{h_l} |\langle s_Q, h_Q; s_l, h_l | s, h_s \rangle|^2. \qquad (16)$$

In eq. (16) $P_{Q \to s_l}$ is the probability for the heavy quark to fragment into the doublet with spin of the light degrees of freedom s_l. p_{h_l} is the probability for the helicity of the light degrees of freedom to be $h_l = h_s - h_Q$, given that the heavy quark fragments to this doublet. Parity invariance of the strong interactions implies that

$$p_{h_l} = p_{-h_l}, \qquad (17)$$

and the definition of a probability implies that

$$\sum_{h_l} p_{h_l} = 1. \qquad (18)$$

The constraints in eqs. (18) and (17) imply that there are $s_l - 1/2$ independent probabilities p_{h_l}.

For the $c\bar{q}$ ground state meson doublet $p_{1/2} = p_{-1/2} = 1/2$ and the relative fragmentation probabilities are

$$\begin{array}{cccc} P^{(D)}_{1/2 \to 0} & : P^{(D^*)}_{1/2 \to 1} & : P^{(D^*)}_{1/2 \to 0} & : P^{(D^*)}_{1/2 \to -1} \\ \frac{1}{4} & : \frac{1}{2} & : \frac{1}{4} & : 0 \end{array} \qquad (19)$$

For the excited $s_l^{\pi_l} = \frac{3}{2}^+$ doublet the relative fragmentation probabilities can be expressed using eq. (16) in terms of $w_{3/2}$. This parameter is defined by $p_{3/2} = p_{-3/2} = (1/2) w_{3/2}$ and $p_{1/2} = p_{-1/2} = (1/2)(1 - w_{3/2})$.

In the charm system eq. (19) does not work well. While the experimental value for the relative probabilities to fragment to longitudinal and transverse D^* helicities agrees with eq. (19), the experimental values for the probabilities to fragment to D and D^* are approximately equal instead of in the ratio

1:3 that eq. (19) predicts. This discrepancy is probably due to the D^*-D mass difference which suppresses fragmentation to the D^*. Recent LEP data shows that predictions for fragmentation based on heavy quark symmetry work better in the b-quark case [8]. The experimental value for the probabilities to fragment to the B and B^* are in the ratio 1:3.

Experimental information on D^{**} production provides the bound, $w_{3/2} <$ 0.24 [7]. It would be very interesting to have an experimental determination of the Falk-Peskin fragmentation parameter $w_{3/2}$.

Heavy quark spin symmetry also makes predictions for the alignment of quarkonia produced by gluon fragmentation. At leading order v/c the gluon fragments to $Q\bar{Q}$ in a color singlet configuration. Two hard gluons occur in the final state to conserve color and charge conjugation, giving a fragmentation probability to $^3S_1$ quarkonia of order $(\alpha_s(m_Q)/\pi)^3(v/c)^3$. However, a term higher order in v/c is much more important because it is lower order in $\alpha_s(m_Q)/\pi$. The gluon can fragment to the $Q\bar{Q}$ pair in a color octet with two soft propagating NRQCD gluons in the final state (each with typical momentum of order $m_Q v(v/c)$ in the quarkonium rest frame). This color octet process [9] gives a contribution to the $^3S_1$ fragmentation probability of order $(\alpha_s(m_Q)/\pi)(v/c)^7$. The fragmenting gluon has large energy (compared with m_Q) and is almost real. Real gluons are transversely aligned. Because the leading interactions of the NRQCD propagating gluons preserve spin symmetry the final state $^3S_1$ quarkonium is also transversely aligned [10]. (There are $\alpha_s(m_Q)$ and v/c corrections [11] that reduce this alignment.) It may be possible to test this prediction in the $Q = c$ case from large $p_\perp$ data on J/ψ and ψ' production at the Tevatron [12].

$B \to D_1(2420) e \bar{\nu}_e$ AND $B \to D_2^*(2460) e \bar{\nu}_e$ DECAY

Semileptonic B decays have been extensively studied. The semileptonic decays $B \to D e \bar{\nu}_e$ and $B \to D^* e \bar{\nu}_e$ have branching ratios of $(1.8 \pm 0.4)\%$ and $(4.6 \pm 0.3)\%$ respectively [13]. They amount to about 60% of the semileptonic decays. The differential decay rates are determined by matrix elements of the $b \to c$ weak axial-vector and vector currents. These matrix elements are usually written in terms of Lorentz scalar form factors and the differential decay rates are expressed in terms of them. For comparisons with the predictions of HQET it is convenient to write the form factors in terms of $w = v \cdot v'$. In the limit $m_Q \to \infty$ heavy quark spin symmetry implies that all six form factors can be written in terms of a single function of w [2]. Furthermore, heavy quark flavor symmetry implies that this function is normalized to unity [2,14] at zero recoil, $w = 1$. The success of these predictions [15] indicates that in this case treating the charm quark mass as large is a reasonable approximation. At order $1/m_{c,b}$ several new functions occur but the normalization of the zero recoil matrix elements is preserved.

In the $m_Q \to \infty$ limit zero recoil matrix elements of the weak axial vector and vector currents from the B-meson to any excited charmed meson vanish because of heavy quark spin symmetry. Since most of the phase space for such decays is near zero recoil (e.g., for B decay to the $s_\ell^{\pi_\ell} = \frac{3}{2}^+$ mesons $D_1(2420)$ and $D_2^*(2460), 1 < w < 1.3$) the $\Lambda_{QCD}/m_{c,b}$ corrections are very important.

The decay $B \to D_1 e \bar{\nu}_e$ has been observed. CLEO and ALEPH, respectively, find the branching ratios [16] $Br(B \to D_1 e \bar{\nu}_e) = (0.49 \pm 0.14)\%$ and $(0.74 \pm 0.16)\%$. For $Br(B \to D_2^* e \bar{\nu}_e)$ there are only upper limits.

The form factors that parametrize the $B \to D_1$ and $B \to D_2^*$ matrix elements of the weak currents $V^\mu = \bar{c}\gamma^\mu b$ and $A^\mu = \bar{c}\gamma^\mu \gamma_5 b$ are defined by

$$\frac{\langle D_1(v',\varepsilon)|V^\mu|B(v)\rangle}{\sqrt{m_{D_1}m_B}} = f_{V_1}\varepsilon^{*\mu} + (f_{V_2}v^\mu + f_{V_3}v'^\mu)(\varepsilon^* \cdot v),$$

$$\frac{\langle D_1(v',\varepsilon)|A^\mu|B(v)\rangle}{\sqrt{m_{D_1}m_B}} = if_A \varepsilon^{\mu\alpha\beta\gamma} \varepsilon_\alpha^* v_\beta v'_\gamma,$$

$$\frac{\langle D_2^*(v',\varepsilon)|A^\mu|B(v)\rangle}{\sqrt{m_{D_2^*}m_B}} = k_{A_1}\varepsilon^{*\mu\alpha}v_\alpha + (k_{A_2}v^\mu + k_{A_3}v'^\mu)\varepsilon^*_{\alpha\beta}v^\alpha v^\beta,$$

$$\frac{\langle D_2^*(v',\varepsilon)|V^\mu|B(v)\rangle}{\sqrt{m_{D_2^*}m_B}} = ik_V \varepsilon^{\mu\alpha\beta\gamma}\varepsilon^*_{\alpha\sigma}v^\sigma v_\beta v'_\gamma. \tag{20}$$

The form factors f_i and k_i are functions of w. In the $m_{c,b} \to \infty$ limit they can be written in terms of a single function $\tau(w)$ [17],

$$\sqrt{6}f_A = -(w+1)\tau, \quad k_V = -\tau,$$
$$\sqrt{6}f_{V_1} = (1-w^2)\tau, \quad k_{A_1} = -(1+w)\tau,$$
$$\sqrt{6}f_{V_2} = -3\tau, \quad k_{A_2} = 0,$$
$$\sqrt{6}f_{V_3} = (w-2)\tau, \quad k_{A_3} = \tau.$$
$$\tag{21}$$

Only the form factor f_{V_1} contributes at zero recoil. Surprisingly one can predict its value [18]

$$\sqrt{6}f_{V_1}(1) = -\frac{4(\bar{\Lambda}' - \bar{\Lambda})\tau(1)}{m_c}, \tag{22}$$

in terms of the $m_{c,b} \to \infty$ Isgur–Wise function τ and the difference between the mass of the light degrees of freedom in the excited $s_\ell^{\pi_\ell} = \frac{3}{2}^+$ doublet $\bar{\Lambda}'$ and the mass of the light degrees of freedom in the ground state doublet $\bar{\Lambda}$. Experimentally the difference $\bar{\Lambda}' - \bar{\Lambda} \simeq 0.39$ GeV. (It can be expressed in terms of measured hadron masses.) A detailed discussion of the $1/m_{c,b}$ corrections to these decays can be found in Refs. [18]. They enhance the rate to $B \to D_1 e \bar{\nu}_e$ (compared with the $m_{c,b} \to \infty$ limit) and lead to the expectation that its branching ratio is greater than that for $B \to D_2^* e \bar{\nu}_e$. This may explain why semileptonic decays to the D_2^* have not been observed.

REFERENCES

1. Eichten, E., and Hill, B., *Phys. Lett.* **B234**, 511 (1990); Georgi, H., *Phys. Lett.* **B240**, 447 (1990).
2. Isgur, N., and Wise, M.B., *Phys. Lett.* **B232**, 113 (1989); *Phys. Lett.* **B237**, 527 (1990).
3. Eichten, E., and Hill, B., *Phys. Lett.* **B243**, 427 (1990); Falk, A., et al., *Nucl. Phys.* **B357**, 185 (1991).
4. Isgur, N., and Wise, M.B., *Phys. Rev. Lett.* **66**, 1130 (1991).
5. Bodwin, G.T., et al., *Phys. Rev.* **D51**, 1125 (1995).
6. Grinstein. B., and Rothstein, I.Z., hep-ph/9703298 (1997) unpublished.
7. Falk, A.F., and Peskin, M.E., *Phys. Rev.* **D49**, 3320 (1994).
8. G. Eigen, Talk Presented at the 7th International Symposium on Heavy Flavor Physics, Santa Barbara, CA July 7-11 (1997).
9. Braaten, E., and Fleming, S., *Phys. Rev. Lett.* **74**, 3327 (1995).
10. Cho, P., and Wise, M.B., *Phys. Lett.* **B346**, 129 (1995).
11. Beneke, M., and Rothstein, I.Z., *Phys. Lett.* **B372**, 157 (1996).
12. Beneke, M., and Kramer, M., *Phys. Rev.* **D55**, 5269 (1997); Leibovich, A., hep-ph/9610381 (1996) unpublished.
13. Particle Data Group, Barnett, R.M., et al., *Phys. Rev.* **D54**, 1 (1996).
14. Nussinov, S., and Wetzel, W., *Phys. Rev.* **D36**, 130 (1987); Shifman, M., and Voloshin, M., *Sov. J. Nucl. Phys.* **47**, 511 (1988).
15. CLEO Collaboration, Duboscq, J.E., et al., *Phys. Rev. Lett.* **76**, 3898 (1996); CLEO Collaboration, Athanas, M., et al., CLEO 97-12 (1997) unpublished.
16. ALEPH Collaboration, Buskulic, D., et al., *Z. Phys.* **C73**, 601 (1997); CLEO Collaboration, Browder, T.E., et al., CLEO Conf. 96-2, ICHEP96 PA05-077 (1996) unpublished.
17. Isgur, N., and Wise, M.B., *Phys. Rev.* **D43**, 819 (1991).
18. Leibovich, A.K., et al., *Phys. Rev. Lett.* **78**, 3995 (1997); hep-ph/9705467 (1997) unpublished.

CKM PARAMETERS, *B* MIXING, AND *CP* VIOLATION

B Mixing on the Lattice: f_B, f_{B_s} and Related Quantities

C. Bernard,[a] [1] T. Blum,[b] T. DeGrand,[c] C. DeTar,[d]
Steven Gottlieb,[e] U. M. Heller,[f] J. Hetrick,[a] C. McNeile,[d]
K. Rummukainen,[g] R. Sugar,[h] D. Toussaint,[i] and M. Wingate,[c]

[a] *Department of Physics, Washington University, St. Louis, MO 63130, USA*
[b] *Department of Physics, Brookhaven National Lab, Upton, NY 11973, USA*
[c] *Physics Department, University of Colorado, Boulder, CO 80309, USA*
[d] *Physics Department, University of Utah, Salt Lake City, UT 84112, USA*
[e] *Department of Physics, Indiana University, Bloomington, IN 47405, USA*
[f] *SCRI, Florida State University, Tallahassee, FL 32306-4130, USA*
[g] *Department of Physics, University of Bielefeld, D-33501 Bielefeld, Germany*
[h] *Department of Physics, University of California, Santa Barbara, CA 93106, USA*
[i] *Department of Physics, University of Arizona, Tucson, AZ 85721, USA*

Abstract. The MILC collaboration computation of heavy-light decay constants is described. Results for f_B, f_{B_s}, f_D, f_{D_s} and their ratios are presented. These results are still preliminary, but the analysis is very close to being completed. Sources of systematic error, both within the quenched approximation and from quenching itself, are estimated, although the latter estimate is rather crude. At present, the largest source of error comes from the extrapolation to the continuum. The chiral extrapolation errors are almost as large in a few cases. A sample of our results is: $f_B = 153 \pm 10\ ^{+36}_{-13}\ ^{+13}_{-0}$ MeV, $f_{B_s}/f_B = 1.10 \pm 0.02\ ^{+0.05}_{-0.03}\ ^{+0.03}_{-0.02}$, and $f_B/f_{D_s} = 0.76 \pm 0.03\ ^{+0.07}_{-0.04}\ ^{+0.02}_{-0.01}$, where the errors are statistical, systematic (within the quenched approximation), and systematic (of quenching), respectively.

INTRODUCTION

B_d-$\bar{B}_d$ mixing offers a way to determine the CKM matrix element V_{td}. Indeed, the mixing parameter x_d, defined as $\Delta M_{B_d}/\Gamma_{B_d}$, is given in the Standard Model by [1]

$$x_d = \tau_{B_d} \frac{G_F^2 M_W^2}{6\pi^2} \eta_B S(x_t) M_{B_d} \xi^2_{B_d} |V_{td}|^2 , \qquad (1)$$

where $\eta_B S(x_t)$ is perturbatively known, $V_{tb} \cong 1$ is assumed, and

---

[1] presented by C. Bernard

$$\xi_{B_d} \equiv f_B \sqrt{B_B}, \qquad (2)$$

with f_B the decay constant of the B_d meson and B_B the corresponding "bag parameter." Since x_d has been measured experimentally, an evaluation of the nonperturbative quantity ξ_{B_d} will determine V_{td}. Similarly, a measurement or bound on the corresponding mixing parameter x_s for B_s mesons will give information about V_{ts} if ξ_{B_s} is known, or about $|V_{td}/V_{ts}|$, given ξ_{B_s}/ξ_{B_d}.

The lattice offers a way to compute quantities like f_B and B_B from first principles. Here, we present a nearly completed computation by the MILC collaboration of the decay constants f_B, f_{B_s}, f_D, f_{D_s}, and their ratios. A calculation of the quantities B_B and B_{B_s}/B_B, in the static limit for the heavy quark, is also in progress; some preliminary results are described in Ref. [2].

SOURCES OF ERROR

A key issue in any lattice computation is the reliability of systematic error estimates. In order to make these estimates, we use a range of lattices, both quenched (*i.e.*, negecting virtual quark loop effects) and unquenched. The lattice spacing and volume are varied over as wide a range as practical. Lattices used are shown in Table 1. We discuss the sources of error in turn.

TABLE 1. Lattices used. $\beta = 6/g^2$, with g the lattice coupling. The quark mass in virtual loops (m_{vir}) is given in lattice units; where it is absent the lattices are quenched. a is the lattice spacing, and ℓ is the spatial box size. Lattice Q is used for a finite size check only, and does not appear in the rest of the analysis. Lattice G was generated by HEMCGC; lattice F, by the Columbia group.

Name	β	m_{vir}	size	# confs.	a^{-1}(GeV)	ℓ(fm)
A	5.7	–	$8^3 \times 48$	200	1.3	1.2
B	5.7	–	$16^3 \times 48$	100	1.3	2.5
E	5.85	–	$12^3 \times 48$	100	1.7	1.4
C	6.0	–	$16^3 \times 48$	100	2.0	1.6
Q	6.0	–	$12^3 \times 48$	235	2.0	1.2
D	6.3	–	$24^3 \times 80$	100	3.1	1.6
H	6.52	–	$32^3 \times 100$	60	4.2	1.5
L	5.445	0.025	$16^3 \times 48$	100	1.3	2.5
N	5.5	0.1	$24^3 \times 64$	100	1.4	3.6
O	5.5	0.05	$24^3 \times 64$	100	1.5	3.2
M	5.5	0.025	$20^3 \times 64$	100	1.5	2.6
P	5.5	0.0125	$20^3 \times 64$	100	1.7	2.4
G	5.6	0.01	$16^3 \times 32$	200	2.1	1.5
F	5.7	0.01	$16^3 \times 32$	49	2.4	1.3

Statistics

The basic objects of interest are the quantum mechanical amplitudes for the propagation of mesons. Such amplitudes can be written as a functional integral, which in this case means the weighted sum over all possible paths for the quarks and over all possible configurations of the gluon field. Since it is impossible, for practical reasons, to include all such "paths," one resorts to a statistical sampling, and the answers therefore have statistical errors. Equivalently, one can say that we are doing a very large dimensional ($\sim 10^8$) integral by Monte Carlo importance-sampling methods. In practice, about 100 to 250 gluon field configurations are needed to reduce the statistical errors to well below the present systematic errors.

Isolating the State of Interest

Since we do not know, *a priori*, the B meson wave function, we cannot create a B directly on the lattice. Instead, we operate on the vacuum with an "interpolating field" which need only have the same quantum numbers as the B. This creates a superposition of all states with B quantum numbers: the B itself, radial excitations, B+ glueballs, *etc*. However, on a Euclidean space lattice states develop in time as e^{-Et}, rather than e^{-iEt}. Therefore, after sufficient Euclidean time, all the higher energy states will have died away, leaving a pure B state.

The question then becomes: how long a time is "sufficient?" A standard approach is to define the "effective mass," m_{eff}, as the instantaneous exponential rate of fall of the B meson propagator. As the higher energy states die away, the propagator will begin to fall like a pure exponential, and m_{eff} will approach the energy (or mass, for zero 3-momentum) of the B. Figure 1 shows a typical plot of m_{eff} vs. time for two propagators. (The propagator G_{sl} is needed because one wants to compute f_B, which is proportional to the amplitude for annihilating the B with an axial current.) For $t \gtrsim 12$, these effective masses show little systematic variation with time. However, because of statistical fluctuations, as well as possible lingering systematic effects, one gets slightly different results depending on which intervals (with $t \gtrsim 12$) are chosen to fit the propagators. The variation over the intervals (what we call the "fitting error") is added in quadrature with the naive statistical error before any further analysis.

Quark Mass Extrapolation/Interpolation

- Light-Quark Extrapolation. We compute with light quark masses (m_q) in the range $m_s/3 \lesssim m_q \lesssim 2m_s$. This is because using physical m_u, m_d would (a) require too much computer time, (b) require too large a lattice, and (c) introduce spurious quenching effects [3].

The interpolation in m_q to the strange quark mass m_s introduces little systematic error (assuming one knows how to fix m_s on the lattice). However, the extrapolation in m_q to physical $m_{u,d}$ (the chiral extrapolation) is a significant source of error. That is because one doesn't know *a priori* the correct functional form of the extrapolation. For example, lowest order chiral perturbation theory predicts that m_π^2 is a linear function of quark mass. However, at the current statistical level, one sees small but significant deviations from linearity that are not well understood. In Fig. 2(a), the linear fit has very poor confidence level (taking into account the correlations in the data), despite the fact that it looks good to the eye.

The deviations from linearity could be due to unphysical effects such as the finite lattice spacing or the residual contamination by radial B excitations. Even the more "physical" cause (chiral logs or higher order analytic terms in chiral perturbation theory) are a source of spurious effects because quenched chiral logs are in general different from those in the full theory [3]. When we include virtual quark loops, the theory is at present only "partially quenched," since the sea quark mass is not tuned to be equal to the valence mass, and again spurious chiral logs will be present [4].

We thus fit quantities like m_π^2 to their lowest order chiral form, despite the poor confidence levels. The systematic error is estimated by repeating the analysis with quadratic (constrained) fits, as shown in Fig. 2(a). The systematic error thus determined is $\leq 10\%$ for decay constants on all quenched data sets used to extrapolate to the continuum; usually it is $\lesssim 5\%$. (After

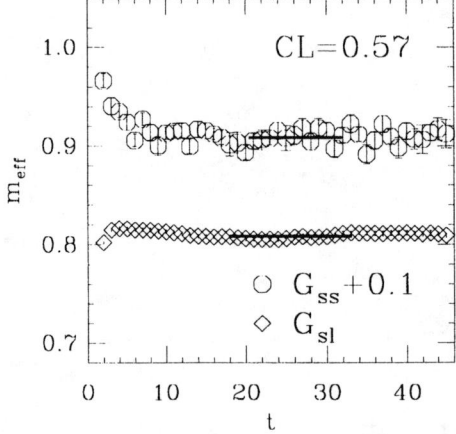

FIGURE 1. The effective mass m_{eff} vs. Euclidean time for two meson propagators (on lattice H) created by the same interpolating field. In G_{sl}, the meson is then annihilated by the axial current; in G_{ss}, by a second interpolating field. For G_{ss}, m_{eff} has been increased by 0.1 for clarity. The two propagators have been fit simultaneously in the ranges shown by the solid lines. The hopping parameters are 0.1474 and 0.125.

extrapolation to the continuum, the error is larger: 7% to 15% — see below.) For lattice F, this systematic error is very large ($\gtrsim 50\%$), which may be due to the small volume, exacerbated by virtual quark loop effects. Lattice F is dropped from further analysis.

• Heavy-Quark Interpolation. Currently practical lattice spacings are in the range 1 GeV$\lesssim a^{-1} \lesssim$ 4 GeV. Thus $m_B a \gtrsim 1$, and it is difficult to simulate a B meson since its Compton wavelength is smaller than the lattice spacing. Our approach to this problem is to interpolate between heavy-quark masses that can be better simulated: infinite mass quarks treated by the static method [5] and normal, propagating quarks with masses lighter than the b.

For the propagating quarks, we correct for the gross lattice artifacts caused by $m_B a \sim 1$ with some of the techniques of Ref. [6]. We use the "EKM norm" and also shift from pole to kinetic heavy-quark mass at tadpole-improved tree level. There are further (but numerically less important) artifacts which we do not correct for; we expect them to be eliminated (or at least drastically reduced) by the continuum extrapolation ($a \to 0$) at the end.

One estimate of the systematic error of this approach is obtained at fixed lattice spacing by comparing two different mass ranges of propagating quarks: "lighter heavies," which give meson masses in the range 1.25 to 2 GeV, and "heavier heavies," which give meson masses in the range 2 to 4 GeV. Figure 2(b) shows the interpolations between the static result and the two propagating mass ranges. Other estimates of the systematic error are available after one takes the $a \to 0$ limit — see below.

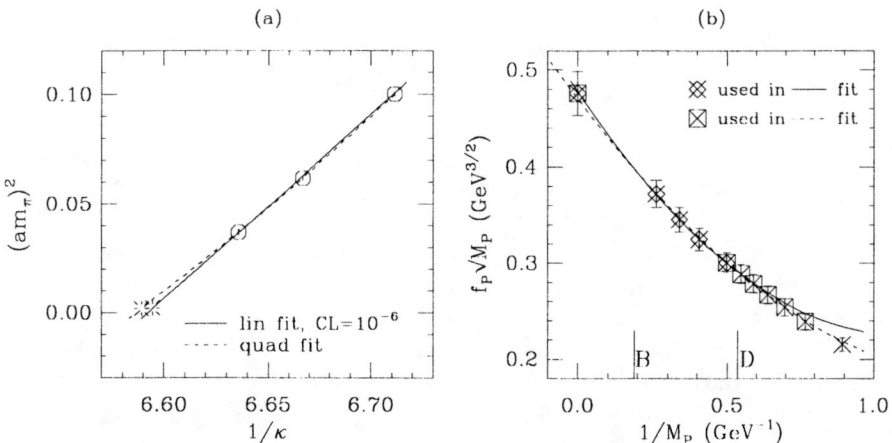

FIGURE 2. (a) The lattice m_π^2 vs. inverse hopping parameter ($1/\kappa \approx m_q + $ const.) on lattice D. Such fits are used to determine the value of κ which corresponds to the physical up or down quark mass. (b) $f_P M_P^{\frac{1}{2}}$ vs. $1/M_P$ on lattice D, where M_P is the mass of a heavy-light pseudoscalar meson (with light quark mass $\approx m_u, m_d$), and f_P is its decay constant. Two different interpolations to the B and D masses are shown.

Perturbation Theory

The axial current A_μ with a lattice cutoff is not the same as the continuum A_μ, but the difference is perturbatively calculable, since it comes from physics at the cutoff scale. At present, we use a (mass independent) one loop matching, with the scale ("q^*") of the coupling estimated along the lines of Ref. [7]. For propagating Wilson quarks, the result is, after tadpole improvement, $q^* = 2.32/a$ [8]. We estimate the systematic error by changing q^* by a factor of 2 and reanalyzing. The error is rather small ($\lesssim 3\%$).

Finite Volume

We compare results on lattice A ($\ell = 1.2$ fm) with those on lattice B ($\ell = 2.5$ fm). All other systematic effects on these two lattices are the same. A is smaller than all other lattices used for our central value; B, much larger. Therefore the difference should give a conservative bound on the finite volume error. This error is ~ 3–4% on decay constants and ~ 1–2% on ratios.

Extrapolation to the Continuum ($a \to 0$)

For any physical quantity Q computed here, we expect (Wilson fermions)

$$Q(a) = Q_{a=0}(1 + a\mathcal{M}_1 + a^2\mathcal{M}_2 + \cdots) , \qquad (3)$$

but we do not know $\mathcal{M}_1$ *a priori*, nor how it compares to $\mathcal{M}_2$. In practice, we find the slope to be quite large for the decay constants ($\mathcal{M}_1 \sim 300$–650 MeV), with f_{B_s} the worst offender. This leads to rather large extrapolation errors (~ 12–27%). The ratios of decay constant are much better behaved, with $\mathcal{M}_1 \sim 100$ MeV, and an error of ~ 4–5%.

Figure 3(a) shows several fits of f_B vs. a used to compute the central value and estimate the two largest sources of systematic error. The central value is obtained from a linear fit to the diamonds, which in turn use linear chiral fits, a lattice scale set by f_π, and the "EKM" corrections described above. The error of the continuum extrapolation is estimated by comparing the central value with the result of (1) a constant fit to the three diamonds with smallest values of a, (2) a linear fit to the squares, which use a different mass shift (the magnetic — m_3 — mass minus the pole mass) in the EKM corrections, and (3) a linear fit to the octagons, which use the "2κ" norm — *i.e.*, no EKM corrections. The continuum extrapolation error is defined as the largest of these three differences. The difference of the extrapolation of the crosses (which have a quadratic chiral extrapolation) and the central value determines the chiral extrapolation error.

Effects of Quenching

The quenched approximation has one great advantage: it saves an enormous amount of computer time. However, it is not a true approximation, since there is no perturbative expansion of the full theory for which the quenched approximation is the first term. Thus one should think of it only as a model. It is a good model since it has confinement and chiral symmetry breaking built in, is completely relativistic, seems to get the low lying hadronic spectrum right to ~ 5 to 10%, and can be shown to have small errors in a few cases [9]. But it is a model, nonetheless. Therefore one must try to estimate its errors, and ultimately to move beyond it.

To this end we have repeated our computations on lattices with virtual quark loops included. However, we emphasize that such computations are not yet "full QCD." This is because (1) the virtual quark mass is fixed and not extrapolated to physical up or down mass (the theory is partially quenched [4]), (2) the virtual quark data is not yet good enough to extrapolate to $a = 0$, and (3) we have two light flavors, not three. Thus the virtual quark simulations are used at this point only for systematic error estimation.

Figure 3(b) shows how the quenching error is estimated. One estimate is obtained by comparing the smallest-a virtual quark simulation (lattice G, the cross at $a = 0.47$ (GeV)$^{-1}$) with the quenched simulations, interpolated to the same value of a. Another estimate can be found by fixing the lattice scale

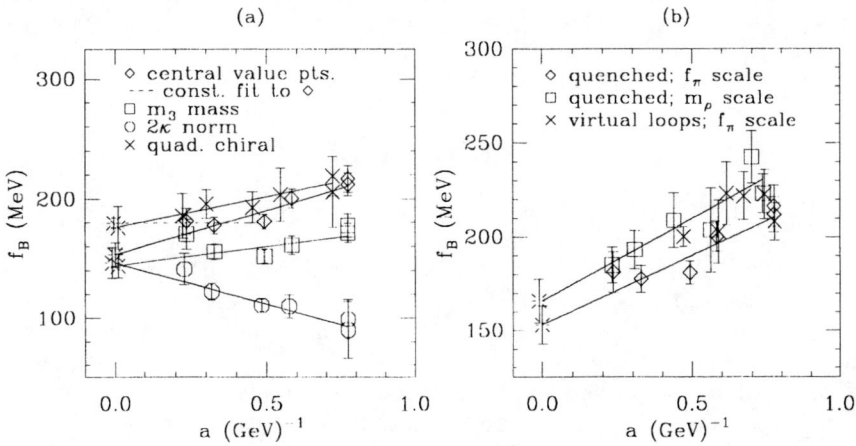

FIGURE 3. (a) Quenched results for f_B as a function of lattice spacing. The linear fit to all the diamonds gives the central value; other points and fits are used for estimating systematic errors — see text. (b) Results for f_B as a function of lattice spacing used for estimating the effects of quenching. The diamonds are the same as in (a); while the squares have the scale fixed by the mass of the ρ meson. The crosses (virtual quark loops included) are not extrapolated to the continuum.

in the quenched simulations by using m_ρ, instead of f_π. In principle, both methods of setting the scale should be equally good. However, quenching can affect m_ρ differently from f_π, so the difference is an estimate of the systematic error.

We emphasize that our quenching error estimate is quite crude at present. The difference between our virtual quark simulations and full QCD must be kept in mind. Further, the comparison of f_π and m_ρ scale results tests the quenched approximation only under the assumption that other systematic errors are well controlled. The errors of our continuum extrapolation are large enough to make this last assumption rather shaky.

RESULTS

We find:

$$f_{B_s} = 164 \pm 9 \,^{+47}_{-13} \,^{+16}_{-0} \text{ MeV} \qquad f_{B_s}/f_B = 1.10 \pm 0.02 \,^{+0.05}_{-0.03} \,^{+0.03}_{-0.02}$$
$$f_B = 153 \pm 10 \,^{+36}_{-13} \,^{+13}_{-0} \text{ MeV} \qquad f_{D_s}/f_D = 1.09 \pm 0.02 \,^{+0.05}_{-0.01} \,^{+0.02}_{-0.0}$$
$$f_{D_s} = 199 \pm 8 \,^{+40}_{-11} \,^{+10}_{-0} \text{ MeV} \qquad f_{B_s}/f_{D_s} = 0.83 \pm 0.02 \,^{+0.06}_{-0.03} \,^{+0.03}_{-0.00}$$
$$f_D = 186 \pm 10 \,^{+27}_{-18} \,^{+9}_{-0} \text{ MeV} \qquad f_B/f_{D_s} = 0.76 \pm 0.03 \,^{+0.07}_{-0.04} \,^{+0.02}_{-0.01} \qquad (4)$$

The errors shown are statistical (plus "fitting"), systematic (within the quenched approximation), and systematic (of quenching), respectively. We note that as experimental measurements of f_{D_s} improve, the ratios f_B/f_{D_s} and f_{B_s}/f_{D_s} may ultimately provide the best way to determine f_B and f_{B_s}.

We thank the Center for Computational Sciences (Oak Ridge), Indiana University, SDSC (San Diego), PSC (Pittsburgh), CTC (Cornell), MHPCC (Maui), and CHPC (Utah) for computing resources. This work was supported in part by the DOE and NSF.

REFERENCES

1. See for example A. J. Buras and M. K. Harlander, in *Heavy Flavors*, edited by A. J. Buras and M. Lindner, World Scientific, Singapore, 1992, p. 58.
2. C. Bernard, talk given at the *Seventh International Symposium on Heavy Flavor Physics*, Santa Barbara, July 7-11, 1997, to be published in the proceedings.
3. S. Sharpe, Phys. Rev. **D41**, 3233 (1990); C. Bernard and M. Golterman, Phys. Rev. **D46**, 853 (1992).
4. C. Bernard and M. Golterman, Phys. Rev. **D49**, 486 (1994).
5. E. Eichten, Nucl. Phys. B (Proc. Suppl.) **4**, 170 (1988).
6. A. El-Khadra, A. Kronfeld, and P. Mackenzie, Phys. Rev. **D55**, 3933 (1997).
7. G.P. Lepage and P.B. Mackenzie, Phys. Rev. **D48**, 2250 (1993).
8. C. Bernard, M. Golterman, and C. McNeile, in preparation.
9. See, for example, C. Bernard and M. Golterman, Nucl. Phys. B (Proc. Suppl.) **30**, 217 (1993). However, in most cases the errors of quenching cannot be unambiguously computed in chiral perturbation theory.

Beautiful CP Violation

Isard Dunietz

Fermi National Accelerator Laboratory, P.O. Box 500, Batavia, IL 60510

Abstract. CP violation is observed to date only in K^0 decays and is parameterizable by a single quantity ϵ. Because it is one of the least understood phenomena in the Standard Model and holds a clue to baryogenesis, it must be investigated further. Highly specialized searches in K^0 decays are possible. Effects in B decays are much larger. In addition to the traditional $B_d \to J/\psi K_S, \pi^+\pi^-$ asymmetries, CP violation could be searched for in already existing inclusive B data samples. The rapid $B_s - \overline{B}_s$ oscillations cancel in untagged B_s data samples, which therefore allow feasibility studies for the observation of CP violation and the extraction of CKM elements with present vertex detectors. The favored method for the extraction of the CKM angle γ is shown to be unfeasible and a solution is presented involving striking direct CP violation in charged B decays. Novel methods for determining the B_s mixing parameter Δm are described without the traditional requirement of flavor-specific final states.

I INTRODUCTION

More than thirty years after its discovery, CP violation remains a mystery. Our entire knowledge about it can be summarized by the single parameter ϵ [1]. CP violation is not just a quaint tiny effect observed in K^0 decays, but is one of the necessary ingredients for baryogenesis [2]. Within the CKM model, it is connected also to the quark-mixing and hierarchy of quark masses. A successful theory of CP violation will have far-reaching ramifications in cosmology and high energy physics.

At present, we are not able to answer even the question raised by Wolfenstein more than 30 years ago: Is CP violation due to a new superweak interaction, which would show up essentially only in mixing-induced phenomena? Or are there direct CP violating effects? There exists a multitude of scenarios for CP violation, all consistent with ϵ. What is needed is the observation of many independent CP violating effects. This would be invaluable in directing us toward a more fundamental understanding of CP violation, in analogy to the history of parity violation. There a variety of measurements guided us to the successful $V - A$ theory [3].

Searches for (direct) CP violation in K and hyperon decays are important [1,4]. Because the expected effects are either tiny for processes with sizable BR's or could be large but then involve tiny BR's $\mathcal{O}(10^{-11})$, ingenious experimental techniques are being developed to overcome those handicaps.

A whole class of additional independent CP measurements can be obtained from studies of b-hadron decays. Although CP violation may not be (entirely) due to the CKM model, that model serves here as a guide. Decays of b-hadrons can access large CKM phases and thus large CP violation, because the b-quark is a member of the third generation. There are many proposed methods that involve large CP violating effects [5]. This talk focuses on recently discussed phenomena, some of which can be studied with presently existing data samples.

First, (semi-)inclusive B decays are expected to exhibit CP violation and CKM parameters can be extracted [6–8]. Even the B_s mixing-parameter Δm could be determined from such flavor-nonspecific final states, in addition to the conventional methods [9,10]. Second, untagged B_s data samples are predicted to exhibit CP violation and permit the extraction of CKM parameters, as long as the B_s width difference is significant [11]. The far-reaching physics potential of the $B_s \to J/\psi\phi$ process is touched upon. The third topic explains why the favorite method for determining the CKM angle γ, pioneered by Gronau-London-Wyler (GLW) [12], is unfeasible. The CKM parameter can be cleanly extracted [13], however, when one incorporates the striking, direct CP violating effects in $B \to D^0/\overline{D}^0$ transitions [14], which were not considered by GLW.

II EXCLUSIVE AND INCLUSIVE B DECAYS

Traditional methods involve exclusive modes such as $J/\psi K_S$ [15], $\pi^+\pi^-$ [16–18], and study the rate-asymmetry between

$$B_d(t) \to J/\psi K_S, \pi^+\pi^- \neq \overline{B}_d(t) \to J/\psi K_S, \pi^+\pi^- . \qquad (1)$$

The effective BR is tiny $\sim 10^{-5}$, but the asymmetries are large $\mathcal{O}(1)$. How does this large asymmetry come about? The unmixed B_d could decay into $J/\psi K_S$ directly, $B_d \to J/\psi K_S$. The CP conjugated process is the direct decay, $\overline{B}_d \to J/\psi K_S$. To excellent accuracy, those two direct decay rates are equal. The B_d could mix first into a $\overline{B}_d$ and then decay to $J/\psi K_S$, $B_d(t) \to \overline{B}_d \to J/\psi K_S$. The CP conjugated process is the mixing-induced $\overline{B}_d(t) \to B_d \to J/\psi K_S$ transition. Again, to excellent accuracy, the magnitudes of the two mixing-induced amplitudes are the same. The large CP violation predicted in the CKM model occurs because of the interference of the direct and mixing-induced amplitudes. To form the asymmetry, it is not sufficient to reconstruct the final state $J/\psi K_S$. One must be able to distinguish those reconstructed events as originating from an initial B_d versus $\overline{B}_d$ (referred to as tagging).

Initially (at $t = 0$) the neutral B meson has no time to mix. At $t = 0$ there is no mixing-induced amplitude and thus no CP violation. There is almost no loss in measuring the asymmetry by not considering $J/\psi K_S$ events within the first B_d lifetime or so. While the rate is largest during that time-interval, the asymmetry is tiny and needs large proper times to build itself up [18,19]. Triggering on detached vertices is thus more efficient for such CP violation studies than one might think naively.

Inclusive B samples are many orders of magnitude larger than the exclusive ones and can be accessed by vertexing. The time-dependent, totally inclusive asymmetry,

$$I(t) \equiv \frac{\Gamma(B^0(t) \to \text{all}) - \Gamma(\overline{B}^0(t) \to \text{all})}{\Gamma(B^0(t) \to \text{all}) + \Gamma(\overline{B}^0(t) \to \text{all})}, \qquad (2)$$

is CP violating [7,8]. That appears to be rather puzzling, especially because the CPT theorem guarantees that the totally inclusive width is the same for particle and antiparticle. That CPT stranglehold is removed, because $B^0 - \overline{B}^0$ mixing provides an additional amplitude and thus novel interference effects. The totally inclusive CP asymmetry $I(t)$ is related to the wrong-sign asymmetry [20,21]

$$\frac{\Gamma(B^0(t) \to W) - \Gamma(\overline{B}^0(t) \to \overline{W})}{\Gamma(B^0(t) \to W) + \Gamma(\overline{B}^0(t) \to \overline{W})} = -a = -Im\frac{\Gamma_{12}}{M_{12}}, \qquad (3)$$

where W denotes "wrong-sign" flavor-specific modes that come only from $\overline{B}^0 \to W$ and never from $B^0 \to W$, such as $W = \ell^- X$ and $W = D_s^+ \{\pi^-, \rho^-, a_1^-\}$ for B_s decays [$W = D^{(*)}D_s^{(*)-}, D\overline{D}\,\overline{K}X, J/\psi \overline{K}^*$ for B_d decays].

The data samples for the $I(t)$ asymmetries exist already. For instance, the SLD collaboration determined the lifetime ratio of neutral to charged b-hadrons by an inclusive topological vertex analysis [22]. The polarization of Z^0 provides a large forward-backward asymmetry of b production and thus an effective initial flavor-tag [23] and it is clear that SLD can study inclusive asymmetries. Similarly, the LEP experiments are able to study $I(t)$ by using their b-enriched samples and optimal flavor-tagging algorithms. CDF has several million high P_T-leptons, which are highly enriched in b content. The data sample of detached vertices on the other hemisphere allows CDF to study $I(t)$. The newly installed vertex detector at CLEO permits meaningful studies, because the $I(t)$ asymmetry becomes significant only after a few B_d lifetimes, see Eq. (4) below.

For $\Delta\Gamma = 0$, the explicit time dependence is [7]

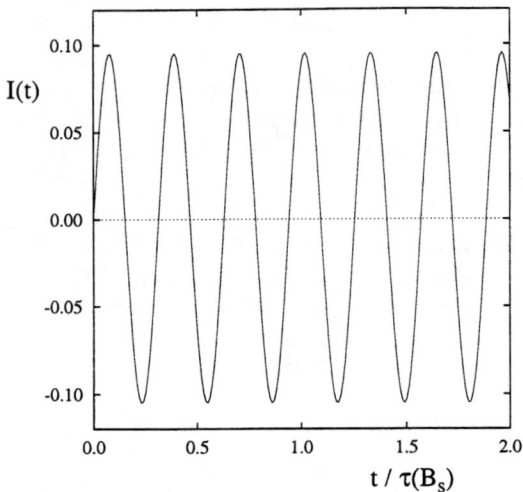

FIGURE 1. The totally inclusive CP asymmetry of $B_s(t) \to$ all, with $a = 0.01, \Delta\Gamma = 0$ and $x = 20$ (see Eqs. (2),(4)).

$$I(t) = a\left[\frac{x}{2}\sin\Delta mt - \sin^2\frac{\Delta mt}{2}\right], \qquad (4)$$

where $x \equiv \Delta m/\Gamma$. The observable a can thus be extracted from a study of $I(t)$.

For B_s mesons, that extraction offers a significant statistical gain over the conventional method [Eq. (3)]. The factor of $x/2$ enhances $I(t)$ over a by an order of magnitude, which corresponds to a statistical gain of $\mathcal{O}(10^2)$. There is another gain, because all B_s decays are used rather than flavor-specific B_s modes that must be efficiently distinguished from B_d modes. The distinction involves stringent selection criteria. The reason is that the wrong-sign asymmetry [Eq. (3)] is time-independent, and the wrong-sign B_d asymmetry is an order of magnitude larger than the B_s one, within the CKM model. Thus, for instance, the high-p (-P_T) leptons must originate from B_s decays and not from B_d decays. This can be achieved by either studying wrong-sign B_s modes at very short proper times [24], or by inferring the existence of a D_s, or by observing such primary kaons that significantly enrich the B_s content, or by a combination of the above. In contrast, the unique time-dependence of $I(t)$ provides automatic discrimination. For the B_s meson at least, the time-dependent inclusive asymmetry may be more effective in extracting the CP violating observable a than the conventional wrong-sign asymmetry.

Figure 1 shows what to expect for the choice $x = 20$ and where New Physics is allowed to enhance $a = |\Gamma_{12}/M_{12}| \sim 0.01$. The observation of a non-vanishing $I(t)$ proves CP violation and in addition allows a determination of the $B_s - \overline{B}_s$ mixing parameter Δm from flavor-nonspecific final states. The traditional methods for extracting Δm require flavor-specific final states and tagging [9,10]. We will mention later on additional ways to extract Δm with flavor-nonspecific final states.

Within the CKM model, the totally inclusive asymmetries are tiny $\mathcal{O}(10^{-3})$ for B_d and $\mathcal{O}(10^{-4})$ for B_s [25,26]. The ability to select specific quark transitions enhances the asymmetries by orders of magnitude, at times to the $\sim (10-20)\%$ level [7]. Such selections permit extractions of CKM phases and to conduct the study in either a time-integrated or time-dependent fashion.[1] Those analyses should be pursued whenever feasible. There exist unitarity constraints, which allow systematic cross-checks. Future B detectors will be able to more fully explore the potential with such semi-inclusive data samples.

III PHYSICS WITH (UNTAGGED) B_S MESONS

One conventional way to determine the CKM angle γ is the time-dependent study of tagged $\overset{(-)}{B_s}(t) \to D_s^{\pm} K^{\mp}$ processes [27], and in the neglect of penguin amplitudes $\overset{(-)}{B_s}(t) \to \rho^0 K_S, \omega K_S$ transitions [17,18,28]. It requires flavor-tagging and the ability to trace the rapid Δmt-oscillations. The requirements are problematic:

(a) Flavor-tagging is at present only a few percent efficient at hadron accelerators [29].[2]

(b) Resolution of Δmt-oscillations is feasible for $x \lesssim 20$ with present vertex technology [9], but LEP experiments reported [10],

$$x \gtrsim 15 . \tag{5}$$

Though Δmt-oscillations may be too rapid to be resolved at present, such large Δm may imply a sizable width difference $\Delta\Gamma$ [31]. Non-perturbative effects may further enhance $\Delta\Gamma$ considerably [32]. Perhaps $\Delta\Gamma$ will be the first observable $B_s - \overline{B}_s$ mixing effect [11], which would circumvent problems (a) and (b). The Δmt-terms cancel in the time-evolution of untagged B_s [11],

$$f(t) \equiv \Gamma(B_s(t) \to f) + \Gamma(\overline{B}_s(t) \to f) = a e^{-\Gamma_L t} + b e^{-\Gamma_H t} , \tag{6}$$

which is governed by the two exponentials $e^{-\Gamma_L t}$ and $e^{-\Gamma_H t}$ alone. That fact permits many non-orthodox CP violating studies and extractions of CKM parameters [11]:

(1) Consider final states with definite CP parity, f_{CP}, such as $\rho^0 K_S, \omega K_S, \ldots$. If the untagged time-evolution $f_{CP}(t)$ is governed by both exponentials $e^{-\Gamma_L t}$ and $e^{-\Gamma_H t}$, then CP violation has occured [11]. The measurement of $f_{CP}(t)$

---

[1] For B_s mesons, Δm could be extracted from such more refined studies.
[2] Though, in principle almost all B-decays could be flavor-tagged [30].

allows even the extraction of CKM parameters [11,33]. The physics of the $J/\psi\phi$ final state is very instructive. The time-evolution of untagged $J/\psi\phi$ could show CP violating effects [33]. The $\overset{(-)}{B_s} \to J/\psi\phi$ has CP-even and CP-odd amplitudes, $\overset{(-)}{A}_+$ and $\overset{(-)}{A}_-$ respectively. Angular correlations [34] allow to measure the interference terms between CP-even and CP-odd amplitudes, which for untagged data samples is proportional to [33],

$$\left(e^{-\Gamma_H t} - e^{-\Gamma_L t}\right)\theta^2 2\eta, \quad \text{where} \quad \theta \approx 0.22. \tag{7}$$

The observation of such a non-vanishing term would prove CP violation and would permit the extraction of the CKM parameter η. Note that the observable depends optimally on the width difference.

Those interference terms once tagged allow the measurement of Δm, even though $J/\psi\phi$ is a flavor-nonspecific final state [34]. To demonstrate the point most sharply, neglect CP violation and set $\Delta\Gamma = 0$. Then $A_+(t) \sim e^{-im_L t}$ and $A_-(t) \sim e^{-im_H t}$. The observable $A_+(t)A_-^*(t) \sim e^{i\Delta m t}$ depends on $\Delta m \equiv m_H - m_L$. Ref. [35] describes yet another method for measuring Δm without flavor-specific final states.

(2) After several B_s lifetimes, the long-lived $B_s^H \sim B_s - \overline{B}_s$ will be significantly enriched over the short-lived B_s^L. Consider then final states f that can be fed from both B_s and $\overline{B}_s$, and that are non-CP-eigenstates. CP violation is proven if the time evolution of untagged $f(t)$ differs from untagged $\overline{f}(t)$,

$$f(t) \neq \overline{f}(t) \Rightarrow CP\ violation. \tag{8}$$

Furthermore, the CKM angle γ can be extracted from time-dependent studies of $D_s^\pm K^\mp(t)$, $\overset{(-)}{D^0}\phi(t)$ [11].[3] CP violating effects and CKM extractions can be enhanced by studying $D_s^{(*,**)\pm}K^{*\mp}(t)$ [36]. In summary, neither flavor-tagging nor exquisite tracing of Δmt-oscillations are necessary, only a large $\Delta\Gamma$.

IV DIRECT CP VIOLATION AND EXTRACTING CKM ANGLES

The favorite method (particularly at $\Upsilon(4S)$ factories) for determining γ has been developed by Gronau, London and Wyler (GLW) [12] and requires the measurements of the six rates $B^\pm \to D^0 K^\pm, \overline{D}^0 K^\pm$ and $D_{CP}^0 K^\pm$. Here D_{CP}^0 denotes that the D^0 is seen in CP eigenstates with either CP-even

---

[3] The determination of γ from $\overset{(-)}{D^0}\phi(t)$ and $D_{CP}^0\phi(t)$ as presented in Ref. [11] must include the effect of doubly-Cabibbo suppressed $\overset{(-)}{D^0}$ decay-amplitudes [14,13].

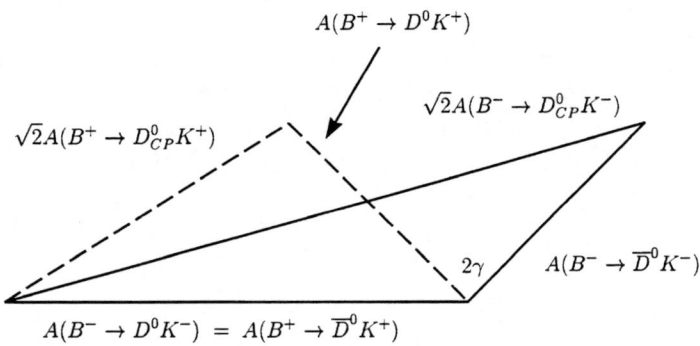

FIGURE 2. The traditional GLW method for extracting the CKM angle γ.

($K^+K^-, \pi^+\pi^-, ...$) or CP-odd ($K_S\phi, K_S\pi^0, ...$) parity. The GLW method focuses on the CP violating rate difference of $B^+ \to D^0_{CP}K^+$ versus $B^- \to D^0_{CP}K^-$ [37], which can reach at best the 10% level and is probably significantly smaller.

In principle, the GLW method is a great idea. However, new CLEO data indicate that the method is unfeasible, and that the largest CP violating effect has been overlooked [14,13]. Once the effect has been incorporated, the CKM angles can be cleanly extracted [13].

Let us review the original GLW method, point out the problem, and show how it can be overcome. Consider CP even D^0_{CP}, for which

$$D^0_{CP} = \frac{1}{\sqrt{2}}(D^0 + \overline{D}^0) \,. \qquad (9)$$

Then

$$\sqrt{2}A(B^- \to D^0_{CP}K^-) = A(B^- \to D^0 K^-) + A(B^- \to \overline{D}^0 K^-), \qquad (10)$$

and that amplitude triangle is shown in Figure 2. The weak phase difference of the two interfering amplitudes is γ. GLW argued that the magnitudes of each of the sides of the triangle can be measured (being proportional to the square roots of the respective rates), and thus claimed that the amplitude triangle can be fully reconstructed.

Figure 2 has not been drawn to scale. The $B^- \to \overline{D}^0 K^-$ amplitude is an order of magnitude smaller than the $B^- \to D^0 K^-$ one, which can be seen as follows [13]. The CKM factors suppress the amplitude ratio by about

1/3. The $\overline{D}^0 K^-$ is color-suppressed while $D^0 K^-$ is also color-allowed, yielding another suppression factor of about 1/4.

Nothing changes when the CP conjugated final states are considered, except that the CKM elements have to be complex conjugated. Apparently, the CP-conjugated triangle can also be determined, see Figure 2. The $A(B^+ \to D^0 K^+)$ is rotated by 2γ with respect to $A(B^- \to \overline{D}^0 K^-)$, and apparently the angle γ can be extracted. Note that the only CP violation in all these processes occurs in

$$\Gamma(B^+ \to D^0_{CP} K^+) \neq \Gamma(B^- \to D^0_{CP} K^-) \tag{11}$$

while there is no CP violation in

$$\Gamma(B^+ \to \overline{D}^0 K^+) = \Gamma(B^- \to D^0 K^-) \text{, and} \tag{12}$$

$$\Gamma(B^+ \to D^0 K^+) = \Gamma(B^- \to \overline{D}^0 K^-) . \tag{13}$$

In principle this argument is correct, but in practice the largest <u>direct</u> CP violating effects (residing in those processes) will be seen in [14,13]

$$B^+ \to D^0 K^+ \neq B^- \to \overline{D}^0 K^- . \tag{14}$$

The $\overline{D}^0$ produced in the $B^- \to \overline{D}^0 K^-$ process is seen in its non-leptonic, Cabibbo-allowed modes f, such as $K^+\pi^-, K\pi\pi$. It was assumed that the kaon flavor unambiguously informs on the initial charm flavor. This assumption overlooked the doubly-Cabibbo-suppressed $D^0 \to f$ process which leads to the same final state $B^- \to D^0[\to f]K^-$. Further, CLEO has measured [38]

$$\left|\frac{A(D^0 \to f)}{A(\overline{D}^0 \to f)}\right| \sim 0.1 , \tag{15}$$

which maximizes the interference,

$$\left|\frac{A(B^- \to K^- D^0[\to f])}{A(B^- \to K^- \overline{D}^0[\to f])}\right| \sim 1 , \tag{16}$$

$$A(B^- \to K^-[f]) = A(B^- \to K^- D^0[\to f]) + A(B^- \to K^- \overline{D}^0[\to f]) . \tag{17}$$

The conditions are ideal for striking direct CP violating effects. They require that the interfering amplitudes be comparable in size (Eq. (16)), that the weak phase difference be large (γ in our case), and that the relative final-state-phase difference be significant. It is an experimental fact that large final state phases occur in many D decays [39]. This enables us to engineer large

CP violating effects by optimally weighting relevant sections of generalized Dalitz plots.

The traditional focus on CP eigenmodes of D^0_{CP} automatically excludes this so potent source of final-state interaction phases. The orthodox method [37,12] accesses only the final-state phase difference residing in $B^- \to D^0 K^-$ versus $B^- \to \overline{D}^0 K^-$, which is expected to be significantly more feeble [40]. The CKM angle γ can be cleanly extracted once one incorporates the findings of this section [13], because penguin amplitudes are absent. The extraction of γ and the observation of CP violation is optimized by combining detailed (experimental) investigations of D^0 decays with $B^{\pm}$ decays to $\overset{(-)}{D^0}$ [13]. This provides yet another reason for accurate measurements of D^0 decays. Note also that observation of direct CP violation (as advocated in this section) would rule out superweak scenarios as the only source for CP violation.

V CONCLUSION

CP violation has been observed only in K^0 decays and is parameterizable by a single quantity ϵ. It is one of the necessary ingredients for baryogenesis [2], and within the CKM model is related to the quark-mixing and hierarchy of quark masses. It is one of the least understood phenomena in high energy physics and a very important one. Just as the successful $V-A$ theory of parity violation [3] emerged from a synthesis of many independent parity violating measurements, so a more fundamental understanding of CP violation will profit from many independent observations of CP violation.

This talk thus emphasized that CP violation should not only be searched in traditional exclusive $B_d \to J/\psi K_S, \pi^+\pi^-$ rate asymmetries. Observable CP violating effects could be present in (semi-)inclusive B decays, and could be searched for with existing data samples. The time-evolutions of untagged B_s data samples have no rapid Δmt-oscillations. Still CP violation could be observed and CKM parameters extracted as long as $\Delta\Gamma$ is sizable. Many striking direct CP violating effects in B decays are possible. The observation of CP violation and CKM extraction are optimized by detailed studies of D decays.

VI ACKNOWLEDGEMENTS

This work was supported in part by the Department of Energy, Contract No. DE-AC02-76CH03000.

REFERENCES

1. B. Winstein and L. Wolfenstein, Rev. Mod. Phys. **65**, 1113 (1993).

2. A.D. Sakharov, JETP Lett. **5**, 24 (1967).
3. R.P. Feynman and M. Gell-Mann, Phys. Rev. **109**, 193 (1958); E.C.G. Sudarshan and R. Marshak, Phys. Rev. **109**, 1860 (1958).
4. G. Buchalla, hep-ph/9612307.
5. For a review see, for instance, A.J. Buras and R. Fleischer, hep-ph/9704376.
6. I. Dunietz and R.G. Sachs, Phys. Rev. **D 37**, 3186 (1988); (E) ibid. **D 39**, 3515 (1989).
7. M. Beneke, G. Buchalla and I. Dunietz, Phys. Lett. **B393**, 132 (1997).
8. L. Stodolsky, hep-ph/9612219.
9. Proceedings of the Workshop on B Physics at Hadron Accelerators, Snowmass, Co., June 21 - July 2, 1993, edited by P. McBride and C. Shekhar Mishra.
10. V. Andreev et al. (The LEP B Oscillations Working Group), "Combined Results on B^0 Oscillations: Update for the Summer 1997 Conferences," LEP-BOSC 97/2, August 18, 1997.
11. I. Dunietz, Phys. Rev. **D52**, 3048 (1995).
12. M. Gronau and D. London, Phys. Lett. **B253**, 483 (1991); M. Gronau and D. Wyler, Phys. Lett. **B265**, 172 (1991).
13. D. Atwood, I. Dunietz and A. Soni, Phys. Rev. Lett. **78**, 3257 (1997).
14. I. Dunietz, Z. Phys. **C56**, 129 (1992); I. Dunietz, in B Decays, Revised 2nd Edition, edited by S. Stone (World Scientific, Singapore, 1994), p. 550.
15. I.I. Bigi and A.I. Sanda, Nucl. Phys. **B193**, 85 (1981).
16. L. Wolfenstein, Nucl. Phys. **B246**, 45 (1984).
17. D. Du, I. Dunietz and Dan-di Wu, Phys. Rev. **D34**, 3414 (1986).
18. I. Dunietz and J.L. Rosner, Phys. Rev. **D34**, 1404 (1986).
19. I. Dunietz and T. Nakada, Z. Phys. **C36**, 503 (1987).
20. A. Pais and S.B. Treiman, Phys. Rev. **D12**, 2744 (1975); T. Altomari, L. Wolfenstein and J.D. Bjorken, Phys. Rev. **D 37**, 1860 (1988); M. Lusignoli, Z. Phys. **C41**, 645 (1989).
21. H. Yamamoto, Phys. Lett. **B401**, 91 (1997).
22. K. Abe et al. (SLD Collaboration), Phys. Rev. Lett. **79**, 590 (1997).
23. W.B. Atwood, I. Dunietz and P. Grosse-Wiesmann, Phys. Lett. **B216**, 227 (1989); W.B. Atwood, I. Dunietz, P. Grosse-Wiesmann, S. Matsuda and A.I. Sanda, Phys. Lett. **B232**, 533 (1989).
24. M. Jimack, private communication.
25. M. Lusignoli, Z. Phys. **C41**, 645 (1989).
26. G. Buchalla, private communication.
27. R. Aleksan, I. Dunietz and B. Kayser, Z. Phys. **C54**, 653 (1992).
28. Ya.I. Azimov, N.G. Uraltsev and V.A. Khoze, JETP Lett. **43**, 409 (1986).
29. B. Wicklund, in the proceedings of the b20 conference, June 29 - July 2, 1997, Illinois Institute of Technology, Chicago, Illinois.
30. I. Dunietz, FERMILAB-PUB-94/163-T, hep-ph/9409355.
31. M. Beneke, G. Buchalla and I. Dunietz, Phys. Rev. **D54**, 4419 (1996).
32. I. Dunietz, J. Incandela, F.D. Snider, and H. Yamamoto, FERMILAB-PUB-96-421-T (hep-ph/9612421), to be published in Z. Phys. **C**.

33. R. Fleischer and I. Dunietz, Phys. Rev. **D55**, 259 (1997).
34. A.S. Dighe, I. Dunietz, H.J. Lipkin and J.L. Rosner, Phys. Lett. **B369**, 144 (1996).
35. Ya. Azimov and I. Dunietz, Phys. Lett. **B395**, 334 (1997).
36. R. Fleischer and I. Dunietz, Phys. Lett. **B387**, 361 (1996).
37. I.I.Y. Bigi and A.I. Sanda, Phys. Lett. **211B**, 213 (1988).
38. H. Yamamoto, Harvard University report, HUTP-96-A-001, January 1996 [hep-ph/9601218]; D. Cinabro et al. (CLEO Collab.), Phys. Rev. Lett. 72, 1406 (1994).
39. See, for instance, P.L. Frabetti (E687 Collaboration), Phys. Lett. **B331**, 217 (1994); G. Bonvicini et al. (CLEO Collaboration), contributed paper to the 28th International Conference on HEP, Warsaw, Poland, July 1996, PA05-090 [CLEO CONF 96-21].
40. R.N. Cahn and M. Suzuki, hep-ph/9708208.

EPR in B Physics and Elsewhere

Harry J. Lipkin [1]

Department of Particle Physics Weizmann Institute of Science, Rehovot 76100, Israel

School of Physics and Astronomy, Raymond and Beverly Sackler Faculty of Exact Sciences, Tel Aviv University, Tel Aviv, Israel

High Energy Physics Division, Argonne National Laboratory, Argonne, IL 60439-4815, USA

Abstract. The application of Einstein-Podolsky-Rosen correlations in $\Upsilon(4s) \to B\bar{B}$ decays to research in CP violation is the first and probably only *use* of EPR as a technique for research in new physics. Elsewhere highly sophisticated EPR projects *question* EPR and test its predictions to look for violations of quantum mechanics, hidden variables, Bell's inequalities, etc.

In 1995 an international conference was held celebrating 60 years of EPR [1]. Nathan Rosen (the R of EPR) was still active and told us about Einstein's view of EPR. My paper on applications to kaon and B physics (ϕ and B factories) [2] was the only talk on the program suggesting that EPR might be *useful*. All the others were on tests of quantum mechanics and included very beautiful photon-interferometry experiments using the latest solid-state technology to split a photon in a crystal into two photons with different energies and show that the split photons with only a piece of the original energy still exhibited the EPR correlations.

- **Elsewhere:** EPR questioned and used to check Quantum Mechanics.
  - Hidden Variables, Bell's Inequality and Bohmian QM.
  - Sophisticated photon-interferometry experiments confirm QM.
- **B physics:** EPR not only accepted but also crucial for BaBar.
  - Lepton asymmetry observed in decay of one B of pair from $\Upsilon(4S)$ "entangled" with ψK_S decay observed from other B.

---

[1] Supported in part by grant No. I-0304-120-.07/93 from The German-Israeli Foundation for Scientific Research and Development and by the U.S. Department of Energy, Division of High Energy Physics, Contract W-31-109-ENG-38.

The Problem of Entangled Wave Functions

A wave function for a pair of spin-1/2 particles moving apart with momenta $\vec{k}$ and $-\vec{k}$ coupled to total spin $S = 0$ must have their spins "entangled" to conserve total spin even if they are far apart. No matter which axis is chosen to measure the spin projections, the two spins must be opposite:

$$\Psi_{(S=0)} = \left|-\vec{k}(\uparrow); \vec{k}(\downarrow)\right\rangle - \left|-\vec{k}(\downarrow); \vec{k}(\uparrow)\right\rangle = \left|-\vec{k}(\leftarrow); \vec{k}(\rightarrow)\right\rangle - \left|-\vec{k}(\rightarrow); \vec{k}(\leftarrow)\right\rangle.$$

If the particles with momentum $-\vec{k}$ are detected and used to trigger a detector for observing the coincident beam with momentum $\vec{k}$, triggering on events with (Spin $\uparrow$) and momentum $-\vec{k}$ creates a polarized (Spin $\downarrow$) beam with momentum $\vec{k}$. But triggering on events with (Spin $\leftarrow$) and momentum $-\vec{k}$ creates a polarized (Spin $\rightarrow$) beam with momentum $\vec{k}$. What does this mean?

- **Elsewhere:** How is this possible? How can the particle with momentum $\vec{k}$ know what is measured at momentum $-\vec{k}$? Something must be wrong with Quantum Mechanics!

- **B physics:** Great! We can make polarized beams with momentum $\vec{k}$ and arbitrary polarizations by choosing the right triggers with momentum $-\vec{k}$!

Entangled Wave Functions in B Physics at $\Upsilon(4S)$ [3]

Both B^o and $\bar{B}^o$ decay into ψK_S. Define B_1 and B_2 as the linear combinations of B^o and $\bar{B}^o$ such that $B_1 \to \psi K_S$, but B_2 does not decay to ψK_S; i.e. the B^o and $\bar{B}^o$ contributions to $B_2 \to \psi K_S$ exactly cancel one another.

The $\Upsilon(4S)$ decays to an "entangled" neutral-B pair. If one is B^o the other must be $\bar{B}^o$. But if one is B_1 the other must be B_2:

$$\Upsilon \to \left|B^o(-\vec{k}); \bar{B}^o(\vec{k})\right\rangle - \left|\bar{B}^o(-\vec{k}); B^o(\vec{k})\right\rangle = \left|B_1(-\vec{k}); B_2(\vec{k})\right\rangle - \left|B_2(-\vec{k}); B_1(\vec{k})\right\rangle.$$

Triggering on events with B^o decay observed with momentum $-\vec{k}$ creates a $\bar{B}^o$ beam with momentum $\vec{k}$.

Triggering on events with $B \to \psi K_S$ decay observed with momentum $-\vec{k}$ creates a B_2 beam with momentum $\vec{k}$.

If CP is conserved, B_1 and B_2 are both mass eigenstates. Then B_2 decays have no time dependence; no lepton asymmetry.

If CP is violated, B_1 and B_2 may not be mass eigenstates. Define B^L and B^H as the linear combinations of B^o and $\bar{B}^o$ that are mass eigenstates. A B_2 beam can be a linear combination of B^L and B^H with a time-dependent relative phase producing $B^o - \bar{B}^o$ oscillations. Time-dependent lepton-asymmetry oscillations can then be observed in B_2 decays. This is the physics underlying the building of an asymmetric B-factory to observe CP violation.

"B-spin," EPR, and CP Violation for Pedestrians [4]

We can see this "entangled" B physics pictorially by defining a fictitious "B-spin" space where B° and $\bar{B}^\circ$ are eigenstates of B_z with spin up and spin down in the z direction. The x direction is defined to make the mass eigenstates B^H and B^L eigenstates of B_x with spin up and spin down in the x direction. The "entangled" pair of B's from $\Upsilon(4S)$ decay have opposite B-spins with respect to any axis. A meson in a state $|B(0)\rangle$ at time $t=0$ develops in time as

$$|B(t)\rangle = e^{-\frac{\Gamma}{2}t} e^{-i\omega\sigma_x(t/2)} |B(0)\rangle = e^{-\frac{\Gamma}{2}t} \cdot \{\cos(\frac{\omega t}{2}) - i\sigma_x \sin(\frac{\omega t}{2})\} |B(0)\rangle ,$$

where Γ is the decay width and ω the $B^H - B^L$ mass difference.

The difference between decay probabilities into B or $\bar{B}$ allowed modes, e.g. a lepton asymmetry A_{lep}, is given by the B-spin polarization in the z direction:

$$A_{lep} \equiv |\langle B^\circ | B(t)\rangle|^2 - |\langle \bar{B}^\circ | B(t)\rangle|^2 = e^{-\Gamma t} \cdot \langle B(0)| e^{i\omega\sigma_x(t/2)} \sigma_z e^{-i\omega\sigma_x(t/2)} |B(0)\rangle$$
$$= e^{-\Gamma t} \cdot [\langle B(0)| \sigma_z |B(0)\rangle \cos(\omega t) + \langle B(0)| \sigma_y |B(0)\rangle \sin(\omega t)] .$$

The state B_2 which does not decay into ψK_S is initially an equal mixture of B° and $\bar{B}^\circ$ and described by a B-spin vector in the xy plane at some angle θ with respect to the x axis. If CP is conserved, B_1 and B_2 are CP eigenstates and mass eigenstates, and $\theta = 0$. When CP is violated, θ is related in the standard model to an angle of the unitarity triangle. The states B_1 and B_2 precess around the x axis out of the xy plane, with a frequency ω given by the $B^H - B^L$ mass difference. They are no longer equal mixtures of B° and $\bar{B}^\circ$ and have an oscillating B_z component. If $B_1 \to \psi K_S$ is observed at $t=0$ in one decay from $\Upsilon(4S)$, the "entangled" semileptonic B_2 decay shows a time-dependent lepton asymmetry given by this oscillating B-spin polarization:

$$\langle B_2| \sigma_z |B_2\rangle = 0; \quad \langle B_2| \sigma_y |B_2\rangle = \sin(\theta); \quad A_{lep}(B_2) = e^{-\Gamma t} \sin(\theta) \sin(\omega t).$$

Most B-decay CP-violation experiments can be simply described as measurements of such polarizations and polarization correlations in this B-spin space.

REFERENCES

1. *The Dilemma of Einstein, Podolsky and Rosen – 60 Years Later. An International Symposium in Honour of Nathan Rosen – Haifa, March 1995.* Edited by A. Mann and M. Revzen, *Ann. Israel Phys. Soc.*, Volume 12. Institute of Physics Publishing, Bristol (1996).
2. Harry J. Lipkin, in ref. 1, pp 49-56.
3. Harry J. Lipkin, *Physics Letters* B219 (1989) 474.
4. Harry J. Lipkin, In "Proceedings of the Workshop on B Physics at Hadron Accelerators," Snowmass, Colorado, June 21-July 3 (1993). Edited by Patricia McBride and C. Shekhar Mishra, SSCL-SR-1225 and Fermilab-CONF 93/267 (1993) p. 411.

HADROPRODUCTION EXPERIMENTS

B Physics at CDF

A. B. Wicklund

High Energy Physics Division
Argonne National Laboratory
Argonne, IL. 60439

Abstract. The CDF experiment at the Fermilab Tevatron has proven to be well suited for precision studies of b physics. Thanks to the excellent performance of the Tevatron Collider and the detector, CDF has accumulated very large data samples and roughly a decade of experience with b physics in $p\bar{p}$ collisions. With the much higher luminosities expected for the Main Injector era, the next decade promises to be an even more fruitful period for CDF. Here we offer a brief overview of issues in hadron-collider b physics and a summary of CDF's accomplishments and future plans.

OVERVIEW

Although b physics was not mentioned in the original 1981 CDF technical design report [1,2], several features were incorporated in the CDF design that made precision b physics possible. These included: a large solenoidal magnetic tracking volume; a well segmented electron/photon calorimeter outside of the tracking region; and a relatively thin muon filter that allowed muon detection down to ~ 1.5 GeV/c transverse momentum (p_T). While not intended for b physics, these detector capabilities allowed CDF to study basic features of b production in the first official physics run (so-called "Run 0", 1988–1989, 4 pb^{-1}), and to develop strategies for a more systematic program of b physics in the second physics run ("Run I", 1992–1996, 120 pb^{-1}), using a silicon vertex detector [3]. In turn, the experience gained from Run I now provides a baseline for planning the CDF b physics program in the Main Injector era ("Run II" and beyond) [4], including detector upgrades that will significantly expand on the present CDF capabilities.

The main motivation for pursuing b physics at hadron colliders is well known. The production cross sections measured at CDF at $\sqrt{s} = 1.8$ TeV imply total event yields of $10^{11} B\bar{B}$ pairs per fb^{-1} integrated luminosity, much higher than the yields expected for e^+e^- B-factories. This rate advantage is significant because once product branching ratios are taken into account, almost all interesting B-decay modes are quite rare. For example, for the "sin 2β" mode, $B^0 \to \psi K_S^0, \psi \to \mu^+\mu^-, K_S^0 \to \pi^+\pi^-$, the combined branching ratio is 1.7×10^{-5}, and the production yields would be 1.4×10^6 events for $1\,\mathrm{fb}^{-1}$ at the Tevatron, compared with 500 events for $30\,\mathrm{fb}^{-1}$ at BaBar (nominal BaBar year). Even though trigger and event selection cuts on p_T and η reduce the geometrical efficiency at CDF to around 1% for this mode, the potential advantage is still rather large.

In addition to high rates, the Tevatron offers other features that are worth noting. First, $p\bar{p}$ is a CP-symmetric initial state, and so we expect equal rates for B and $\bar{B}$ hadrons, at least in the central region. Second, the p_T spectrum for B hadrons scales like the B mass and is significantly harder than that for light hadrons. As a result, the B hadrons are Lorentz-boosted at all rapidities, including the central region where the production rate is highest; at $y=0$, the average $p_T(B)$ is around $3.5\,\mathrm{GeV}/c$. Thus, with p_T cuts on the B decays, one can take advantage of the long B lifetime to identify B decays and to exploit the time dependence of mixing and CP violation signatures. Third, the hard $p_T(B)$ spectrum is also a useful tool in improving signal to background; whereas $B\bar{B}$ production makes up $\sim 0.2\%$ of the $p\bar{p}$ inelastic cross section, at high p_T the ratio of b jet to inclusive jet production is measured to be around 2%. Finally, the Tevatron (like LEP) produces all species of b hadrons, including $B_s, \Lambda_b, \Xi_b, B_c$, and B^* and B^{**} excitations; in this respect, b physics at the Tevatron complements that at the $\Upsilon(4S)$ B factories.

In order to take advantage of the potentially high yields at the Tevatron, efficient trigger schemes are needed. For Runs 0 and I, CDF relied on single and dilepton triggers to collect very large samples of semileptonic and J/ψ decays, with typical trigger thresholds at $p_T \sim 8\,\mathrm{GeV}/c$ (single leptons) and at $p_T \sim 2\,\mathrm{GeV}/c$ (dileptons). From silicon vertex-based analyses, the inclusive J/ψ sample has a B fraction of $\sim 20\%$, the remainder coming from prompt sources. Similarly, the inclusive lepton samples are found to be typically 40% from B decays, the remainder coming from $c\bar{c}$ production or fake leptons. Silicon vertex cuts can be used to further improve the B sample purities, but to start with, the signal to background in these samples is comparable to the b fractions that are produced in e^+e^- collisions at the $\Upsilon(4S)$ or the Z^0 poles ($\sim 20\%$).

Combining the high purity and large yields for these lepton trigger samples, the CDF Run I data provide the largest single sample of exclusive decays for J/ψ modes, ~ 1800 total events in $B \to \psi K_S^0, \psi K^+, \psi K^{*0}, \psi\phi$, and $\psi\Lambda$. CDF also has the largest single sample of quasi-exclusive semi-leptonic decays, $\sim 10,000$ total events in the modes $B \to l^+\nu D^0, D^+, D^{*+}, D_s$, and Λ_c. Figure 1 shows examples of the $B^0 \to \psi K_S^0$ and ψK^{*0} and $B^+ \to \psi K^+$ signals from Run I. Figure 2 shows reconstructed charm peaks from semileptonic B^0 and B^+ decays. These correspond to total samples of order 10^5 $B \to J/\psi X$ and 10^6 $B \to l^+\nu X$ inclusive events. For comparison, typical $B^0 \to \psi K_S^0$ yields at LEP are ~ 8 events per 3.6 million Z^0 decays [5], and at CLEO ~ 46 events

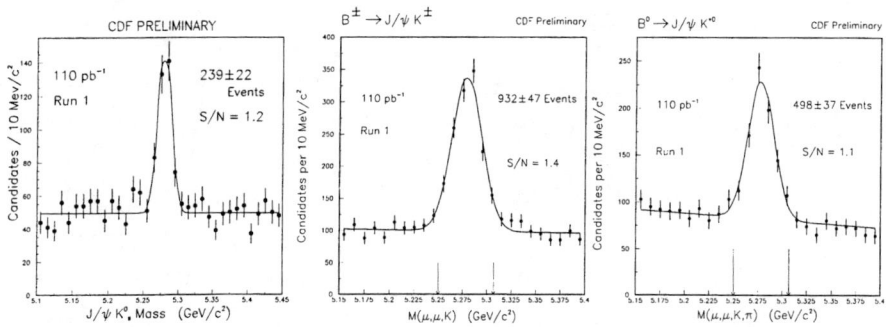

FIGURE 1. (left to right) ψK_S^0, ψK^+, and ψK^{*0} mass peaks from CDF Run I.

per $3.1\,\mathrm{fb}^{-1}$ integrated luminosity [6]. Thus, for modes where CDF can exploit an efficient trigger, it is demonstrably possible to take advantage of the very large b production cross section in the central region.

While lepton and J/ψ triggers open up a broad palette of b physics, there are at least two further challenges that are not *easily* met at a hadron collider. The first is flavor tagging– identifying the flavor of a neutral B hadron at birth– needed for mixing and CP violation studies. This can be done by identifying the associated b jet in a $b\bar{b}$ final state, or by measuring the charge of the parent $\bar{b}$ jet that produced the B^0 or B_s^0 ("self-tagging" or "same-side tagging"). These place demands on the rapidity coverage for tracking and particle identification and on the ability to reject particles from associated gluon jets or underlying event debris. CDF has used both tagging methods to make competitive measurements of $B^0 - \overline{B^0}$ time-dependent mixing. Using the experience gained in Run I, CDF will improve the tagging efficiency in Run II by extending the tracking coverage in η, improving both muon and electron coverage, and possibly adding kaon identification by time-of-flight.

The second challenge is to trigger on all-hadronic B decays, such as $B^0 \to \pi^+\pi^-$, $\pi^+\pi^+\pi^-\pi^-$, and $B_s^0 \to D_s^-\pi^+$, $D_s^\pm K^\mp$. Taking branching ratios into account, a single-lepton trigger (*e.g.*, triggering on $\overline{B} \to l$ and then searching for associated B decay to hadrons) would not be a viable way to get adequate statistics. Instead, a silicon-based trigger on secondary vertices, coupled with trigger-level tracking cuts, appears to be a good way to capture large numbers of hadronic B decays, and CDF plans to deploy a silicon vertex trigger in Run II. A key feature that makes this workable in the central region is the very small transverse spread of the beam at the interaction point (IP). In Run I the typical beam spot size was ± 22 microns in both transverse dimensions. Thus, it is not necessary to reconstruct the primary vertex event by event in the trigger, but only to find tracks with large projected impact parameter with respect to the IP. For forward detectors (LHC-B and BTeV), the task will be more challenging because of the longitudinal spread of the interaction region.

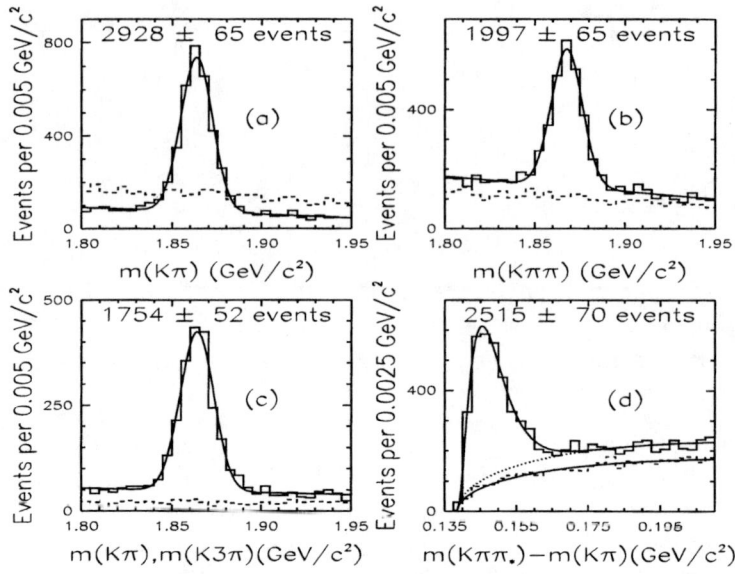

FIGURE 2. (Left) Charm peaks in $B^+ \to l^+\nu \overline{D^0} X$ and $B^0 \to l^+\nu D^{(*)-} X$: (a) $\overline{D^0} \to K^+\pi^-$; (b) $D^- \to K^+\pi^-\pi^-$; (c) $D^{*-} \to \overline{D^0}\pi^-, \overline{D^0} \to K^+\pi^-$ or $K^+\pi^+\pi^-\pi^-$; (d) $D^{*-} \to \overline{D^0}\pi^-, \overline{D^0} \to K^+\pi^-\pi^0$.

Below, we summarize physics topics that CDF has examined, including b production, spectroscopy, rare decays, lifetimes, and mixing properties. We then conclude with a brief overview of future plans.

CDF B PHYSICS RESULTS

Run 0

The CDF detector, as configured at the start of Run 0, is described in Ref. [7]. The most relevant components for B physics are the 3-meter-diameter by 3-meter-long central drift chamber, which covered the region $|\eta| \leq 1.2$; the central calorimeter, which featured fine-grained electromagnetic shower detection; and muon chambers located outside the central calorimeter at $5\lambda_{abs}$.

The first physics run in 1988–89 yielded samples of about 1000 $J/\psi + \psi'$, 40,000 inclusive leptons, and 900 $e\mu$ dilepton pairs. The $e\mu$ sample produced the first CDF publication on b physics, a measurement of $\overline{\chi}$, the species- and time-averaged mixing parameter [8].

Clear signals were seen for exclusive B hadron production in the modes $B^+ \to \psi K^+$ [9], $B^0 \to \psi K^{*0}$ [10], and $B \to e^+\nu D^0 X$ [11]. These had a major impact on CDF's planning for B physics in Run I. The observed signals corresponded to a substantially larger B hadron cross section than predicted by NLO QCD [12]. Likewise, the b quark inclusive cross sections, as determined with inclusive electron [11], muon [13], J/ψ, and ψ' [14] samples, also indicated higher than expected cross sections. In order to extract the b cross section from the charmonium samples, it was assumed that B's were the main source of ψ' production and that B's and $\chi_c \to \psi\gamma$ were the main sources of J/ψ production. It proved possible to measure the χ_c contribution, using the excellent CDF calorimeter segmentation to identify the soft photon in χ_c decay [15,16], and thus, by subtraction, infer the $b \to J/\psi$ contribution. With the assumption that *direct* J/ψ production is negligible, this measurement implied that ~60% of J/ψ's originate from B decay. The high cross section was good news from an engineering standpoint (*i.e.*, better for a future b physics program), but seemed inconsistent with the older UA1 results at $\sqrt{s}=630$ GeV [17], and generated considerable interest in the theory community [18].

Run I

During 1989–1992 major improvements were made to both the Tevatron Collider and the CDF detector. The most important CDF upgrade component was the four-layer silicon vertex detector (SVX), located very close (3 to 8 cm radii) to the IP, yielding typical impact parameter resolutions of $13 + 40/p_T$ μm [3]. This device quickly resolved the cross section issues raised in Run 0. The SVX data allowed model-independent separation of B production in $B \to J/\psi$, ψ', $e\nu X$, $\mu\nu X$... from backgrounds such as prompt charmonium production, $c\bar{c}$, and misidentified leptons. In particular, it was easy to show that a relatively small fraction (~20%) of J/ψ and ψ' production comes from B decay [19]. Also, the fraction of inclusive leptons coming from B decays was somewhat overestimated in the Run 0 analyses; the SVX measurements allowed precise determinations of the sample composition, including fake rates. The unexpectedly large cross sections for *direct* ψ and ψ' production are very interesting in their own right [20], but remain a background in the B physics industry.

Currently all Run I CDF B cross section measurements are based either on the SVX impact parameter data [21–23] or exclusive reconstruction [24,25]. Figure 3 shows examples of the exclusive and inclusive cross section measurements, compared with NLO QCD, using MRSD0 structure functions. Both the single b quark and B hadron, and the correlated $b\bar{b}$ cross sections in the central region are consistently 2-3 times higher than the nominal QCD predictions. To check the earlier discrepancy with UA1 data, CDF interleaved data at $\sqrt{s}=0.63$ TeV with $\sqrt{s}=1.8$ TeV at the end of Run I; this permitted a direct comparison of the b cross sections at the two energies with the same apparatus, the same decay channels (inclusive muons), and minimal systematic bias. The *ratio* of experimental cross sections at the two energies agrees well with

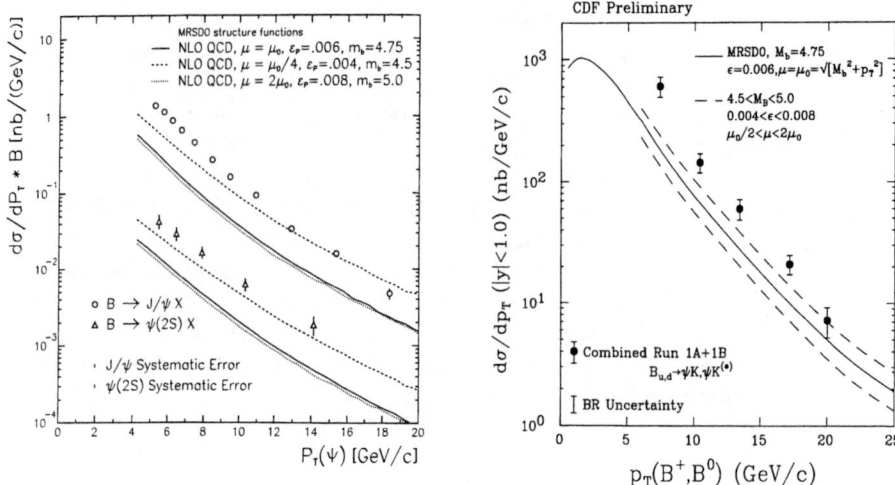

FIGURE 3. (left) Inclusive J/ψ and ψ' cross sections from B decay, compared with NLO QCD; (right) B^+, B^0 differential cross sections compared with NLO QCD.

NLO QCD expectations [26]; however, the absolute cross sections (including UA1's) are systematically higher than the QCD predictions at both energies [27]. Thus, CDF cross section measurements show that the absolute yields and also the probability of finding the second b jet for tagging are favorable for a b physics program in the central region.

In addition to the SVX detector, other improvements to CDF augmented the b physics capability in Run I. First, the trigger strategies were optimized, using the experience gained in Run 0. By lowering thresholds, tightening trigger-matching cuts, and extending the muon coverage in η, CDF increased the yield of J/ψ's per pb^{-1} from $\sim$230 in Run 0 to $\sim$4500 in Run I. The purity of the single muon triggers was improved using additional absorber with read-out chambers at $8\lambda_{abs}$; this reduced hadron punch-through backgrounds by a factor of twenty. The inclusive electron trigger purity was improved by matching the shower-maximum detector signal to the electron track [28]. Electron identification was also improved by addition of preshower detectors at $1X_o$. Relativistic rise dE/dx information was obtained from the outer 54 layers of the central drift chamber; with $\sim 9\%$ resolution, this gave $\sim 2\sigma$ separation between electrons and minimum ionizing particles, and allowed statistical separation of $\pi^\pm$, $K^\pm$, and $p, \bar{p}$. For slow particles, $\frac{1}{\beta^2} dE/dx$ was available from both the drift chamber and the SVX. Finally, it should be noted that the SVX provides not only precise impact parameter information; by matching tracks to secondary vertices, the SVX also allows clean reconstruction of multiparticle charm and bottom decays, such as $D^0 \to K^-\pi^+\pi^-\pi^+$, which would be otherwise buried in combinatorial background. Overall, the CDF detector improvements in Run I allowed much larger bandwidths for single and dilepton triggers and better B purity than in Run 0.

Properties of b Hadrons

Masses

The exclusive J/ψ decay modes provide very straightforward signatures to measure b hadron masses. The J/ψ and ψ' decays themselves provide a built-in calibration for tracking systematics. With samples of 32 $B_s^0 \to \psi\phi$ and 20 $\Lambda_b \to \psi\Lambda$ events, CDF measurements dominate the world averages for B_s^0 [29] and Λ_b masses [30]. With much higher statistics in Run II, we can look forward to precision mass measurements for a variety of states using the J/ψ sample: B_s, Λ_b, Σ_b, Ξ_b, B_c, and strong-decay excitations (B^{**}, Σ_b^*, Λ_b^*).

Branching Ratios and Decay Distributions

CDF has used the exclusive J/ψ sample to establish relative branching ratios for the decays $B^+ \to \psi K^+$, $\psi' K^+$, ψK^{*+}, $B^0 \to \psi K^{*0}$, $\psi' K^{*0}$, ψK^0, and $B_s^0 \to \psi\phi$ [31–33], as well as the Cabibbo-suppressed decay $B^+ \to \psi\pi^+$ [34]. The decay angular distributions are used to study the CP-composition in $B_s^0 \to \psi\phi$ and $B^0 \to \psi K^{*0}$ [35]; this information is potentially important for CP violation studies with these modes [36]. The published CDF results on the longitudinal polarization fraction in $B^0 \to \psi K^{*0}$ are compatible with the recent CLEO analysis: $\Gamma_L/\Gamma = 0.65 \pm 0.10 \pm 0.04$ (CDF), $0.52 \pm 0.07 \pm 0.04$ (CLEO) [6]. A value $\Gamma_L/\Gamma = 1$ would signal a pure CP-even final state; the observed value is consistent with an admixture of even and odd partial waves, and a full angular distribution analysis is needed to separate the CP-odd P wave from the CP-even S and D waves. The published CDF result on $B_s^0 \to \psi\phi$, $\Gamma_L/\Gamma = 0.56 \pm 0.21^{+0.02}_{-0.04}$, is also consistent with an admixture of even and odd waves. The current CDF results are based on 19 pb^{-1}, and with the full statistics of Run I and eventually Run II, it will be possible to carry out a precise determination of the full angular distribution for both ψK^{*0} and $\psi\phi$ modes, similar to the CLEO analysis on $B^0 \to \psi K^{*0}$ [6].

Rare Decays

CDF has set competitive limits on rare decays involving non-resonant dilepton final states, $B^+ \to K^+\mu^+\mu^-$ and $B^0 \to K^{*0}\mu^+\mu^-$, as well as $B^0 \to \mu^+\mu^-$ and $B_s^0 \to \mu^+\mu^-$ [37]. The limits on the first two modes are currently only an order of magnitude above the standard model expectations; CDF should have large enough dimuon samples in Run II to observe of order 100 or more events in the $K^+\mu^+\mu^-$ and $K^{*0}\mu^+\mu^-$ modes in Run II. These decays involve loop diagrams and are potentially sensitive to physics beyond the standard model.

Radiative Decays

CDF has also done a feasibility study on rare radiative decays. The relative branching ratios for $B^0 \to \rho^0\gamma$ to $B^0 \to K^{*0}\gamma$ or $B_s^0 \to K^{*0}\gamma$ to $B_s^0 \to \phi\gamma$ are proportional to the ratio of CKM matrix elements V_{td}^2/V_{ts}^2, up to hadronic corrections of order unity. CDF implemented a photon-plus-two-charged-particle trigger for about $23\,\mathrm{pb}^{-1}$ integrated luminosity in Run I. Using standard photon identification and isolation cuts, and impact parameter cuts on the charged particles, the preliminary CDF analysis finds 1(0) signal candidate events with expected physics signals of 0.95 ± 0.51 (0.34 ± 0.18) events in the $K^{*0}\gamma$ ($\phi\gamma$) channels. Preliminary limits on the $B_s^0 \to \phi\gamma$ decay branching ratio are found to be 3.9×10^{-4} at 90% confidence; so far, the only published limit on this mode is 7.0×10^{-4} from DELPHI [38]. A second experimental method is to use photon conversions into e^+e^- pairs in place of the photon in these final states. This would improve the mass resolution and the background rejection, using an electron rather than a photon trigger. The conversion radiator is supplied by the CDF inner detector (about 12% X_o in Run II). Combining the conversion efficiency with the product branching ratios, the radiative decay signal, $B^0 \to K^{*0}\gamma$, would be around 3% of the "known" signal $B^0 \to K^{*0}\psi, \psi \to e^+e^-$, for which CDF expects several 1000's in Run II. CDF already has clean B peaks using $J/\psi \to e^+e^-$, and has shown the utility of conversion photons with cleanly resolved peaks in $\chi_c \to \psi\gamma$ [30].

Search for the B_c

The spectroscopy of $\bar{b}c$ states can best be studied at hadron colliders. The mass of the weakly decaying B_c^+ is predicted to be 6.24–6.31 GeV [39], and its production rate is predicted to be around 4×10^{-3} relative to the B^+ meson [40]. The decay rate is expected to be the sum of $\bar{b}$ and c decays plus the $\bar{b}c$ annihilation process; one of the more interesting experimental questions is whether the $\bar{b}$ or c-quark decays first. Predictions for the lifetime range from 0.4 to 1.4 ps, depending on whether the decay is c or $\bar{b}$ dominated [41]. Predictions are given in Ref. [42] for branching ratios into states involving J/ψ. Combining the theoretical assumptions, we get the following estimates for ratios of B_c^+ to B^+ production:

$$\frac{B_c^+ \to \psi l^+ \nu}{B^+ \to \psi K^+} \simeq 0.09 \times \frac{\tau}{0.5ps} \times \epsilon \tag{1}$$

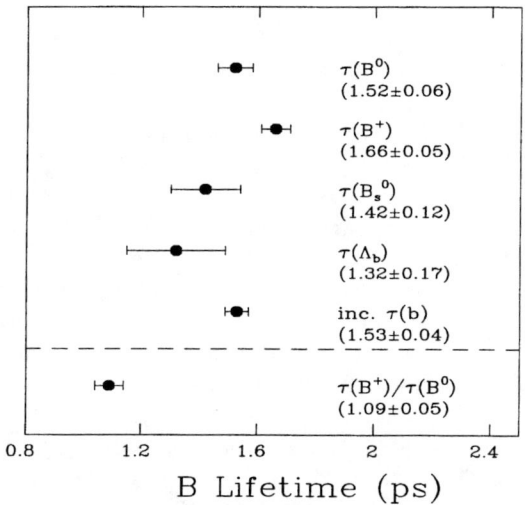

FIGURE 4. Exclusive B hadron lifetimes measured in CDF, combining J/ψ, ψ' and semileptonic modes. Also shown are the B^+/B^0 lifetime ratio and the species-averaged lifetime, the latter based on inclusive J/ψ and ψ' production. Values shown for B^+, B^0, and B_s^0 are preliminary.

$$\frac{B_c^+ \to \psi\pi^+}{B^+ \to \psi K^+} \simeq 0.009 \times \frac{\tau}{0.5ps} \times \epsilon, \quad (2)$$

where τ is the B_c^+ lifetime and ϵ is the relative detection efficiency. CDF has published limits on the $\psi\pi^+$ mode [34], and has candidates in the semileptonic modes. The important point here is that CDF has a *large* $B^+ \to \psi K^+$ signal in the denominator (c.f., Figure 1), and so the B_c^+ signal should be detectable, depending on $\tau(B_c^+)$. ALEPH has reported a clean candidate event in the $\psi\mu^+\nu_\mu$ final state [43]; as noted earlier, the total yield of B hadron decays to J/ψ's at LEP is substantially lower than the Run I CDF yields. Stay tuned!

Lifetime Measurements

In 1990 the PDG species-averaged lifetime for B hadrons was 1.18 ± 0.11 ps. The current average values are 1.538 ± 0.019 ps from LEP semileptonic decays and 1.533 ± 0.036 ps from CDF inclusive J/ψ decays. This underscores the dramatic impact that CDF, SLD, and LEP experiments have had on time-dependent measurements such as B lifetimes and mixing parameters. Current world averages on the individual B hadron lifetimes now approach accuracies of around 5%. The effects of non-spectator contributions have been calculated with the heavy quark expansion technique [44,45]. The expected pattern for $\tau(\Lambda_b) : \tau(B_s^0) : \tau(B^0) : \tau(B^+)$, ~0.9:1.0:1.0:~1.05 appears to match the data fairly well, 0.79:0.98:1.00:1.06 [46], although the short Λ_b lifetime is not understood.

CDF has measured the species lifetimes using large samples of semileptonic and exclusive J/ψ decays [47–53]. The systematic uncertainties are larger for the semileptonic samples, since the B hadron is not fully reconstructed (missing neutrino) and there are uncertainties associated with feeddown from excited charm states (e.g., D^{**}'s). For the exclusive J/ψ modes, typical systematic uncertainties are on the order of 1%, mainly due to modeling of the $c\tau$ resolution function. Since the semileptonic channels have higher statistics, the errors are comparable for both modes, but it is likely that after Run II all of the B hadron lifetimes, like the mass measurements, will be dominated by the J/ψ data. Figure 4 shows a summary of the CDF results with semileptonic and J/ψ modes combined. The values are in good agreement with the LEP results and the world averages [46].

Two additional lifetime measurements are accessible at hadron colliders, namely the B_c^+ lifetime discussed above and the lifetime difference between the short and long-lived B_s^0 states (expected to be the predominantly CP-even and odd $B_s^0 \pm \overline{B_s^0}$ mixing eigenstates, respectively). Theoretical estimates give $\Delta\Gamma/\Gamma \sim 0.16$ [54,55], with phenomenological upper limits, based on $b \to c\bar{c}s$ transition rates, of $\Delta\Gamma/\Gamma \leq 0.44 \pm 0.06$ [54]. With the present sample of 420 $B_s^0 \to l\nu D_s X$ events, CDF has probed the sensitivity of a double exponential fit to separate the two lifetimes; extrapolating from this, CDF should be able to measure $\Delta\Gamma/\Gamma$ to ± 0.02–0.03 in Run II. If $\Delta\Gamma/\Gamma$ is indeed large, it would open up new avenues to CP violation, including the elusive phase γ [54]. In the limit of large $\Delta\Gamma/\Gamma$, the B_s^0 system would be similar to the $K_L - K_S$ system; the separability of the two eigenstates by lifetime acts like a flavor tag, and allows one to compare the CP properties in the decay of each eigenstate. In addition, if the difference is large, the relation between ΔM and $\Delta\Gamma$ can be used to estimate Δm_s for the B_s system [54].

Time-Dependent Mixing

Time-dependent measurements of $B^0\overline{B^0}$ oscillations at LEP, SLD, and CDF have yielded precise values for the mass difference, $|\Delta m_d|$, between the neutral B_d eigenstates. Eventually, combined with accurate measurements of the mass difference for the B_s eigenstates, this information should lead to relatively model-independent determinations of $|V_{td}|/|V_{ts}|$.

The current CDF analyses of B_d-mixing are based on semileptonic decays, using both the high-p_T single lepton and lower-p_T dilepton trigger samples. In all cases, one of the leptons is combined with other charged particles in the same b jet to measure the decay vertex in the transverse plane and to estimate $\beta\gamma$ and the proper decay time. The analyses further split depending on whether the charged particles accompanying the lepton are reconstructed as charmed D^0, D^+, and D^{*+} or are treated inclusively. In both cases, the B hadron flavor at decay time is tagged by the decay lepton. In the reconstructed charm sample, illustrated in Figure 2, the B_d^0 is identified by its semileptonic decay (with a few % cross-talk from charged B decays). In the inclusive samples, the B_d^0 time-dependent oscillation must be extracted from backgrounds from B^+, B_s^0, and Λ_b decays.

The flavor of the B hadron at birth is deduced either from the associated "away-side" b jet or from the fragmentation products in the "same-side" $\bar{b}$-jet. For the dilepton trigger samples, the away-side jet is tagged using the second trigger lepton. For the inclusive single lepton triggers, both away-side tags, based on soft leptons or jet charge, and same-side (SST) tags, based on the fragmentation charge, are used.

To date, these combinations have led to five CDF measurements of Δm_d, which are shown in Figure 5. These are:

(1) single lepton-plus-charm with SST ($\sim$ 9K events- c.f., Figure 2)

(2) single inclusive lepton with away-side tag on soft lepton or jet charge ($\sim$ 250K events)

(3,4) inclusive lepton from both $e\mu$ and $\mu\mu$ dilepton samples ($\sim$ 20K events)

(5) lepton-plus-charm from dilepton sample ($\sim$ 0.5K events)

The lepton-plus-charm samples are essentially pure B, after sideband subtraction on the mass peaks. The inclusive lepton samples are also very pure in B content, after selecting lepton plus charged particle jets having displaced vertices; the b fractions are estimated using the lifetime distributions, the jet mass, and the lepton $p_T(\text{rel})$, and are typically of order 90%.

The time-dependent oscillation is given by

$$B_{tag}^0(0) \to B^0(t) = \frac{e^{-\Gamma t}}{2}[1 + D \cos \Delta m t] \tag{3}$$

$$B_{tag}^0(0) \to \overline{B^0}(t) = \frac{e^{-\Gamma t}}{2}[1 - D \cos \Delta m t]; \tag{4}$$

the equations give the relative probabilities for a neutral B, tagged as a B^0 at birth ($t = 0$) to decay as a B^0 or a $\overline{B^0}$ at time t; Δm is the oscillation frequency; and D is the flavor tagging "dilution" ($D = 2R - 1$, where R is the probability for a right-sign tag). The known variation of D as a function of the event variables (for example, $p_T(\text{rel})$ for lepton tags, or charge sum for jet-charge tags) is input to the maximum likelihood fits for each event. The fits then determine the overall magnitude of D from the amplitude of the mixing oscillation and Δm_d from the phase.

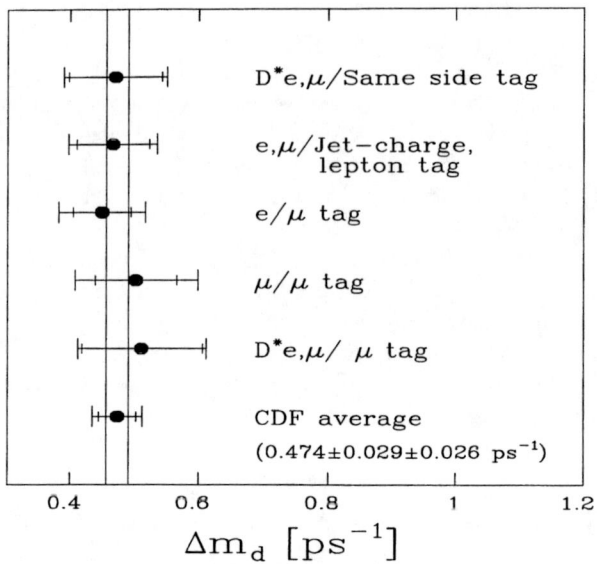

FIGURE 5. Summary of CDF measurements (preliminary) of Δm_d in the $B^0 - \overline{B^0}$ system. The band shows the current world average from LEP, SLD, and CDF measurements. The labels give the B_d decay signature/ tagging method.

If we denote the overall tagging efficiency, including right- and wrong-sign tags, as ϵ, then the statistical accuracy of a sample of N signal events corresponds to that of a tagged sample of size $N \times \epsilon D^2$. From the inclusive lepton analyses, the figure of merit for the different tagging methods, ϵD^2, is found to be:

$\epsilon D^2 \simeq 0.8\%$ (away-side jet charge)

$\epsilon D^2 \simeq 1.1\%$ (away-side lepton tags)

$\epsilon D^2 \simeq 2.4\%$ (SST on B^0)

$\epsilon D^2 \simeq 5.2\%$ (SST on B^+)

The away-side tagging methods (jet-charge and lepton tags) are handicapped by several effects: mixing of the away side jet, which gives an intrinsic $D_{mix} = 1 - 2\overline{\chi}$; sequential decays for leptons; the limited η coverage of the present CDF tracking system; and background tags from gluon jets at small η. The same-side tags do not suffer from these effects and so yield a higher efficiency. In the CDF SST algorithm, prompt tracks from the $b \to B$ fragmentation or from B^{**} decay [56] are selected, and the parent B hadrons are tagged according to the expected correlations: $B^0 \pi^+$, $\overline{B^0} \pi^-$, $B^+ \pi^-$, and $B^- \pi^+$. As an example, Figure 6 shows the time dependence of the $B \to \overline{B}$ transition using SST tags;

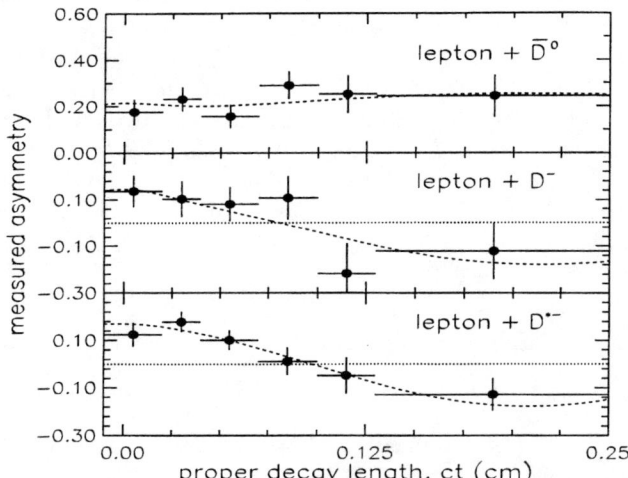

FIGURE 6. Measured asymmetry, $(B_{tag}(0) \to B) - (B_{tag}(0) \to \overline{B})$ for charged (top) and neutral B's (middle and bottom). The curves show the mixing fits; the amplitudes are given by the SST dilution factors.

neutral B's show the expected mixing oscillation, while charged B's, used as a check on the method, do not [57].

THE FUTURE

With the main injector, the Tevatron is expected to deliver luminosities of order 2×10^{32}, and both the CDF and D0 experiments expect an integrated luminosity of $2\,\text{fb}^{-1}$ in Run II. To handle the higher luminosities and shorter (396 ns) bunch spacing, the CDF detector is undergoing a major renovation [4]. The silicon vertex-tracking system will be upgraded from the present single-sided 4-layer device to a double-sided 7-layer system having both $X-Y$ and $R-Z$ readout; the new device will cover 90% of the luminous region in Z, compared with $\sim 60\%$ in the Run I device, and will be capable of standalone track reconstruction to $\eta=2$. These features should improve the signal-to-noise for B reconstruction and will increase the tagging efficiency for all tagging methods. In addition, the central drift chamber will be replaced, and high-p_T tracks ($p_T \geq 1.5\,\text{GeV}/c$) will be available for the level-1 decision. The trigger itself will be pipelined to accomodate a 50 kHz rate out of level-1; this bandwidth requirement is driven primarily by the all-hadronic B trigger. At the second trigger level, drift-chamber tracks found in level-1 will be matched to the silicon vertex-detector hits, allowing offline-quality information on the track impact parameters. Fast processors in level-2 can then be used to form track-based triggers with secondary-vertex cuts, as well as the traditional lepton-based triggers. The third level trigger, which will perform event reconstruction using the full event readout, will select inclusive lepton, dilepton, and J/ψ samples as in Run I, and also the all-hadronic triggers needed for studies of CP violation ($B^0 \to \pi^+\pi^-$) and B_s^0 mixing ($B_s^0 \to D_s^-\pi^+, D_s^-\pi^+\pi^+\pi^-$.)

We have already alluded to some of the B physics goals that can be met with very high statistics lepton and J/ψ samples in Run II, for example:

- Precision masses and lifetimes for fully reconstructed B's
- Determination of $\frac{\Delta\Gamma}{\Gamma}$ for B_s^0, using $B_s^0 \to l^+\nu D_s^-$, $\psi\phi$
- Observation of rare decays, $B^+ \to \mu^+\mu^- K^+$, $B^0 \to \mu^+\mu^- K^{*0}$
- Radiative decay branching ratios, e.g., $B_s^0 \to \phi\gamma$, $B_s^0 \to K^{*0}\gamma$
- Detailed studies of the B_c meson

In addition, CDF expects to:

- Observe CP violation in the channels $B_d^0 \to \psi K_S^0$ and $B_d^0 \to \pi^+\pi^-$
- Search for CP violation in $B_s^0 \to \psi\phi$ (=0 in S.M.)
- establish Δm_s using all-hadronic B_s triggers.

CP Violation

We conclude with a brief discussion of CP violation studies in Run II. The general form for the time evolution for B^0 decays to CP-eigenstates like ψK_S^0 and $\pi^+\pi^-$ is given by

$$a_{CP}(B_d \to f; t) = A_{CP}^{dir} \cos \Delta(m_d t) + A_{CP}^{mix} \sin \Delta(m_d t), \quad (5)$$

where the CP-violating asymmetry is defined by

$$a_{CP}(B_d \to f; t) = \frac{(B_d^0(t) \to f) - (\overline{B_d^0(t)} \to f)}{(B_d^0(t) \to f) + (\overline{B_d^0(t)} \to f)}. \quad (6)$$

Here, "$B_d^0(t) \to f$" is the probability for a neutral B, produced as a B^0 at $t = 0$, to decay to the CP-eigenstate f at time t; A_{CP}^{dir} denotes direct CP violation in the decay, while A_{CP}^{mix} denotes CP violation due to interference between mixing and decay processes [58]. For the final state $f = \psi K_S^0$, the first term in Eq. (5) is expected to be small, and the mixing induced term A_{CP}^{mix} is given by the quantity $-\sin(2\beta)$. For the final state $f = \pi^+\pi^-$, the first term could be large, depending on penguin contributions; the mixing induced term A_{CP}^{mix} is given by $-\sin(2\alpha)$ plus possible additional contributions from penguin amplitudes.

As with mixing measurements, the observed asymmetry is reduced by the tagging dilution factor D:

$$a_{CP}^{obs.} = D \times a_{CP}(B_d \to f; t) \tag{7}$$

Thus, it is necessary to calibrate the dilution accurately. For the away-side tagging methods, D can be calibrated directly using the high-statistics sample of $B^+ \to \psi K^+$ [59]. For "same side tagging", since B^+ and B^0 have different charge correlations with the fragmentation tracks, D must be calibrated using the mixing oscillation signal from $B_d^0 \to \psi K^{*0}$, supplemented by mixing measurements using B_d^0 semileptonic decays.

Extrapolating from the observed event yields in Run I, taking into account lower trigger thresholds and better tagging coverage with the upgraded detector, CDF expects to obtain up to 15,000 ψK_S events in Run II ($2\,\mathrm{fb}^{-1}$), with overall tagging efficiency $\sim 5.4\%$ [4]. A simple time-averaged asymmetry measurement would yield an uncertainty

$$\delta \sin(2\beta) \simeq \frac{1+x_d^2}{x_d} \frac{1}{\sqrt{\epsilon D^2 N}} \sqrt{\frac{S+B}{B}}, \tag{8}$$

where the dilution factor $(1+x_d^2)/x_d = 2.13$ arises from time averaging the $\sin \Delta m_d t$ dependence, and S/B is the signal to background. With these input assumptions, $\delta \sin(2\beta) = 0.09$. In practice, it will be necessary to fit the time dependence. This verifies the $\sin(\Delta m_d t)$ dependence expected for A_{CP}^{mix} and reduces $\delta \sin(2\beta)$; it also improves the effective signal to background, since the combinatorial background from prompt J/ψ production occurs at $t=0$, where the CP asymmetry should vanish.

Assuming a nominal $BR(B_d^0 \to \pi^+\pi^-)$ of 1×10^{-5}, the all-hadronic trigger designed for CDF in Run II would yield approximately 10,000 events in this mode [4]. Ignoring A_{CP}^{dir}, Monte Carlo studies indicate that an error $\delta \sin(2\alpha) \sim 0.12$ can be achieved, with the same ϵD^2 as for ψK_S^0 above. Again, it will be necessary to fit the time dependence, and here the penguin-induced $\cos(\Delta m_d t)$ oscillations may turn out to be large. In that case, the interpretation of "$\sin(2\alpha)$" is more complicated; some strategies for this case are discussed in Ref. [60]. In addition to the possible complications from penguin diagrams, backgrounds are expected from $B_d^0 \to K^+\pi^-$ and $B_s^0 \to K^+\pi^-, K^+K^-$, which overlap the $\pi^+\pi^-$ mass distribution within $\sim \pm 2\sigma$. Neither of these backgrounds would contribute to the $\sin(\Delta m_d t)$ oscillation, but they introduce an overall dilution factor in the observed asymmetry [61]. This overall dilution can be determined by measuring the total background from $K^+\pi^-$ and K^+K^- production using relativistic-rise dE/dx measurements from the central drift chamber; this purely statistical separation can be done on the full signal sample before tagging. Since the time-dependence of the CP asymmetry is especially important for $B_d^0 \to \pi^+\pi^-$, it is worth noting that the proper time resolution is quite good in this mode due to the large opening angle between the decay pions. With the all-hadronic trigger selections and just the Run I silicon vertex resolution, the average resolution on the decay proper time would be around 6% of the B_d^0 lifetime; this should improve with 3D vertexing in Run II.

CONCLUSION

CDF has shown that it is possible to take advantage of the high B hadron production rates in the central region at the Tevatron, using selected triggers. The very high yields and the mix of B hadron flavors make the hadron-collider B program complementary to that at e^+e^- B factories. With the upgraded CDF detector, it should be possible to increase the present samples of J/ψ and lepton triggers by factors of fifty. CDF also plans to deploy all hadronic silicon-based triggers for studies of CP violation and B_s^0 mixing. Thus, the CDF collaboration is optimistic that after ten years of experience doing B physics at the Tevatron, the best is still yet to come.

REFERENCES

1. D. Ayres et al., *Design Report for the Fermilab Collider Detector Facility*, 1981 (unpublished).
2. There was, nevertheless, appreciation of the potential for b physics at hadron colliders. See for example: R. Diebold and J. Sauer, *Electron Identification in Jets*, CDF Note 72 (1980); L. L. Chau et al., *Heavy Quark Jets*, Proc. 1982 Summer Study on Elementary Particle Physics and Future Facilities, Snowmass (1982), p. 510.
3. D. Amidei et al., *Nucl. Instrum. Methods* **A 269**, 93 (1988).
4. The CDF II Collaboration, *The CDF II Detector Technical Design Report*, FERMILAB-Pub-96/390-E.
5. OPAL Collaboration, *Z. Phys.* **C70**, 197 (1996).
6. CLEO Collaboration, *Measurement of the Decay Amplitudes and Branching Fractions of $B \to J/\psi K^{*0}$ and $B \to J/\psi K$ Decays*, CLNS 96/1455.
7. CDF Collaboration, F. Abe et al., *Nucl. Instrum. Methods Phys. Res. Sect.* **A271**, 387 (1988).
8. CDF Collaboration, F. Abe et al., *Phys. Rev. Lett.* **67**, 3351 (1991).
9. CDF Collaboration, F. Abe et al., *Phys. Rev. Lett.* **68**, 3403 (1992).
10. CDF Collaboration, F. Abe et al., *Phys. Rev. D* **50**, 4252 (1994).
11. CDF Collaboration, F. Abe et al., *Phys. Rev. Lett.* **71**, 500 (1993).

12. P. Nason, S. Dawson, and R. K. Ellis, *Nucl. Phys.* **B303**, 607 (1988); *ibid* **327**, 49 (1989); *ibid* **B335**, 260 (1990).
13. CDF Collaboration, F. Abe *et al.*, *Phys. Rev. Lett.* **71**, 2396 (1993).
14. CDF Collaboration, F. Abe *et al.*, *Phys. Rev. Lett.* **69**, 3704 (1992).
15. CDF Collaboration, F. Abe *et al.*, *Phys. Rev. Lett.* **71**, 2537 (1993).
16. CDF Collaboration, F. Abe *et al.*, *Phys. Rev. Lett.* **79**, 578 (1997).
17. N. Ellis and A. Kernan, Phys. Rep. 195, 23 (1990).
18. E. Berger, R. Meng, and W. K. Tung, *Phys. Rev. D* **46**, 1859 (1992); M.L. Mangano, *Z. Phys.* **C58**, 861 (1992); J. Smith and W. K. Tung, *Proceedings of the Workshop on B Physics at Hadron Accelerators* Snowmass, Co. (1993), P. McBride and C. S. Mishra, ed.
19. CDF Collaboration, F. Abe *et al.*, *Phys. Rev. Lett.* **71**, 3421 (1993).
20. R. K. Ellis, W. J. Stirling, and B. R. Webber, *QCD and Collider Physics*, Cambridge Univ. Press, 1996, ch. 10.
21. CDF Collaboration, F. Abe *et al.*, *Phys. Rev. D* **53**, 1051 (1996).
22. CDF Collaboration, F. Abe *et al.*, *Phys. Rev. D* **55**, 2546 (1997).
23. CDF Collaboration, F. Abe *et al.*, *Phys. Rev. Lett.* **79**, 572 (1997).
24. CDF Collaboration, F. Abe *et al.*, *Phys. Rev. Lett.* **75**, 1451 (1995).
25. The CDF Collaboration, *Measurement of the B-Meson Differential Cross-Section in $p\bar{p}$ Collisions at $\sqrt{s}=1.8$ TeV*, FERMILAB-CONF-96/198-E.
26. CDF Collaboration, *Measurement of the Ratio of b Quark Production Cross-Sections at $\sqrt{s}=630$ GeV and $\sqrt{s}=1800$ GeV*, FERMILAB-CONF-96/176-E.
27. To avoid confusion, note that the older cross section comparisons used DFLM structure functions in the NLO QCD predictions; more modern structure functions such as MRSD0 and MRSA predict significantly smaller b cross sections.
28. K. Byrum *et al.*, *Nucl. Instrum. Methods* **A364** 144 (1995).
29. CDF Collaboration, F. Abe *et al.*, *Phys. Rev. D* **53**, 3496 (1996).
30. CDF Collaboration, F. Abe *et al.*, *Phys. Rev. D* **55**, 1142 (1997).
31. CDF Collaboration, F. Abe *et al.*, *Phys. Rev. Lett.* **76**, 2015 (1996).
32. CDF Collaboration, F. Abe *et al.*, *Phys. Rev. D* **54**, 6596 (1996).
33. CDF Collaboration, *Branching Fractions of $B^+ \to \psi(2S)K^+$ and $B^0 \to \psi(2S)K^{*0}$ Decays at CDF*, FERMILAB-CONF-96/160-E.
34. CDF Collaboration, F. Abe *et al.*, *Phys. Rev. Lett.* **77**, 5176 (1996).
35. CDF Collaboration, F. Abe *et al.*, *Phys. Rev. Lett.* **75**, 3068 (1995).
36. The self-conjugate decay modes, $B_s^0 \to \psi\phi$ and $B_d^0 \to \psi K^{*0}$, $K^{*0} \to K_S^0 \pi^0$ have $CP = +1$ (S, D waves), $CP = -1$ (P wave). Experimentally, the decay mode $K^{*0} \to K^+\pi^-$ is used to determine the partial wave structure for the ψK^{*0} channel.
37. CDF Collaboration, F. Abe *et al.*, *Phys. Rev. Lett.* **76**, 4675 (1996).
38. DELPHI Collaboration, *Z. Phys.* **C72**, 207 (1996).
39. E. Eichten and C. Quigg, *Phys. Rev. D* **49**, 5845 (1994); W. Kwong and J. Rosner, *Phys. Rev. D* **44**, 212 (1991).
40. E. Braaten, K. Cheung, and T. C. Yuan, *Phys. Rev. D* **49**, R5049 (1996).
41. M. Beneke and G. Buchala, *Phys. Rev.* **D53**, 4991 (1996); I. I. Bigi, *Phys. Lett.* **B371**, 105 (1996).

42. C. H. Chang and Y. Q. Chen, *Phys. Rev. D* **49**, 3399 (1994).
43. ALEPH Collaboration, *Search for the B_c meson in hadronic Z decays*, **CERN-PPE**/97-026.
44. M. B. Voloshin and M. A. Shifman, *Sov. Phys. JETP* **64**, 698 (1986); I. I. Bigi et al., *B Decays, 2nd ed.*, S. Stone (ed.), World Scientific, Singapore (1994); I. I. Bigi, *UND-HEP-95-BIG02*; G. Bellini, I. I. Bigi, and P. J. Dornan, *Phys. Rep.* **289**, 1 (1997).
45. M. Neubert and C. T. Sachrajda, *Nucl. Phys.* **B 483**, 339 (1997).
46. T. R. Junk, *A Review of B Hadron Lifetime Measurements*, 2nd. Int. Conf. on B Physics and CP Violation, Honolulu (1997).
47. CDF Collaboration, F. Abe et al., *Phys. Rev. Lett.* **74**, 4988 (1995).
48. CDF Collaboration, F. Abe et al., *Phys. Rev. Lett.* **72**, 3456 (1994).
49. CDF Collaboration, F. Abe et al., *Phys. Rev. Lett.* **76**, 4462 (1996).
50. CDF Collaboration, F. Abe et al., *Phys. Rev. Lett.* **77**, 1439 (1996).
51. CDF Collaboration, F. Abe et al., *Phys. Rev. Lett.* **77**, 1945 (1996).
52. CDF Collaboration, *B States and Lifetimes at CDF*, FERMILAB-CONF-96/155-E.
53. CDF Collaboration, *Measurement of the Lifetime of the B_s^0 Meson from $D_s^- l^+$ Correlations*, FERMILAB-CONF-96/154-E.
54. M. Beneke, G. Buchalla, and I. Dunietz, *Phys. Rev. D* **54**, 4419 (1996).
55. M. Voloshin, M. Shifman, N. Uraltsev, and V. Khoze, *Sov. J. Nucl. Phys.* **46**, 112 (1987).
56. M. Gronau, A. Nippe, and J. Rosner, *Phys. Rev. D* **47**, 1988 (1993). In the present analysis, CDF does not distinguish B^{**} daughters from other fragmentation tracks.
57. CDF Collaboration, *Observation of $\pi - B$ Charge-Flavor Correlations and Measurement of Time-Dependent $B^0 \overline{B^0}$ Mixing in $p\bar{p}$ Collisions*, FERMILAB-CONF-96/175-E.
58. Y. Nir and H. Quinn, *Ann. Rev. Nucl. Part. Sci.* **42**, 211 (1992); A. Buras and R. Fleischer, *Quark Mixing, CP Violation and Rare Decays After the Top Quark Discovery*, in *Heavy Flavors II*, A. J. Buras and M. Lindner (ed.), World Scientific (1997); I. I. Bigi and A. I. Sanda, *Nucl. Phys.* **B281**, 41 (1987).
59. This assumes that the away-side tagging dilution is the same for B^+ and B^0 hadrons. This is clearly not the case at the $\Upsilon(4S)$, where a B^+ hadron is tagged by an away-side B^-, while a B^0 hadron is tagged by an away-side $\overline{B^0}$. At either a hadron collider or an $\Upsilon(4S)$ B factory, the time-averaged CP asymmetry for $B^0 \overline{B^0}$ states produced in odd-L waves vanishes due to Bose statistics. At the hadron collider, this is exactly compensated by the larger CP asymmetry in the even-L waves. So long as the $B^0 \overline{B^0}$ states are a stochastic mix of even- and odd-L states, the dilution measured with B^+ hadrons is the one needed in Eq. (7) for the B_d^0 CP-asymmetry.
60. F. DeJongh and P. Sphicas, *Phys. Rev. D* **53**, 4930 (1996).
61. The B_s^0 contributions have much faster oscillations than the $\sin(\Delta m_d t)$ term. The $B_d^0 \to K^+ \pi^-$ backgrounds could contribute to the $\cos(\Delta m_d t)$ oscillation term if there is intrinsic CP violation leading to $B_d^0 \to K^+ \pi^- \neq \overline{B_d^0} \to K^- \pi^+$.

B Physics at DØ

Ronald Lipton
for the DØ Collaboration
Fermi National Accelerator Laboratory,
Batavia, Illinois 60510

Abstract. DØ has measured *B* production in the semi-muonic final state at the Fermilab Tevatron. Cross sections for *b*-quark production are found to be a factor of 2-2.5 higher than theory in the central rapidity region and ~4 times higher than theory in the forward region. Data taken at $\sqrt{s}$=630 GeV show similar features. Upgrades to DØ will significantly enhance the ability of the experiment to address the central issues of *b* physics.

INTRODUCTION

QCD-based models make specific predictions of the rapidity, p_T and energy dependence of heavy quark production. The large *b* samples at the Tevatron provide a unique opportunity to test and refine next-to-leading order QCD models [1,2]. In this paper we describe DØ measurements of beauty production in several regions of phase space and compare them to next-to-leading order theory. We also describe future upgrades to the experiment which will enable DØ to address a wide range of beauty physics in Run II.

THE DØ DETECTOR

DØ is a compact detector optimized for measurements of leptons, jets, and missing E_T at the Fermilab Tevatron. The experiment consists of a central tracker, a uranium-liquid argon calorimeter, and a muon system instrumented with drift chambers and toroidal magnets. There is no central magnetic field and the muon system provides the only track momentum measurements. Our B physics results therefore always include a semi-muonic final state. The acceptance of the muon system determines the character of our beauty measurements.

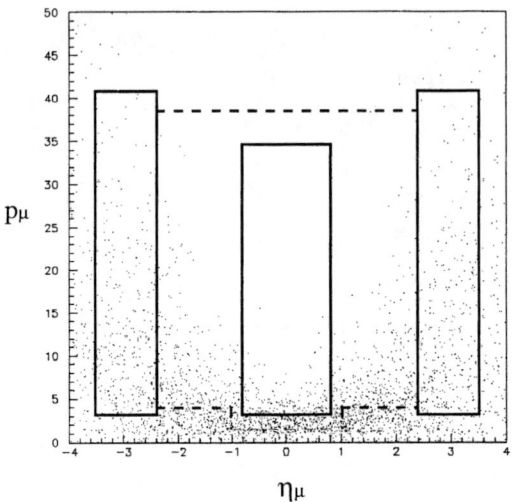

FIGURE 1. Scatterplot of muon momentum vs. rapidity for ISAJET generated muons from $B \to J/\psi\ K_s$, $J/\psi \to \mu\mu$ events. The solid boxes indicate the acceptance for data presented in this paper. The dashed line represents the muon acceptance expected for the upgraded DØ detector.

Figure 1 shows an inclusive muon spectrum from $B \to J/\psi\ K_s$, $J/\psi \to \mu\mu$ generated by ISAJET. The solid boxes indicate the phase space acceptance regions for the data presented in this paper. In the low η region the acceptance is defined by muon energy loss in the calorimeter and toroid/steel absorber. Punchthrough is minimized for high p_T muons at the expense of some loss of acceptance for low p_T muons from B decay. In the high η region we use an independent small-angle muon system (SAMUS) whose acceptance extends to lower transverse momentum.

TECHNIQUE

Muon candidates are required to pass a standard series of cuts including number of hits in the chambers, track fit chi-square and timing, calorimeter energy, and toroid field integral. A reconstructed muon candidate can come from π or K decay, hadronic punchthrough, $b \to c \to \mu$, $c \to \mu$, or $b \to \mu$. In most cases the fraction of $b \to \mu$ can be determined from the data. In our data the $b \to \mu$ signal is extracted using a combination of the p_T, impact parameter, and isolation distributions of candidate muons. Fits are performed with the b-fraction (f_b) and background fractions as free parameters.

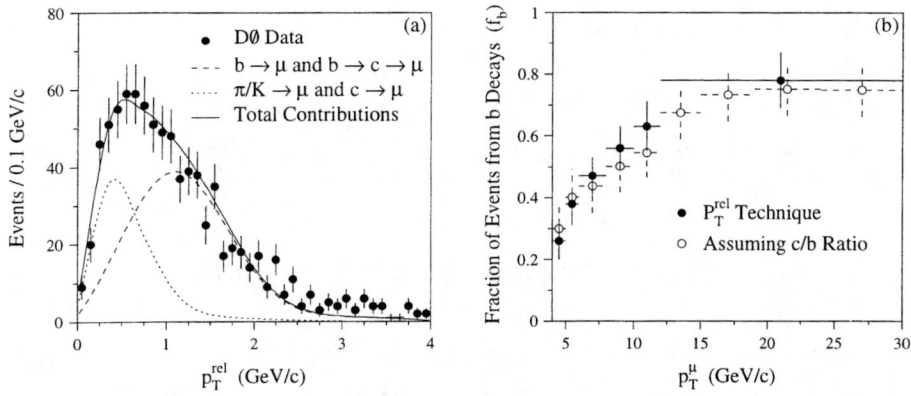

FIGURE 2. a) Distribution for transverse momentum relative to the jet axis for central single muon candidates. b) Fraction of muons from b-quarks extracted from the fit in 2a.

Figure 2a shows the distribution of the p_T of muons relative to the jet axis (p_T^{rel}) for an inclusive single muon sample. In this case a two component fit is used to separate the direct $b \to \mu$ decays from the background. The resulting b-fraction for the inclusive central single muon sample is shown in Figure 2b.

The $B \to J/\psi$ signal is separated from prompt J/ψ production using a different technique. Muons from the decay are identified by their isolation from jets in the calorimeter. The b fraction determined by isolation is confirmed using the impact parameter of the muons in the Vertex Drift Chamber.

RESULTS

DØ B production studies can be separated into three categories with the following signals:
1. Central $\eta<1$, $b\bar{b}$ correlations
   - Inclusive muons
   - Muons + Jets
   - Dimuons + Jets
   - $J/\psi \to \mu\mu$
2. Forward $\eta>2.5$
   - Inclusive muons
   - $J/\psi \to \mu\mu$
3. 630 GeV
   - Inclusive muons

Central Production and Correlations

The central b-quark cross section has been measured by DØ using the four processes listed above [3,4]. Figure 3 shows the inclusive cross section as a function of p_T at $\sqrt{s}=1.8$ TeV. The NLO-QCD prediction is also shown. Theoretical errors are obtained by varying the mass of the b-quark from $4.5<m_b<5.0$ GeV/c^2 and varying the QCD scale between 100 and 187 MeV. Our data consistently lie above the central value of the QCD prediction but the data and predictions lie within 1σ of the combined experimental and theoretical errors.

Measurements of $b\bar{b}$ correlations provide additional information on the mechanisms of beauty production. The lowest order flavor creation diagrams predict pairs which are closely correlated in azimuthal angle ($\Delta\phi$). Higher order effects such as flavor excitation and gluon splitting tend to produce pairs more widely separated in ϕ.

Dimuon events are selected where each muon has an associated jet and the pair mass is between 6 and 35 GeV/c^2. The b-fraction in each $\Delta\phi$ bin is extracted from a maximum

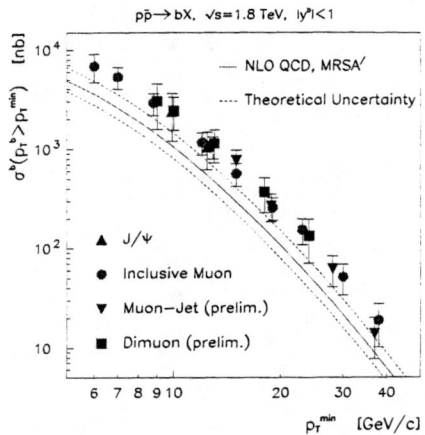

FIGURE 3. Integrated b quark cross section vs. $p_T^{min}(b)$ measured in four modes.

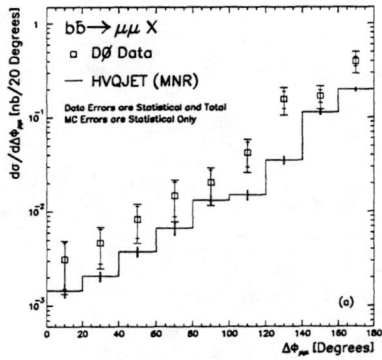

FIGURE 4. The $\Delta\phi_{\mu\mu}$ spectrum for $b\bar{b}$ production and the HVQJET prediction.

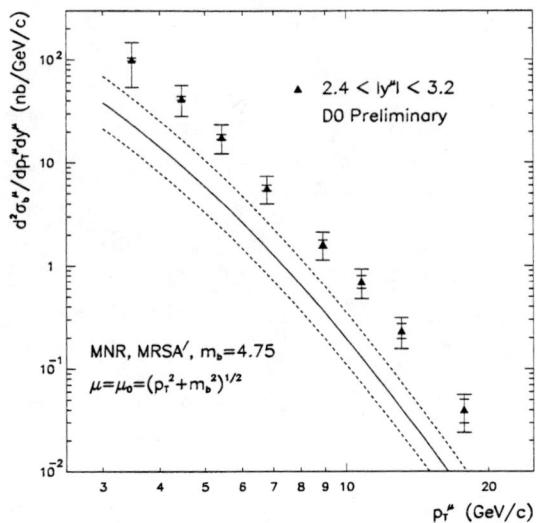

FIGURE 5. Inclusive muon cross section vs. p_T^μ for data in the high rapidity region.

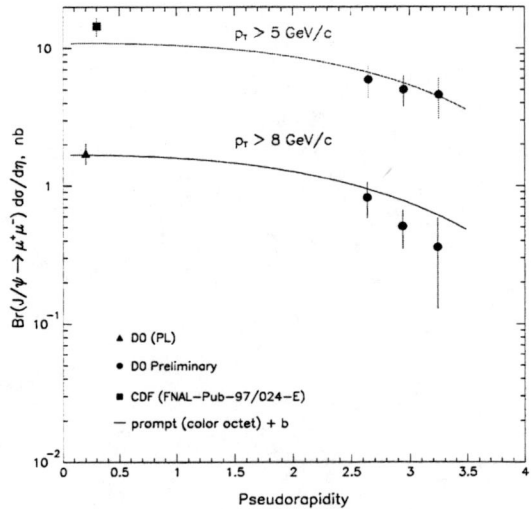

FIGURE 6. J/ψ production cross section as a function of rapidity for two transverse momentum regions. The solid lines are predictions of the Color Octet model.

likelihood fit of the signal and background distributions. The result is shown in Figure 4.

Also shown is the NLO-QCD calculation using the HVQJET [5] Monte Carlo. The data agrees with the shape of the HVQJET prediction and is larger in magnitude, consistent with the inclusive b-quark production results.

Forward production

The primary backgrounds for $b \to \mu$ at low p_T in the forward region are π and K decays. These backgrounds are estimated using the ISAJET inclusive particle spectrum which was verified using the central region charged particle spectrum measured by CDF. The calculation agrees with the measured cross section at the lowest p_T where the data are dominated by decays. We also used the ISAJET model to estimate the relative contribution to the b signal from charm decay. This model was tested using the distribution of p_T relative to jets for those events with reconstructed jets.

Figure 5 shows the resulting $b \to \mu$ cross section for $2.4 < \eta_\mu < 3.2$. Also shown is the NLO-QCD prediction using the MRSA' structure functions. The data in this region are a factor of four above the central value of the predictions. The theoretical and experimental errors in this case do not overlap.

Contributions to inclusive J/ψ production are expected to vary with rapidity. The differential J/ψ cross section as a function of η for two p_T ranges is shown in Figure 6. B meson decay, as predicted from the inclusive muon data, only accounts for a few percent of the J/ψ's in the high η region. Predictions based on color singlet production also fail to account for the observed yield. A Color Octet model, which was adjusted to fit prompt J/ψ production in the central region, does agree with the data.

630 GeV Data

Tevatron running at 630 GeV provides an opportunity to measure b quark production at two energies with the same detector and compare the Tevatron 630 GeV measurements to UA1. Our data consist of single muon triggers with $\eta_\mu < 0.8$ and $4 < p_T < 10$ GeV/c. Cosmic ray backgrounds are more important in this data sample due to the low instantaneous luminosity. The b-quark fraction (f_b) and backgrounds from π and K decays are calculated using an ISAJET model for inclusive production. The result is shown in Figure 7. Theoretical curves are from the next-to-leading order calculation of Mangano, Nason and Rudolfi using the MRSA' parton density function. When compared with theory the 630 GeV inclusive cross section is similar to the 1800 GeV result. Both are a factor of 2-2.5 above the predicted values but have shapes well reproduced by theory.

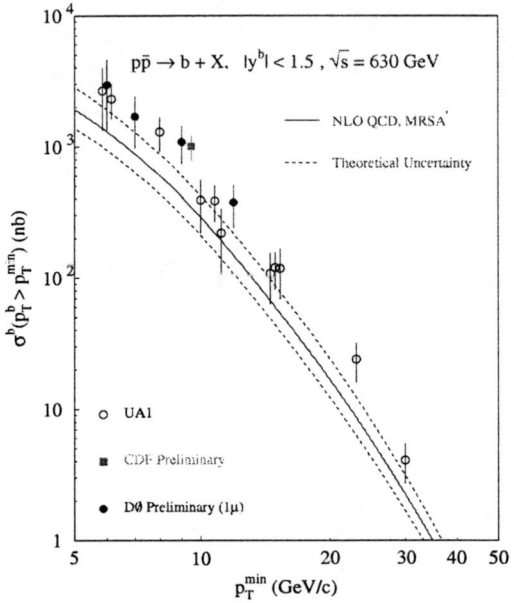

FIGURE 7. Integrated b quark cross section vs. p_T^{min}(b) for $\sqrt{s}$=630 GeV data.

THE DØ UPGRADE

Both the DØ and CDF detectors are being upgraded for the increased luminosity and shorter crossing interval to be provided by the Fermilab Main Injector. Major detector upgrades include:
- 2 Tesla solenoid magnet
- Silicon + scintillating fiber tracker
- Muon system with lower thresholds and improved triggering and tracking

These improvements have especially important consequences for the range of b physics which can be addressed by DØ. The magnet and tracking system will enable DØ to reconstruct hadronic final states. The silicon vertex detector will allow b tagging by impact parameter. The fiber tracker will provide a fast level 1 tracking trigger. Finally the muon system is being rebuilt to provide lower p_T thresholds in the central region and a more powerful, pixel based trigger at higher values of rapidity. The dashed line in Figure 1 shows the coverage of the upgraded muon system.

Studies have been made of the ability of the upgraded DØ detector to study the "flagship" b physics of the next decade. The detector will be able to trigger on dimuons at $p_T > 1.5$ GeV with no prescale. A 2 fb^{-1} run will yield more than 1500 muon tagged $J/\psi K_s$ decays [6]. These events will provide a measurement of $\sin(2\beta)$ with an error less than 0.15. Implementing other tagging methods will further reduce the error.

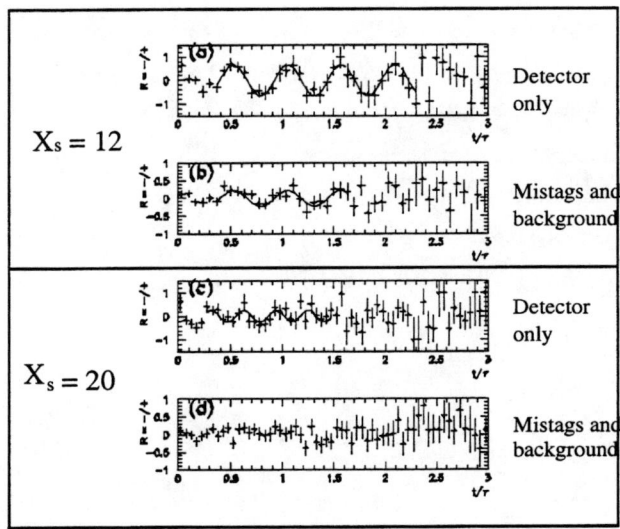

FIGURE 8. Simulated measurements of B_s oscillations for two values of x_s showing the effects of detector resolution as well as mistags and backgrounds.

Figure 8 shows a simulation of measurements of B_s mixing in DØ. The all charged mode $D_s n\pi$ is used to reduce the lifetime error due to momentum reconstruction. A data sample of 1000 tagged events is modeled. With such a sample our x_s reach is between

15 and 20. Achieving this sample size in an all hadronic final state depends on implementing a clean single muon trigger with minimal p_T threshold.

Conclusions

DØ has measured b production in a variety of modes and a large region of phase space at the Tevatron. NLO QCD calculations consistently underestimate production cross sections. The discrepancy is largest in the high η data, where the discrepancy is a factor of four. QCD models are more successful at describing the shapes of the differential cross sections and $b\bar{b}$ correlations. DØ is now building an upgraded detector which will have significantly enhanced capability to study b physics.

---

1. P. Nason, S. Dawson, and R.K. Ellis, *Nucl. Phys.* **B303**,607(1988), B327, 49 (1989).
2. M. Mangano, P. Nason, G Ridolfi, *Nucl. Phys.* **B373**, 295 (1992).
3. D0 Collaboration, S. Abachi et. al., *Phys. Rev.* Lett. **74**, 2632 (1995).
4. D0 Collaboration, S. Abachi et. al., *Phys. Lett.* **B370**, 239 (1996).
5. M.M. Baarmand, D0 note 2517, 1995.
6. G. Lima and A. Maciel, D0 note 3228, 1997.

HERA-B, an experiment to study CP violation at the HERA proton ring using an internal target

Walter Schmidt-Parzefall

University of Hamburg, II. Institute for Experimental Physics, Luruper Chaussee 149, 22761 Hamburg, Germany

Abstract. This paper discusses a dedicated fixed target experiment at the HERA proton ring with the goal of observing CP violation in the $B \to J/\psi K_s^0$ decay channel, using a detector with a highly selective trigger for lepton pairs from the J/ψ decay. The precision of the asymmetry, $sin2\beta$, is expected to reach 0.13 within one year of running.

Introduction

The first evidence for the discovery of the b quark 20 years ago at FNAL, which we are celebrating, is shown in Fig. 1. Only one year later DESY was able to observe the b quark in e^+e^- collisions (Fig. 2) and determined the b-quark's charge $-1/3$. Since then a fruitful B-physics program went on at DESY. From 1982 to 1992 the ARGUS detector was operational. After the end of this project a new B-physics experiment named HERA-B was approved in 1995.

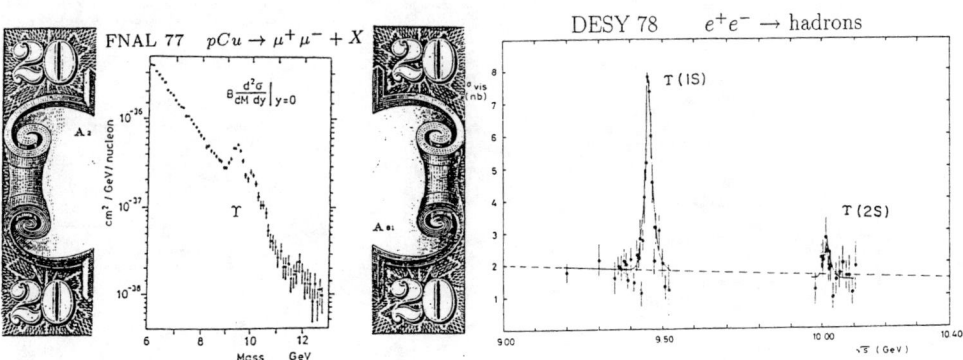

Figure 1: The 20th anniversary of the discovery of the b quark. The mass distribution of μ pairs from pCu collisions shows a peak at 9.46 GeV from the Υ states, being $b\bar{b}$-quark bound states.

Figure 2: The first evidence of the b quark in e^+e^- collisions. The visible cross section as function of the C.M. energy shows two well-resolved peaks from the two lowest-lying Υ states.

CP424, *Twenty Beautiful Years of Bottom Physics*
edited by Burnstein/Kaplan/Rubin
© 1998 The American Institute of Physics 1-56396-745-6/98/$15.00

HERA-B is an experiment to study B physics using an internal target in the proton beam of the HERA collider at DESY. The main physics aims of the experiment are the observation of CP violation in the B system via the reaction $B \to J/\psi K_s^0$, the measurement of $B_s \bar{B}_s$ mixing and the observation of the B_c meson. In addition a rich physics programme will be performed. This physics programme is largely different from the physics at the future e^+e^- B factories. While they run at one fixed energy, where only some B states are produced, at a hadron machine all B states are produced simultaneously.

For a detailed description of the project see Ref. [1] and references therein.

The production of b quarks by the internal target

The basis of the B physics at the HERA-B experiment are the $b\bar{b}$-quark pairs produced from a wire target placed in the halo of the HERA proton beam, see Fig. 3. Two sets of four target wires are foreseen.

To reach the required sensitivity, the HERA-B detector has to run at an event rate of $\dot{N} = 40$ MHz. The bunch frequency of HERA is 10 MHz. Thus at each bunch crossing on average 4 superimposed events must be accepted. These 4 events will originate from different places on the target wires and can thus be disentangled. The thickness of a target wire is of the order of 0.3% of an interaction length. A HERA proton must pass through the target around 300 times before it interacts. Fortunately the beam optics and the proton diffusion speed at HERA are such that this condition is met. The event rate $\dot{N}$ is simply given by

$$\dot{N} = \epsilon \frac{N_p}{\tau} = 40 \text{MHz},$$

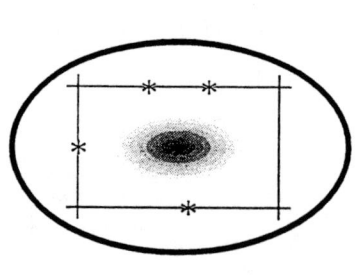

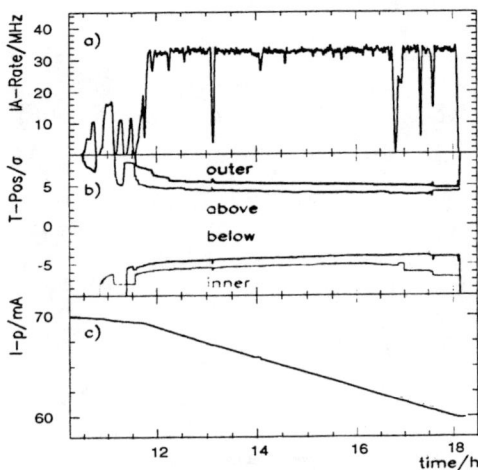

Figure 3: The wire target of the HERA-B experiment. Two sets of four targets are placed around the proton beam, typically at a distance of $5\sigma_{beam}$ from the beam center. On the average four events per bunch-crossing are produced on the eight wires.

Figure 4: The long-time performance of the wire target: a) the event rate, stabilized at 32 MHz, b) the positions of the four wires, c) the proton current.

where $N_p = 1.3 \cdot 10^{13}$ is the number of protons in the beam corresponding to a beam current of 100 mA and $\tau = 50$h is the beam lifetime of HERA.

The target efficiency ϵ is the fraction of those outward-diffusing halo protons which interact in the target. Its value has been determined experimentally with a test target at the HERA-B proton beam. While the wire is moved out of the shadow of a collimator towards the beam, first the wire and the collimator are competing with each other; finally, at a large enough distance from the collimator, most of the protons lost are interacting in the wire. A target efficiency of $\epsilon = 0.6$ has routinely been achieved. The target is operated by a servo system, which moves the wire and thus stabilizes the rate. As Fig. 4 shows, a stable rate above 30 MHz has been achieved over several hours.

At the HERA proton energy of 820 GeV the centre-of-mass energy is $\sqrt{s} = 40$GeV. At this energy the pp cross section $\sigma_{b\bar{b}}$ for the production of $b\bar{b}$-quark pairs is not yet well known. Recent measurements by experiments at FNAL [2] gave

$$E789 \quad \sigma_{b\bar{b}} = (5.7 \pm 1.5 \pm 1.3) \text{ nb}$$
$$E771 \quad \sigma_{b\bar{b}} = (43^{+27}_{-17} \pm 7) \text{ nb}$$

whereas the QCD calculation of Ellis, Dawson and Nason [3] gave as a central value

$$\sigma_{b\bar{b}} = 12 \text{nb}.$$

Since this calculation compares well with the data of several other experiments, it might well be that it is correct also in this case. The rate of b-quark pairs produced in $b\bar{b}$ is then for a copper target with A = 63.55 and $\sigma_{inel} = 0.782$b, assuming $\sigma_{b\bar{b}} = 12$nb,

$$\dot{n}_{b\bar{b}} = \frac{A\sigma_{b\bar{b}}}{\sigma_{inel}} \dot{N} = 37 \text{Hz}.$$

In one year with $10^7 s$ of running about $3.7 \cdot 10^8$ b-quark pairs will be produced. Over the time of the experiment more than 10^9 b quarks will be available as the basis of the physics programme of the HERA-B experiment.

Kinematics

The design of the HERA-B detector has been optimized for the detection of the reaction $B \rightarrow J/\psi K_s^0$ and additional flavour-tagging B-decay particles. Fig. 4 shows the typical momenta of the various B-decay particles. Also shown are the typical decay paths. The velocity of the centre-of-mass system of the collision is $\beta\gamma = 21$. Thus light particles which are emitted at CM-angles Θ_{CM} with $|cos\Theta_{CM}| < 0.95$ are Lorentz-transformed to lab. angles Θ_{LAB} within the range 10mr $< \Theta_{LAB} <$ 210mr.

The angular acceptance range of the HERA-B detector is a close approximation to this ideal angular range: 90% of the tracks are accepted.

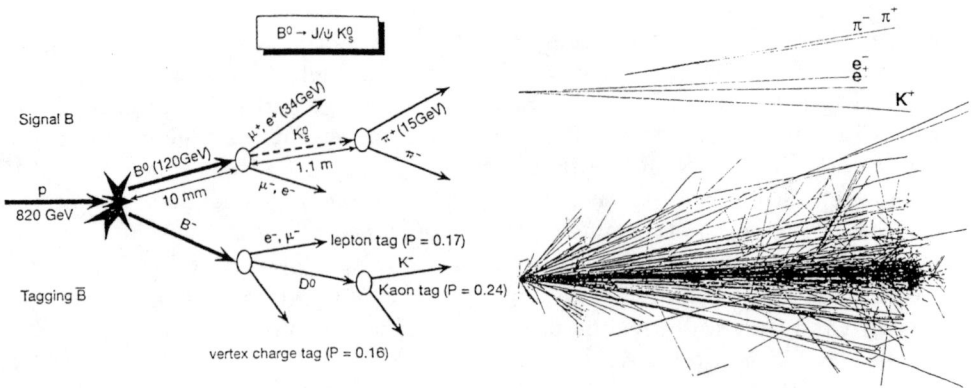

Figure 5: The decay kinematics of the CP-violating reaction $B \to J/\psi K_s^0$ and the second B meson used for tagging.

Figure 6: A GEANT simulation of an event with a decay $B \to J/\psi K_s^0$ and additional 5 superimposed events.

The HERA-B detector

A top view of the HERA-B detector is shown in Fig. 5.

The proton beam passes through the detector through an aluminium beam pipe with 0.3mm thickness which has an approximately conical shape with a 7mr half-opening angle. Thus it allows the detector components to be efficient from 10mr out. The electron beam of HERA also passes through the detector and must be shielded against the magnetic field of the spectrometer magnet.

The spectrometer magnet has a field integral of 2.2 Tm. The shape of the poles was dictated by the requirements for the electron-beam shield.

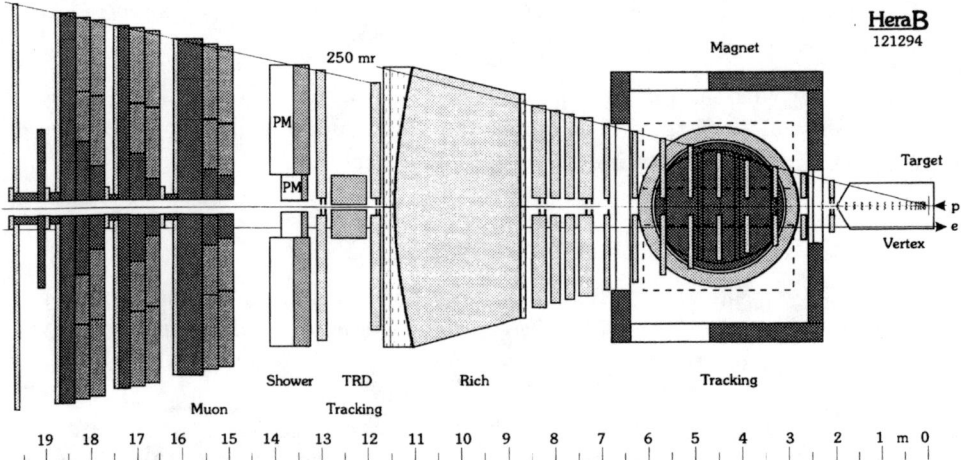

Figure 7: A top view of the HERA-B detector consisting of a wire target, 8 stations of silicon vertex detectors, an inner tracker system of gas microstrip chambers, an outer tracker system of honeycomb straw chambers, a ring-imaging Cherenkov counter with a C_4F_{10} radiator and multianode-PMT photon detection, a transition radiation detector with xenon-filled straw chambers, a shower counter with lead-scintillator and tungsten-scintillator sandwiches read out by wavelength-shifter fibres, and a muon detection system with steel or steel-loaded concrete absorbers with tube chambers and pad chambers.

The physics of the HERA-B experiment

The physics accessible by the HERA-B experiment depends on the feasibility of an appropriate trigger. The lepton-pair trigger will trigger on a J/ψ. However, to extend the physics programme any combination of leptons will be triggered on (possibly with the exception of like-sign electron pairs). As an upgrade, a high-p_T trigger will be added.

The main physics goal planned with the J/ψ trigger is, of course, the CP-violating reaction $B^0 \to J/\psi K_s^0$. But there are many more states involving a J/ψ produced which also allow interesting physics. They are listed in Table 1. This list is still incomplete but it illustrates the rich potential of a J/ψ trigger. All the states listed here are ground states. When combined with additional particles, they open up a new field of spectroscopy of heavy-quark states. The reaction $B_s \to J/\psi \phi$ is analogous to the reaction $B_d \to J/\psi K_s^0$, but in the case of a decaying B_s very little CP violation is expected. This reaction is a stringent test of the Standard Model. The reaction $B^0 \to J/\psi K^{*0}$ cannot have CP violation and can be used as a calibration reaction to recognize and subtract systematic asymmetries. The reactions $B_c \to J/\psi \ell \nu$ and $B_c \to J/\psi \pi$ will allow the discovery of the B_c meson and will be the basis for its spectroscopy. Also included in the J/ψ trigger are the rare $b \to s$ B decays, $B \to \ell^+\ell^- K$ and $B \to \ell^+\ell^- K^*$, which will be studied by HERA-B.

This physics programme is considerably extended by the reconstruction of semileptonic B decays. A detailed study showed that semileptonic B decays like $B \to D^* \ell \nu$ can be reconstructed in the HERA-B environment. Since the neutrino remains undetected, the kinematics of the reaction can only be reconstructed up to a 2-fold ambiguity which, however, can be resolved in many cases.

The most important of the semileptonic B decays is the reaction $B_s \to D_s \ell \nu$. This reaction will enable a measurement of $B_s \bar{B}_s$ oscillations at HERA-B.

Table 1
Number of expected J/ψ events per year

Decay	Reconstructed events
$B^0 \to J/\psi K_s^0$	1440
$B^0 \to J/\psi K^{*0}$	5500
$B^+ \to J/\psi K^+$	5900
$B^+ \to J/\psi K^{*+}$	1400
$B_s^0 \to J/\psi \phi$	1200
$B_c^+ \to J/\psi \ell^+ \nu$	2000
$B_c^+ \to J/\psi \pi^+$	70
$\Lambda_b \to J/\psi \Lambda$	400

Table 2
Efficiencies, dilution factors and statistical powers of the B-meson tags

	Lepton tag	Kaon tag	Charge tag
Cuts			
p	>5 GeV/c	$\in[5,50]$ GeV/c	–
p_t	>0.8 GeV/c	–	–
$\chi^2(v_B,t)$	–	>2	>2
$\delta(v_B,t)$	<1 mm	<1 mm	<1 mm
$\chi^2(v_X,t)$	>4	>4	>5
Tag	sign(q_ℓ)	sign(q_K)	sign($\sum_i q_i \cdot p_i$)
Efficiencies			
ϵ (tag found)	16.1%	46.0%	96.4%
ϵ_c (correct)	11.5%	31.3%	56.0%
ϵ_w (wrong)	4.6%	14.7%	40.4%
Significance			
Dilution D	0.43	0.36	0.16
Power $P = D\sqrt{\epsilon}$	0.17	0.24	0.16
Combined power		$P = 0.31 \pm 0.03$	

The central scientific subject of the HERA-B experiment is a precise determination of the CKM matrix [4] which determines the weak couplings of quarks in the Standard Model:

$$V = \begin{pmatrix} V_{ud} & V_{us} & V_{ub} \\ V_{cd} & V_{cs} & V_{cb} \\ V_{td} & V_{ts} & V_{tb} \end{pmatrix} \approx \begin{pmatrix} 1 - \frac{\lambda^2}{2} & \lambda & \lambda^3 A(\rho - i\eta) \\ -\lambda & 1 - \frac{\lambda^2}{2} & \lambda^2 A \\ \lambda^3 A(1 - \rho - i\eta) & -\lambda^2 A & 1 \end{pmatrix},$$

where $\lambda = sin\theta_c = 0.22$ is the Cabibbo angle. The parameters A, ρ and η are the subject of B physics.

The unitarity condition relevant for B physics reads as follows:

$$V_{ud}V_{ub}^* + V_{cd}V_{cb}^* + V_{td}V_{tb}^* = 0.$$

It can be visualized as a triangle in the complex (ρ, η) plane, the so called unitarity triangle, as shown in Fig. 8. To fully fix the triangle one has to measure the coordinates (ρ, η) of its apex. The measurement of two sides and the three angles allows a test of the consistency of the CKM ansatz, which explains CP violation by an imaginary part in the matrix.

The CP range of HERA-B

The present knowledge of the unitarity triangle is shown in Fig. 8. Ali and London [5] have determined the allowed $\rho - \eta$ space from a simultaneous fit to all available experimental data and theoretical quantities including the t-quark mass.

A full GEANT simulation of the secondary interactions of the typically 4 superimposed events in the detector has been made, see Fig. 6. These events were then reconstructed and the corresponding resolutions and efficiencies for the reaction $B^0 \to J/\psi K_s^0$ determined. They are shown in Table 2. Table 3 gives the sensitivity of the HERA-B experiment to $sin2\beta$ thus obtained.

Table 3
Estimate of the number of reconstructed $B^0 \to J/\psi K_s^0$ events after a running time of 10^7 s at an interaction rate of 40 MHz, and the resulting statistical precision of the measurement $sin(2\beta)$ after tagging. The results are shown for three different values of the $b\bar{b}$ cross-section per nucleon.

$\sigma_{b\bar{b}}$ [nb]	6	12	17
σ_{inel}/A (Cu)	13	13	13
$\sigma_{b\bar{b}}/\sigma_{inel}$ [mb]	4.6×10^{-7}	9.2×10^{-7}	1.3×10^{-6}
Interaction rate [MHz]	40	40	40
$b\bar{b}$ rate [Hz]	18	37	52
Fraction of detected $B^0 \to J/\psi K_s^0$	3.7×10^{-6}	3.7×10^{-6}	3.7×10^{-6}
Number of $B^0 \to J/\psi K_s^0$ per 10^7 s	680	1360	1930
Statistical factor	2.3	2.3	2.3
Tagging power P_{tag}	0.31	0.31	0.31
$\delta sin 2\beta$ after 10^7 s	0.19	0.13	0.11

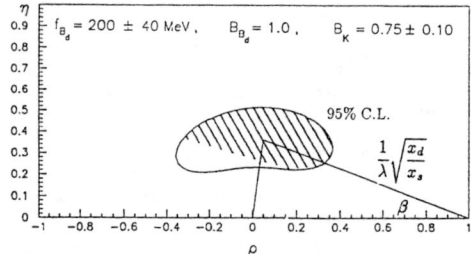

Figure 8: The CP-range of the HERA-B experiment, showing the unitarity triangle and the 95% CL contour for the Standard Modell expectation of its apex (ρ, η). The hatched area is the range where HERA-B can make a 4σ measurement within a year's running time of $10^7 s$ for a cross section of $\sigma_{b\bar{b}} = 12 nb$.

B_s Mixing

A measurement of the frequency x_s of $B_s\bar{B}_s$ oscillations is of similar importance as the observation of CP violation in the B system. It determines the length of one of the sides of the unitary triangle and thus completely determines the whole triangle. After small and known corrections the following relation holds:

$$\frac{x_d}{x_s} = \frac{|V_{td}|^2}{|V_{ts}|^2} = \frac{|V_{td}|^2}{|V_{cb}|^2}.$$

The length of the triangle's side is then

$$\frac{|V_{td}|}{|\lambda V_{cb}|} = \frac{1}{\lambda}\sqrt{\frac{x_d}{x_s}}$$

x_s is at present bound within [6]

$$10 \leq x_s \leq 33$$

At HERA-B the measurement of $B_s\bar{B}_s$ oscillations will be based on the semileptonic decay $B_s \to D_s^* \ell \nu$. The lepton-pair trigger will trigger on the lepton of this semileptonic decay and a lepton from the other B meson of the event, which is used for tagging.

Table 4 shows the corresponding rates and efficiencies.

Table 4
Estimate of the number of reconstructed $B_s \to D_s \ell \nu$ events after a running time of 10^7 s at an interaction rate of 40 MHz after tagging for a $b\bar{b}$ cross-section per nucleon $\sigma_{b\bar{b}} = 12$ nb.

Number of $b\bar{b}$ pairs produced per year	3.7×10^8
$B(b \to B_s) + B(\bar{b} \to B_s)$	0.16
$2B(B_s \to D_s^{(*)}\ell\nu)$	0.126
$B(D_s^+ \to \phi\pi^+)$	0.037
$B(\phi \to K^+K^-)$	0.49
Geometrical efficiency	0.6
Track reconstruction efficiency	0.9
Trigger efficiency	0.054
Cuts on vertices and masses	0.38
Cuts on $m_{D_s\ell}$ and σ_τ	0.27

Number of events per year	
$\phi\pi$	400
$\phi\pi\pi\pi$	90
$K^{*0}K^+$	380
Total	870

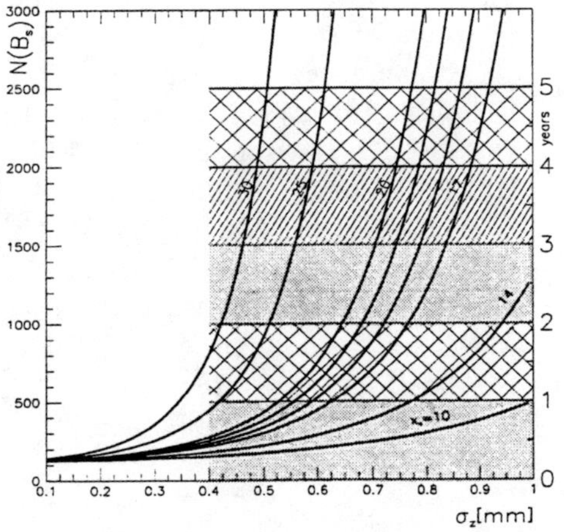

Figure 9: The range of the HERA-B experiment for a measurement of $B_s - \bar{B}_s$ mixing for various mixing parameters x_s. The vertex resolution of the detector under construction is $\sigma_z = 0.82 mm$. By an improvement of the vertex resolution of a factor of two, the whole Standard Model x_s-range can be covered.

In a Monte Carlo study the tracks were followed through the silicon vertex detector, where they undergo multiple scattering, and were then reconstructed. The most frequent background reactions were added. Finally the sensitivity curves as shown in Fig. 9 were obtained. Up to $x_s = 17$ the oscillations can be recognized by the detector under construction. The sensitivity of the detector is mainly limited by the resolution of the vertex detector, $\sigma_z = 0.85$mm. Should it turn out that x_s is larger than about 17, the HERA-B detector could be upgraded with an improved vertex detection system in order to cover the full x_s range allowed by the Standard Model.

Outlook

The construction of the HERA-B detector is a technical challenge. The requirements of occupancies, rates and radiation hardness go to the limit of present-day technologies. The detector will be completed by end 1998, and HERA-B expects to make valuable contributions to B physics.

References

1. H. Albrecht *et al.*, DESY-PRC **92/04** (1992)
   H. Albrecht *et al.*, DESY-PRC **93/04** (1993)
   W. Hofmann, Nucl. Instr. and Meth. **A333**, 153 (1993)
   T. Lohse *et al.*, DESY-PRC **94/02** (1994)
   P. Križan *et al.*, Nucl. Instr. and Meth. **A351**, 111 (1994)
   E. Hartouni *et al.*, DESY-PRC **95/01** (1995)
2. D. M. Jansen *et al.*, Phys. Rev. Lett. **74**, 3118 (1995)
   L. Spiegel (for E771), Frascati Physics Series, Vol.**VII**, 223 (1996)
3. P. Nason, S. Dawson and K.R. Ellis, Nucl. Phys. **B303**, 607 (1988); **B327**, 40 (1989); **B335**, 260 (1990)
   S. Frixione et al., Nucl. Phys. **B412**, 225 (1994)
4. N. Cabibbo, Phys. Rev. Lett. **10**, 531 (1963)
   M. Kobayashi and T. Maskawa, Prog. Theor. Phys. **49**, 652 (1973)
5. A. Ali and D. London, CERN-TH **7248/94**; Z. Phys. **C65**, 431 (1995)
6. F. Abe *et al.*, Phys. Rev. Lett. **73**, 225 (1994)
   F. Abe *et al.*, Phys. Rev. Lett. **D50**, 2966 (1994)

BTeV: A heavy quark experiment at the Tevatron

Patricia McBride

Fermilab, Batavia, IL 60510

Abstract. The Fermilab Tevatron collider is an excellent laboratory for studying the physics of heavy quarks. Large numbers of events containing $c\bar{c}$ pairs and $b\bar{b}$ pairs will be created in $p\bar{p}$ collisions at 2 TeV center-of-mass energy during the next run of the collider. BTeV is an experimental program that has been proposed for the new C0 interaction hall at the Tevatron. The detector, which will have a forward and backward coverage, is designed for the study of rare decays, mixing, and CP violation in the c and b systems. The proposed detector has excellent vertex resolution and will use a Level 1 vertex trigger to collect large samples of interesting events containing charm and beauty decays.

INTRODUCTION

The observation of CP violation in the kaon systems remains one of the most intriguing puzzles of particle physics. The CKM model provides an explanation within the Standard Model for the observed CP asymmetries and predicts large asymmetries in the B sector although CP violation in the B decays has not yet been observed. The BTeV experiment at the Tevatron is designed to measure these predicted CP-violating effects in B decays. BTeV also plans to make high-statistics measurements in the charm sector which will offer an excellent opportunity for observing effects that are outside the Standard Model, since in the Standard Model charm CP violation and mixing are expected to be very small.

There are several experimental facilities which will begin data taking in the next several years that will have the chance to observe CP violation in the B system. It is expected that the first observation will be in the channel $B_d \to J/\psi K_s^0$. The BTeV experiment will thus be a second generation B experiment and will most likely begin data taking after the first observations of CP violation in the B system. The primary goal of the BTeV experiment is to collect large data samples of events with decays of beauty and charm and to make precision measurements of mixing and CP violation and rare decays. The physics goals of a second generation heavy quark experiment are compelling and will yield important information about the CKM matrix [1].

BTeV offers many possibilities for heavy quark physics. BTeV will collect an estimated $10^7 - 10^8$ reconstructed $\overset{(-)}{D^0} \to K^{\mp}\pi^{\pm}$ charm decays which is a factor of 10 to 100 over the presently fixed target samples. BTeV will also collect large statistics for many B decays. Unlike the case at the e^+e^- B factories, all species of B hadrons, B^0, B^+, B_s^0, b-baryons and B_c mesons, are produced at hadron colliders. One of the primary goals of this experiment is to measure the angles α, β, and γ of the unitarity triangle that is derived from the CKM matrix for the B system. The triangle in the $\rho - \eta$ plane is shown in Figure 1. The sides and angles of the triangle represent Standard Model parameters. Constraints on the triangle are shown in the bottom plot of Figure 1. The angle α is accessible through $B_d \to \pi^+\pi^-$, the angle β through $B_d \to J/\psi K_S^0$ and the angle γ through measuring the time dependent asymmetry in the decays $B_s, \bar{B}_s \to D_s^{\pm} K^{\mp}$ [2] and also through the decay $B^+ \to D^0 K^+$ [3] [4]. The measurement of the asymmetry in $B_d \to \pi^+\pi^-$ and the determination of γ will only be possible at experiments with high statistics B samples as well as good particle identification.

The full BTeV detector will have a high resolution vertex detector, a detached vertex trigger, lepton identification and lepton trigger systems, and excellent particle identification. One of the benchmark measurements for the BTeV detector is the precision measurement of B_s oscillations which can be done by measuring the time evolution and determining the mixing parameter $x_s = \Delta M/\Gamma$. The proposed baseline detector has an x_s reach of ≈ 50 in one year of running at a luminosity of $5 \times 10^{31} \text{cm}^{-2}\text{s}^{-1}$.

B-HADRONS AT THE TEVATRON

Following the commissioning of the new Main Injector, the Fermilab collider will produce on the order of 10^{11} b hadrons and 10^{12} c hadrons during each year of running at luminosities of 10^{32} cm^{-2} s^{-1}. Table 1 shows the projections for the running conditions at the C0 interaction region at the Tevatron.

QCD calculations indicate that in the central region at the Tevatron, b's produced are approximately uniformly in pseudorapidity, as seen in plot in the left side of Figure 2. A plot of the $\beta\gamma$ of b quarks produced at the Tevatron is shown on right side Figure 2. Bs produced in the forward region have much larger values of $\beta\gamma$ than those produced with low pseudorapidity near $|\eta| = 0$. That translates into longer decay distances, better vertex separation, higher momentum tracks and hence better time resolution in the forward region. The production angles of the B and $\bar{B}$ hadrons produced in the forward region are highly correlated as shown in in Figure 3. A large number of $b\bar{b}$ pairs can be detected in the angular coverage in the forward (and backward) region which is a great advantage for measurements that require tagging the flavor of the other b.

THE BTEV DETECTOR

The proposed BTeV experiment is a forward collider detector with two arms that cover the angular region from 10 mrad. to 300 mrad. in the forward and backward region. The detector design has been optimized for the new C0 hall which will be constructed at the Tevatron in 1998. The BTeV detector layout for the C0 hall is shown schematically in Figure 4. The baseline design, which is outlined in the BTeV Expression of Interest [5], includes a large central dipole magnet that bends in a vertical plane and a vertex detector containing a series of pixel planes that covers the long interaction region. The dipole

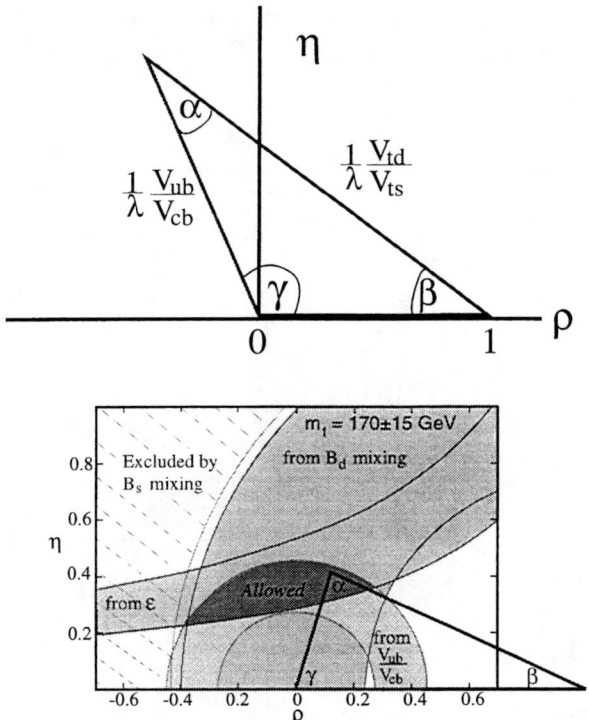

FIGURE 1. *Top plot:* Sketch of the CKM triangle showing the usual conventions for the angle α, β and γ. The angles can be determined from measurements of CP violation in B decays. *Bottom plot:* Drawing of the $\rho-\eta$ plane showing 1 σ constraints from measurements of ϵ, B_d mixing, and V_{ub}/V_{cb} and the region excluded by B_s mixing measurements.

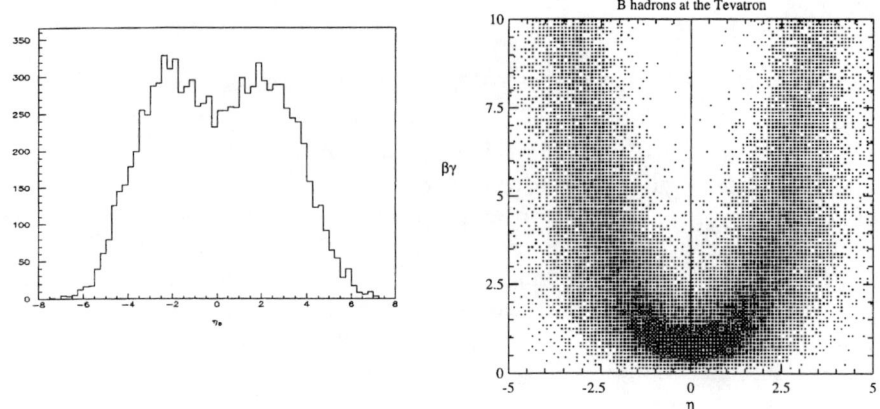

FIGURE 2. Pseudorapidity (η) distribution (left) and $\beta\gamma$ vs pseudorapidity (η) (right) for b hadrons produced at the Tevatron collider.

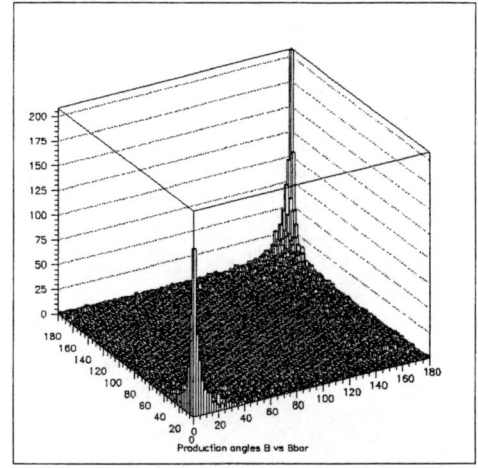

FIGURE 3. The production angle (in degrees) for the b hadron plotted versus the production angle for the the $\bar{b}$ hadron showing the correlation of b and $\bar{b}$ hadrons in the forward region at the Tevatron ($\sqrt{s}$ =2 TeV).

TABLE 1. The Tevatron collider as a b and c source at the C0 interaction region during Run II and beyond.

Luminosity in Run II	$5 \times 10^{31} \text{cm}^{-2}\text{s}^{-1}$
Luminosity (ultimate)	$2 \times 10^{32} \text{cm}^{-2}\text{s}^{-1}$
b cross-section	$100 \mu\text{b}$
# of b's per 10^7 sec (Run II)	10^{11}
b fraction	0.2%
c cross-section	$> 500 \; \mu\text{b}$
Bunch spacing	132 ns
Luminous region length	$\sigma_z = 30$ cm
Luminous region width	$\sigma_x, \sigma_y = \approx 50 \; \mu\text{m}$
Interactions/crossing	$< 0.5 >$

magnet for BTeV was used previously in the Fermilab experiment E605 and is made from soft iron that was originally part of the Nevis Cyclotron. The vertex detector will be made from pixel detectors that will be developed for BTeV. Each detector arm consists of a forward tracking system, a Ring Imaging Cherenkov detector, an electromagnetic calorimeter and a muon system. Lepton, dilepton and detached vertex triggers will be used to trigger on heavy quark events. The full BTeV detector will operate in collider mode, but there is also the possibility to use a thin wire target during the initial phase of running.

The baseline design for the vertex detector is shown in Figure 5. The detector is made from 93 layers of pixel detectors, and the individual pixel size is $30 \mu\text{m} \times 300 \mu\text{m}$. The dimensions are chosen to enhance the resolution and the pattern recognition speed. The planes are grouped in triplets so that small mini-tracks can be formed at each triplet station. This is to help reduce the speed of pattern recognition and permit fast tracking in the vertex detector at the first level of the trigger. In the baseline design the inner edges of the pixel detectors are 6 mm from the beam. While reducing this distance would improve vertex resolution, it also increases radiation damage. The optimization of these effects is under study. The pixel planes will be operating inside the vacuum following a design that was introduced by the P238 Collaboration at the CERN S$p\bar{p}$S [6].

At the Tevatron only a small fraction of interactions will contain a $b\bar{b}$ or $c\bar{c}$ pair. In addition, the interesting branching ratios for the decay modes of interest are very small ($\lesssim 10^{-5}$). Lepton triggers have been shown to work successfully for B physics studies at hadron colliders, however, many of the modes BTeV is interested in studying involve hadronic decays. To be efficient for both b and c decays BTeV must, therefore, have a trigger that is relatively independent of decay mode. The efficiency is optimized if the trigger accepts events that can also be found off-line. The BTeV trigger focuses on the key feature of heavy quark events, the presence of detached vertices, and uses this

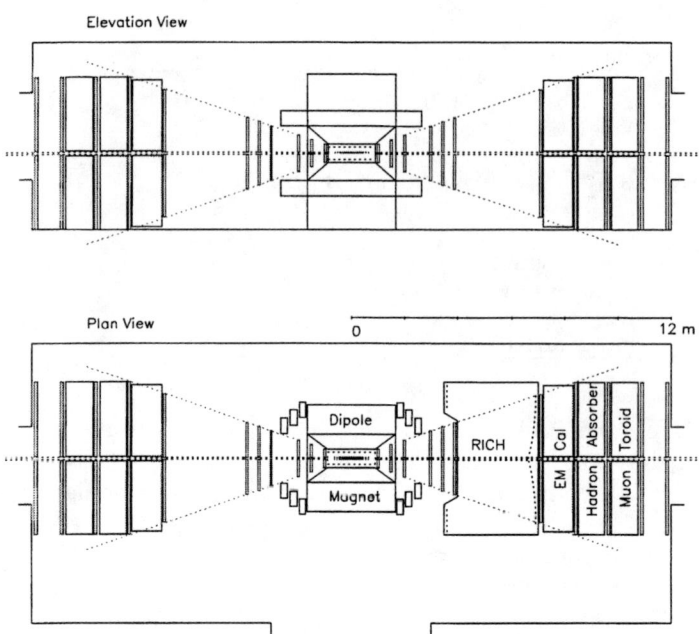

FIGURE 4. Schematic of the BTeV detector showing the central dipole magnet, vertex detector, and the two forward arms each containing a tracking system, a Ring Imaging Cherenkov detector, an electromagnetic calorimeter and a muon system.

at Level 1. Other branches of the trigger will include dimuon triggers and electron and dielectron triggers.

For a heavy quark experiment to operate at the Tevatron, a rejection factor of about 100 for light quark events is needed at Level 1, in order to reduce the data to a manageable amount at Level 2. As stated above, BTeV will use its detached vertex trigger [7] at the first level. One of the key ingredients for the vertex trigger is that the vertex detector have excellent spatial resolution, fast readout and low occupancy. The proposed pixel detector meets these requirements. Sub-event parallelism and a heavily pipelined parallel processing architecture will be used to do the tracking and vertexing in the vertex trigger.

Inexpensive processors and memory are readily available commercially so it is straightforward to buffer the event data while the calculations are performed. The movement of data through the trigger system will be controlled by a switching and control network. A prototype of the vertex trigger processor and switching system is under development at Fermilab.

The reference design for the vertex trigger algorithm is outlined below. Hits in each pixel plane are assigned to a detector sub-unit or ϕ slice. Station hits are formed in a hit processor for each ϕ slice of a triplet pixel detector station. The resulting mini-vectors are sent from the hit processor via a sorting switch to a farm of track processors where the tracking is done. Tracks are then passed to a farm of vertex processors. The primary vertex is found in the vertex processor and tracks with large impact parameter are selected. Momentum information is used in the trigger to exclude low momentum tracks, whose impact parameters are badly smeared by multiple scattering. Preliminary simulation studies show that the reference design for the vertex trigger can achieve efficiencies of approximately 70% for accepted and reconstructed B decays such as $B_d \rightarrow \pi\pi$ and rejections of better than a factor of 100 for light quark events [5].

Particle identification will be critical for studying modes such as $B_d \rightarrow \pi^+\pi^-$ where the backgrounds from other two body B decay modes are expected to be large. It is useful for tagging since kaons from the other B can be used to flavor tag the signal b. To achieve reasonable separation of kaons and pions over the momentum range of 3 GeV/c $< p <$ 70 GeV/c is needed to accomplish these two goals. The BTeV RICH detector will have a C_4F_{10} gas radiator with $\pi/K/p$ thresholds of 1.6/9.0/17.1 GeV/c to achieve reasonable

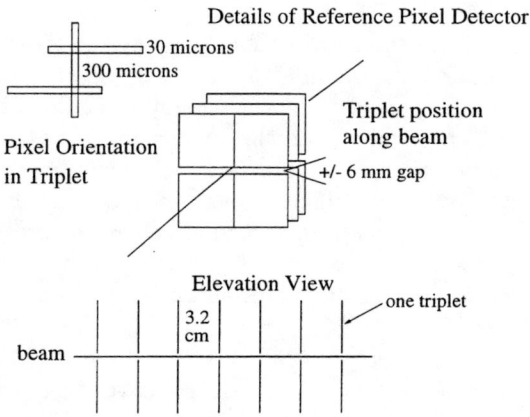

FIGURE 5. Schematic diagram of the BTeV vertex detector. The detector will be made of planes of pixel detectors. In the baseline design, these pixel planes are arranged into a triplet pattern with 31 stations positioned along the interaction region.

separation over this momentum range. Increased capability for identifying low momentum particles could be achieved by adding a thin aerogel radiator at the entrance to the gas RICH. This has been proposed by the LHC-B collaboration for their RICH [8] and will be investigated for use in BTeV.

Downstream tracking chambers are used to provide better momentum measurements for tracks exiting the magnet. They will also provide the only momentum measurement for tracks, such as daughter tracks from K_S^0 decays, that do not pass through the vertex detector. The baseline technology for this detector is planes of straw tubes.

Lepton identification is essential for any heavy quark experiment. The ability to identify electromagnetic final states is an integral part of the BTeV physics program. Electron and photon identification will be done with electromagnetic calorimeters located behind the RICH detectors. Various technologies are under investigation for this detector. Muon identification will be done with a system of iron toroids and muon chambers at the back of each detector arm. Many of the physics goals such as rare decay searches and tagging for CP violation and mixing studies rely on efficient muon identification. The muon system and electromagnetic calorimeter will also be used to provide J/ψ and prompt lepton triggers.

Physics Reach of the BTeV Detector

BTeV will be well positioned to make precision measurements of CP violating and rare decays of b and c hadrons. It will be able to make a competitive precision measurement of the angle β using the "golden" mode $B_d \rightarrow J/\psi K_S^0$. In one year of running at nominal luminosity, the uncertainty on sin2β will be $\delta\sin2\beta = 0.042$ [5] with $J/\psi \rightarrow \mu^+\mu^-$ only.

A summary of the expected yields for B decay modes of interest for one year of running at a luminosity of 5×10^{31}cm^{-1}s^{-2} is given in Table 2. The geometrical acceptance, expected trigger efficiency and reconstruction efficiency for final analysis cuts are included in these estimates. For CP-violating asymmetries, the sensitivity can be characterized by the number of untagged events scaled by the effective tagging efficiency ϵD^2, where D is the dilution due to mistags. The combination of lepton and kaon tags gives a tagging efficiency of 15% with a mistag fraction of about 25%, which yields an effective tagging efficiency $\epsilon D^2 \approx 4\%$. Jet charge and same-side tags will increase the sensitivity by improving ϵD^2 to $\approx 8-10\%$.

An observed CP asymmetry in $B_d \rightarrow \pi^+\pi^-$ provides information on the angle α. The BTeV vertex trigger is an ideal way to trigger on this mode and the RICH provides good separation of the $\pi\pi$ decays from other two body decays of B mesons. The backgrounds in this channel come mainly from $b\bar{b}$ events. Preliminary studies from a sample of 5 million simulated $b\bar{b}$ events indicate that the signal (S) to background (B) ratio ($S/S+B$) will be an

TABLE 2. BTeV event sample sizes after one year of running at luminosity of $5 \times 10^{31} \text{cm}^{-2} \text{s}^{-1}$.

Channel	Observed mode	Untagged events/year	Physics
$B_d \to J/\psi K_s^0$	$\mu^+\mu^-\pi^+\pi^-$	28000	$\sin 2\beta$
$B_d \to \pi^+\pi^-$	$\pi^+\pi^-$	17000	$\sin 2\alpha$
$B_s \to D_s K$	$\phi\pi K; \phi \to K^+K^-$	2350	$\sin \gamma$
$B_s \to D_s K$	$K^{0*}KK$	3000	$\sin \gamma$
$B_s \to D_s \pi$	$\phi\pi^-\pi^+; \phi \to K^+K^-$	17000	x_s
$B_s \to J/\psi \bar{K}^{0*}$	$\mu^+\mu^- K^-\pi^+$	1470	x_s
$B^\pm \to K^\pm \mu^+\mu^-$	$K^\pm \mu^+\mu^-$	300	SM rare decay

important dilution factor if one assumes a branching fraction for $B_d \to \pi^+\pi^-$ of 0.75×10^{-5} [9]. The uncertainty on $\sin 2\alpha$ from one year of running is expected to be $\delta\sin 2\alpha \approx 0.10$ [5] ignoring penguin contributions.

BTeV will have excellent resolution for secondary vertices which translates into good proper time resolution. This makes it an ideal detector for studying B_s mixing since x_s is expected to be large. Current limits on B_s mixing from LEP indicate that $x_s > 15$ [10]. Studies of several decay modes of the B_s [11] have been made in order to understand the sensitivity to B_s oscillations. The decays $B_s \to J/\psi \bar{K}^{0*}$ and $B_s \to D_s \pi$ when fully reconstructed in BTeV give a time proper resolution of ≈ 0.045 ps. For these studies we simulated the baseline BTeV detector using the detector simulation package MCFast [12]. The mixed and unmixed decays of $B_s \to D_s\pi$ for an $x_s = 40$ are shown in the left hand plots of Figure 6. The x_s reach for the mode $B_s \to D_s\pi$ is shown in the right hand plot of Figure 6.

One of the major goals of BTeV will be the measurement of the angle γ. Simulation studies are underway to investigate the reach of the BTeV detector from the time evolution of the decays $B_s, \bar{B}_s \to D_s^\mp K^\pm$ [13]. The sensitivity of the measurement depends on the mixing parameter x_s and on the branching fractions for $B_s \to D_s^\mp K^\pm$ which are not well known. It is clear that good particle identification will be crucial for this measurement in order to separate the signal from the background coming from $B_s \to D_s\pi$.

The physics potential for BTeV is vast and has been more fully documented elsewhere [5] [14] [15]. In addition to the extensive program in the B sector, BTeV plans to collect a high statistics charm decay sample in order to look for CP violation and mixing in the charm sector. The proposed schedule for BTeV calls for first collisions with a partial detector in the C0 hall during Run II. The full detector would then be installed and BTeV would begin running with a luminosity of $5 \times 10^{31} \text{cm}^{-2}\text{s}^{-1}$ soon after.

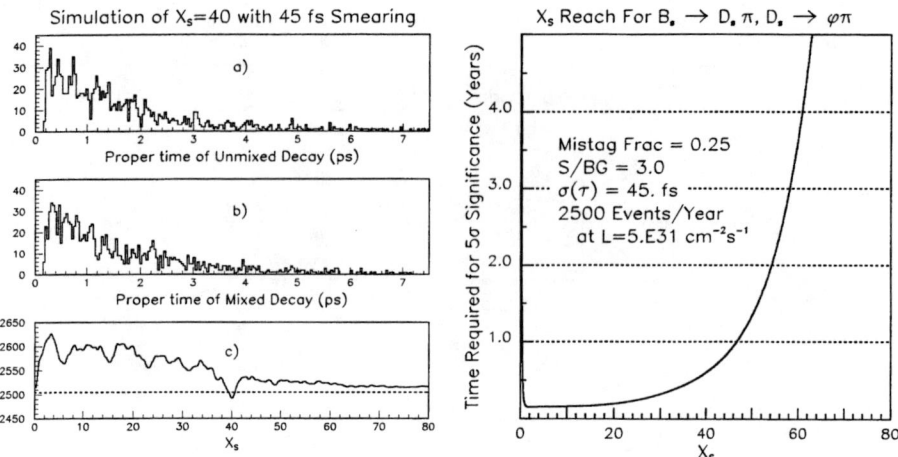

FIGURE 6. Proper lifetime plots of a) unmixed and b) mixed decays for the decay mode $B_s \to D_s \pi$ for one year of running. Part c) shows the corresponding negative log likelihood as a function of x_s. The right hand plot shows the x_s reach of the BTeV detector using the mode $B_s \to D_s \pi$.

Acknowledgements

The author would like to acknowledge the work of Penny Kasper, Rob Kutschke, Joel Butler, Paul Lebrun, Mike Procario, Tomasz Skwarnicki, Vince Smith and Kevin Sterner who produced the simulation results for the BTeV Expression of Interest.

REFERENCES

1. N. Cabibbo, *Phys. Rev. Lett.* **10**, 531 (1963); M. Kobayashi and K. Maskawa, *Prog. Theor. Phys.* **49**, 652 (1973).
2. R. Aleksan, I. Dunietz and B. Kayser, Z. Phys. **C54**, 653 (1992).
3. M. Gronau and D. London, Phys. Lett. **B253**, 483 (1991); M. Gronau and D. Wyler, Phys. Lett. **B265**, 172 (1991).
4. D. Atwood, I. Dunietz and A. Soni, Phys. Rev. Lett. **78**, 3257 (1997).
5. BTeV: An Expression of Interest for a Heavy Quark Program at C0, BTeV-pub-97/2, http://www-btev.fnal.gov/public_documents/c0_eoi_1.ps, May 1997.
6. J. Ellett *et al.*, Nucl. Instrum. Meth. A 317, 28 (1992).
7. D. Husby, P. Chew, K. Sterner, W. Selove, Nucl. Inst. and Meth. A **383** (1996) 193.

8. R. Forty, CERN-PPE/96-176, Sept. 1996 published in Proc. of the 4th Int. Workshop on B-Physics at Hadron Machines, Rome, Italy, June 1996, F. Ferroni, P. Schlein (Eds.) North-Holland, 1996.
9. M. Procario, BTeV-int-97/10.
10. V. Andreev et al. (The LEP B Oscillations Working Group), "Combined Results on B^0 Oscillations: Update for the Summer 1997 Conferences," LEP-BOSC 97/2, August 18, 1997.
11. R. Kutschke, BTeV-int-97/9.
12. P. Avery et al., "MCFast: A Fast Simulation Package for Detector Design Studies," in the Proceedings of the International Conference on Computing in High Energy Physics, Berlin (1977) and http://www-pat.fnal.gov/mcfast/mcfast_docs.html.
13. Penny Kasper, BTeV-int-97/13.
14. J. N. Butler, "Prospects for Heavy Flavor Physics at Hadron Colliders," in proceedings for Frontiers in Contemporary Physics; "Fundamental Particles and Interactions," May 11-16, 1997 at Vanderbilt University, Nashville TN.
15. S. Stone, "The Goals and Techniques of BTeV and LHC-B," in the proceedings of the International School of Physics, "Heavy Flavor Physics - A Probe of Nature's Grand Design," Varenna, Italy, 8-18 Jul 1997.

Studies of CP Violation in B-meson Decays at LHC

Tatsuya Nakada

Paul Scherrer Institute
CH-5232 Villigen-PSI
Switzerland

Abstract. CP violation is one of few remaining open questions in the standard model. Although observed CP violation phenomena in the neutral kaon decays can be accommodated in the framework of the standard model, it is not excluded that they are signs of new physics. A better insight into this problem can be obtained by studying CP violation in B-meson decays in detail. The ultimate opportunity of doing such studies is offered by LHC, in particular with the LHC-B detector which is designed for this purpose.

INTRODUCTION

The standard model can so far describe the world of elementary particles successfully and consistently. The theory has been tested up to the quantum correction level and all the precision tests performed at both high and low energies show no sign of deviation from the standard model predictions, except a few intriguing experimental results in the neutrino sector.

Similar to mass generation and the Higgs particle, CP violation is one of the few remaining untested aspects of the standard model. Currently observed CP violation in the neutral kaon decays can be accommodated within the standard model. However, it does not exclude that a mechanism beyond the standard model is responsible.

CP violation is one of the three necessary conditions to generate matter-antimatter asymmetry in the universe [1]. The standard model however does not seem to be capable of generating a sufficient amount of CP violation to explain the observed dominance of matter in our universe [2]. This calls for new sources of CP violation beyond the standard model.

In the standard model, CP violation is described in the framework of the 3×3 complex unitary mass-mixing matrix (KM-matrix) [3]. Such a matrix is uniquely defined by four parameters, which include a phase. In order to generate CP violation, the KM-matrix elements, V_{ij}, must satisfy

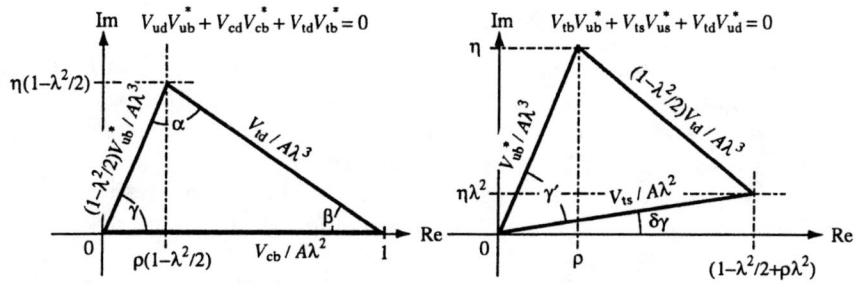

FIGURE 1. Two unitarity triangles under Wolfenstein's parametrization.

$\mathrm{Im}(V_{ij}V_{lk}V_{ik}^*V_{lj}^*) \neq 0$ [4]. One of the most commonly used parametrizations with parameters λ, A, ρ and η was introduced by Wolfenstein [5] where λ is practically identical to the sine of the Cabibbo angle [6]. From the unitarity condition of the KM matrix, six unitarity triangles can be constructed. Only two of them are expected to have the three sides with a similar length. In Figure 1, those two unitarity triangles are drawn. It follows that $\arg V_{ub} = -\gamma$, $\arg V_{td} = -\beta$ and $\arg V_{ts} = \delta\gamma + \pi$. For generating CP violation in the standard model, $\eta \neq 0$ is required.

Due to the large errors on the current measurements of $|V_{ub}|$ and $|V_{td}|$, a non-zero value of η can be obtained only from $\mathrm{Re}\epsilon$, the CP violation parameter observed in K^0-$\overline{K}^0$ oscillations. Both theoretical and experimental uncertainties contribute to the error on $|V_{ub}|$. They will be progressively reduced in coming years using the new data coming from the experiments running at high luminosity e^+e^- storage rings. The errors on $|V_{td}|$ and the standard model description of $\mathrm{Re}\epsilon$ are dominated by the theoretical uncertainties in calculating the strong interaction effect. Their improvement will be slower.

New information is expected to come from CP violation in the neutral B-meson decays [7]. In order to discover an effect due to new physics, it is most important to measure with a high precision CP violation parameters where the standard model predictions have little theoretical uncertainty. It is also important to measure CP violation in many different decay modes, even with a reduced accuracy. Although there are some theoretical uncertainties to interpret in some of those modes, such measurements nevertheless allow a consistency test to be made.

The theoretically cleanest decay modes are $B_s \to D_s K$ where no penguin diagram contributes. By comparing the four proper-time dependent decay rates,

$$B_s \to D_s^+ K^-(t)$$
$$\overline{B}_s \to D_s^+ K^-(t)$$
$$B_s \to D_s^- K^+(t)$$

$$\overline{B}_s \to D_s^- K^+(t)$$

$\gamma' - \delta\gamma$ can be measured [8] without theoretical uncertainty. Both tree and penguin diagrams contribute to the $B_d \to J/\psi K_S$ decay. However, two decay amplitudes have almost the same weak interaction phase. Therefore, $B_d \to J/\psi K_S$ decays give $\sin 2\beta$ [7] with very little theoretical uncertainty.

Since $\gamma' - \delta\gamma$ and β can be measured from those decay modes without limitation from theoretical uncertainties, one should measure them as accurately as possible to test the standard model.

The interpretation of the CP asymmetry measured from the decay mode $B_d \to \pi^+\pi^-$ is more difficult. If the decay is only due to the tree diagram, $\sin 2\alpha$ [7] can be measured. However, the recent CLEO measurements [9] indicate that the contribution of the penguin diagrams to $B_d \to \pi^+\pi^-$ is considerable, i.e. $\sim 20\%$. If the penguin diagram contribution can be determined by studying other decay modes which are dominated by penguin diagrams, α can be extracted from $B_d \to \pi^+\pi^-$ alone [11].

The LHC offers the opportunity to study those decay modes with high statistics. In proton-proton collisions at 14 TeV, the $b\overline{b}$ cross section is expected to be of the order of 500 μb which leads, even for a modest luminosity of 1.5×10^{32} cm^{-2}s^{-1}, to about 10^{12} $b\overline{b}$ pairs in a standard (10^7 s) year of running. Moreover, a sizable fraction of the inelastic interactions consists of events with a $b\overline{b}$ pair ($\approx 5 \times 10^{-3}$). Another advantage is that many different kinds of b-hadrons, e.g. B_u, B_d, B_s, B_c and b-baryons, are produced in pp interactions.

GENERAL PURPOSE DETECTORS AT LHC

ATLAS and CMS [12] are two general purpose collider detectors designed to perform high-p_t physics such as studies of the top quark and searches for the Higgs and supersymmetric particles in pp interactions at LHC in the central region. The b quark is an important tool for high-p_t physics. With the increasing interest in physics of the B-meson itself, the two collaborations include the study of B-meson decays as a part of their physics programme [13].

It is expected to take several years for LHC to reach its design luminosity of 10^{34} cm^{-2}s^{-1} which is required to fully exploit LHC for high-p_t physics. Thus, the physics of b-quarks will be important for ATLAS and CMS during the first few years of the LHC operation. Once LHC achieves the design luminosity, b-quark physics will become exceedingly difficult due to the large background.

Both ATLAS and CMS have an excellent muon detection capability. The muon is used in the first-level trigger, either with a single muon (ATLAS) or a combination of a single and di-muon (CMS). The excellent detection capability for electrons allows ATLAS to reconstruct $B_d \to J/\psi K_S$ decays with $J/\psi \to e^+e^-$. Due to the strong magnetic field of their solenoid (4 T), CMS has difficulty in using the electron channel.

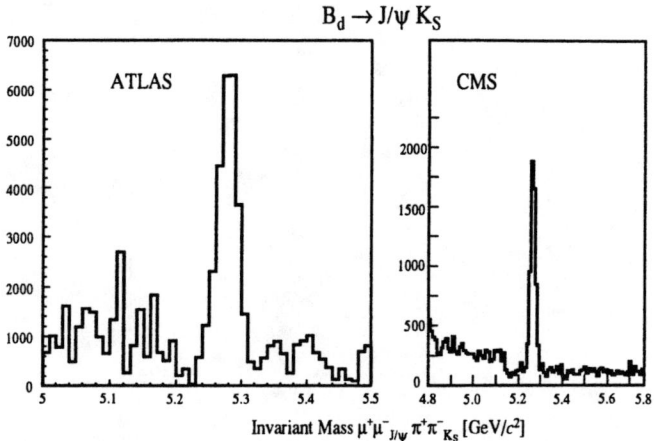

FIGURE 2. Reconstructed invariant mass distribution for $B_d \to J/\psi K_S$ decays with ATLAS and CMS detectors.

Both experiments have improved their vertex detectors by placing their first plane much closer to the beam than the original designs shown in their Letters of Intent. The new designs provide a better impact parameter resolution which reduces the background for the reconstructed B mesons and improves the proper-time resolution. However, radiation damage becomes a serious concern when operating at the nominal LHC luminosity.

From $B_d \to J/\psi K_S$ decays (Figure 2), both general purpose detectors can measure $\sin 2\beta$ very well. However, measuring $\gamma' - \delta\gamma$ from the $B_s \to D_s K$ decays is not possible due to lack of particle identification capability. As shown in Figure 3, the invariant mass distribution of "$D_s K$" without particle identification is totally dominated by $B_s \to D_s \pi$ and $\to D_s^* \pi$ which have much larger branching fractions than $B_s \to D_s K$. Since no CP violation is expected in $B_s \to D_s^{(*)} \pi$ decays from the standard model, to measure CP violation without isolating the $B_s \to D_s K$ decay is very difficult.

Particle identification is also important for reconstructing $B_d \to \pi^+\pi^-$ decays. Without particle identification, other two-body b-hadron decays such as $B_d \to \pi^\pm K^\mp$ and $B_s \to K^+K^-$ cannot be distinguished from $B_d \to \pi^+\pi^-$ decays from the invariant mass distribution.

It is clear that a dedicated experiment to study CP violation in B-meson decays with particle identification capability is necessary.

THE LHC-B EXPERIMENT

The LHC-B [14] experiment shown in Figure 4 is a forward spectrometer dedicated to the study of CP violating B-meson decays at the LHC. Some of

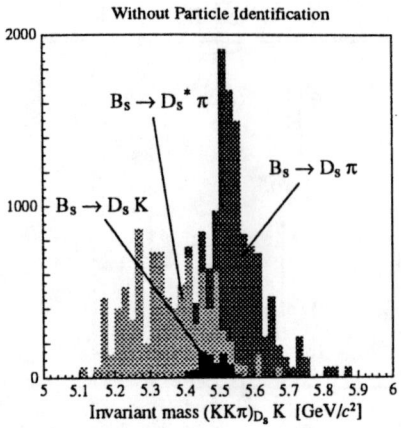

FIGURE 3. Invariant mass distribution for $B_s \to D_s\pi$, $B_s \to D_s^*\pi$ and $B_s \to D_sK$ all reconstructed as "$B_s \to D_sK$" without applying particle identification.

the reasons for choosing the forward geometry are as follows:

- The B-meson production angles are peaked in the forward and backward direction with respect to the beam direction. The produced b and $\bar{b}$ are typically correlated in one unit of rapidity. Therefore, the geometric efficiency is high for detecting all the decay particles from one b-hadron together with a decay particle from the accompanying hadron to be used as a flavour tag.

- A particle identification system of a manageable size covering all of the necessary momentum region can be built based on Ring Imaging Cherenkov Counters (RICH).

- An efficient early-level trigger can be designed based on muons, electrons and hadrons with large transverse momenta. In the forward direction, longitudinal momenta of particles are large. Threshold values for a p_t-cut can therefore be decided on the basis of background suppression rather than detector requirements.

- The large Lorentz boost of accepted B-mesons (corresponding to about 7 mm mean decay distance) allows proper-time measurements to be made with a few percent uncertainty. This is crucial for studying CP violation and oscillations with B_s-mesons because of the expected high oscillation frequency.

- Forward planar detector systems, quite similar to those used in fixed target experiments, are less complicated, easier to install, maintain and upgrade.

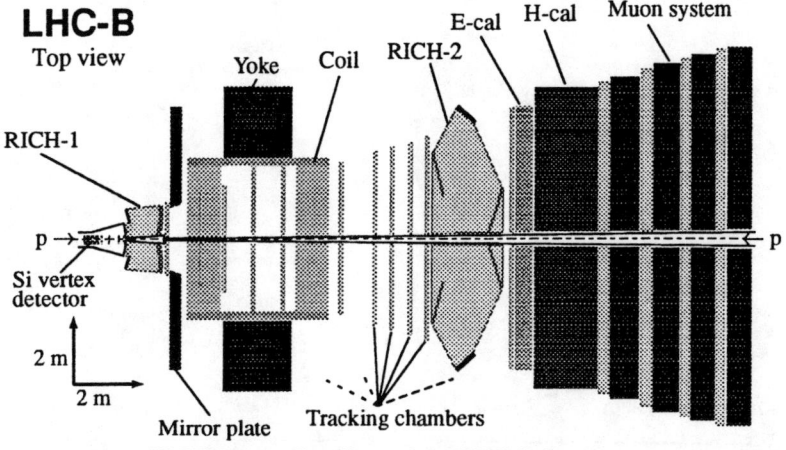

FIGURE 4. Top View of the LHC-B detector.

Due to its open geometry, the detector set-up is relatively simple and intrinsically much cheaper than the general purpose LHC detectors and fills the region of phase space which will not be covered by them.

Due to the efficient trigger, the LHC-B experiment will be able to exploit fully its physics potential at a much lower luminosity (1.5×10^{32} cm^{-2}s^{-1}). This matches well with the expected machine condition in the start-up period of the LHC. Working at a lower luminosity also eases the problem of radiation damage. The experiment can operate in conditions corresponding to the LHC design luminosity as well by locally reducing the luminosity at the LHC-B interaction region.

Detector

As seen from Figure 4, the detector consists of a micro-vertex detector system at the intersection point (placed in Roman pots), a tracking system, aerogel and gas RICH counters, a large-gap dipole magnet, electromagnetic and hadronic calorimeters, and a muon system.

The spectrometer will be installed at IP-8, currently occupied by the DELPHI experiment. Unlike the other interaction points, the pp collisions will not take place at the centre of the hall, but are displaced by ~ 11 m. This arrangement has the advantage that large detector components such as the magnet, calorimeters and muon system can be placed in the already existing hall thus requiring no extra excavation and providing easy installation and maintenance.

One of the most crucial components of the LHC-B detector is the RICH system. It consists of two detectors with three different radiators in order to

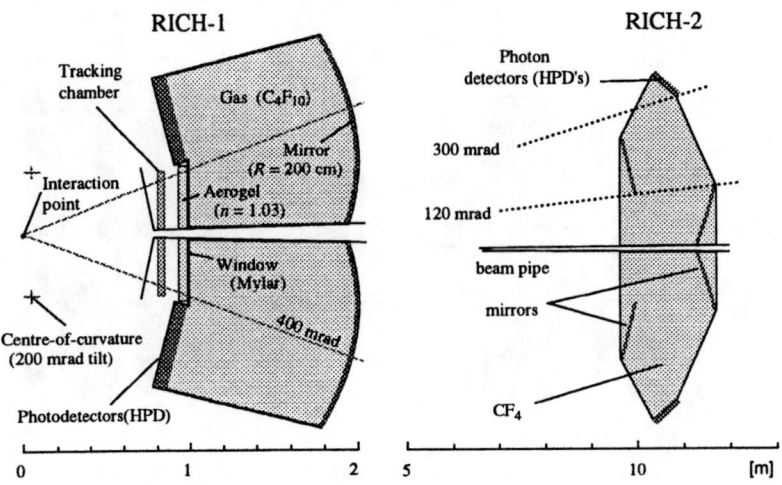

FIGURE 5. The Ring Imaging Cherenkov Counter designed for the LHC-B experiment.

cover the required momentum range, 1 to 150 GeV/c. Figure 5 shows their schematic views. The first RICH placed in front of the dipole magnet uses aerogel, with a refractive index of $n = 1.03$, and C_4F_{10} gas ($n = 1.0014$) as radiators. Cherenkov light of both radiators is detected with planes of Hybrid Photo Diodes (HPD's) placed outside of the spectrometer acceptance. The second gas RICH detector, used for high momentum particles, is placed after the magnet. It uses CF_4 ($n = 1.0005$) as the radiator. Again, planes of HPD's placed outside of the spectrometer acceptance are used to detect Cherenkov photons.

Trigger

The current design of the LHC-B trigger is based on four decision levels. Due to the large mass and the transverse momentum spectrum of the B-meson, decay products of the B-meson have on average higher p_t than particles produced in an ordinary so-called "minimum bias" event. Decay products of the B-meson also originate from vertices which are displaced from the primary interaction point by several mm.

The early levels of the LHC-B trigger exploit those two characteristics. Level-0 decision is based on high-p_t hadrons or electrons found in the calorimeter system or muons found in the muon system. It provides a modest reduction of minimum bias events by ~ 10. High-p_t muon and electron triggers are sensitive to leptons from semileptonic b decays and other decay modes with leptons such as J/ψ. The high-p_t hadron trigger increases the yield of final states which do not have a lepton in them. For example, this trigger signif-

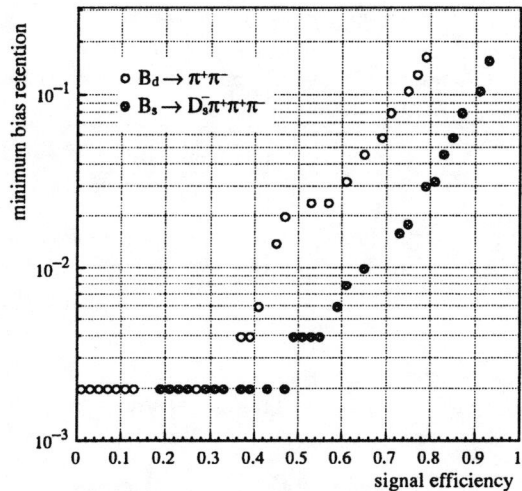

FIGURE 6. The Level-1 vertex trigger efficiencies for events with $B_d \to \pi^+\pi^-$ or $B_s \to \bar{D_s}3\pi$ decays as a function of the minimum bias retention rates. Efficiencies for B-meson events are normalized to those where the B-meson final states under consideration can be reconstructed in the spectrometer.

icantly enhances the yield for $B \to \pi^+\pi^-$ over that obtained from a lepton trigger alone.

At Level-1, those p_t measurements are refined using the tracking system. Thus a further reduction of the event rate can be achieved. In addition, data from the micro-vertex detector are used to select events with multiple vertices. In the forward direction, the impact parameter of a track is given by the r information only and a track is almost in a fixed ϕ plane. The micro-vertex detector therefore consists of layers of Si planes with r-ϕ strip geometry. This simplifies the track finding and impact parameter calculation so that the trigger algorithm can be executed in the limited time available. Figure 6 shows the Level-1 vertex trigger efficiencies for events with $B_d \to \pi^+\pi^-$ or $B_s \to D_s3\pi$ decays as a function of the minimum bias retention rates. Efficiencies for B-meson events are normalized to those where the B-meson final states under consideration can be reconstructed in the spectrometer.

Compared to the trigger scheme described in the Letter of Intent [14], a factor of two or more in the yield of B mesons which are used in the physics analysis is expected.

After a positive decision of the Level-1 trigger, data are read out to an event buffer. Hereafter, all the detector information is in principle available for the trigger decision. At Level-2, a further enhancement of events with b-hadron is achieved, for example by refining the reconstruction of the b-hadron decay vertices. At Level-3, the trigger decision is made by reconstructing

TABLE 1. Expected precisions for the measured angles of the unitarity triangles with the LHC-B detector given in the Letter of Intent with one year (10^7 s) of data taking. The assumed running luminosity is 1.5×10^{32} cm^{-2}s^{-1}. With the improved trigger scheme described in this paper, we anticipate results better than those shown in the table.

angle	decay channel	precision	Important features of the LHC-B experiment
$\sin 2\alpha$	$B_d \to \pi^+\pi^-$	0.04	Particle identification
$\sin 2\beta$	$B_d \to J/\psi K_s$	0.02	
γ	$B_d \to DK^{*0}$	7° to 12°	Particle identification
$\gamma' - \delta\gamma$	$B_s \to D_s K$	6° to 16°	Particle identification and proper time resolution
$\sin 2\delta\gamma$	$B_s \to J/\psi\phi$	0.01 to 0.02	Proper time resolution

the full event. Due to the large b-hadron production rate, not all the events with b-hadrons can be recorded. Therefore, the b-hadron final states are reconstructed to select only the decay modes of interest.

CP Physics Performance

In addition to the capability to collect a large sample of B-meson decays, special features of the LHC-B experiment are its particle identification capability and its excellent proper-time resolution.

The importance of good particle identification can be demonstrated with the decay channel $B \to \pi^+\pi^-$. Here, the most dangerous background comes from other two-body decay modes of B-mesons such as $B \to K^+\pi^-$ and $B_s \to K^+K^-$, which can only be rejected by particle identification. Simulation studies show that a clean reconstruction of $B \to \pi^+\pi^-$ is possible with high efficiency.

The proper-time resolution of the B-meson with the LHC-B detector is found to be ~ 0.042 ps. With this excellent resolution, the B_s oscillation parameter, x_s, can be measured up to 55 or more, which is beyond the upper limit predicted in the standard model.

Table 1 summarizes the expected performance of the LHC-B detector for measuring CP violation parameters given in the Letter of Intent [14]. It assumes a luminosity of 1.5×10^{32} cm^{-2}s^{-1} and data taking for 10^7 s. The errors include an estimated systematic error. With the improved trigger scheme described in this paper, we anticipate results better than those shown in the table.

CONCLUSIONS

B-meson decays provide some of the most accurate tests of the standard model in the aspects of flavour mixing and CP violation. LHC will be the ultimate source of B-mesons where a dedicated experiment, LHC-B, is planned together with two general purpose pp experiments. Particular characteristics of LHC-B are the particle identification, efficient trigger and excellent propertime resolution which allow two angles of the unitarity triangle to be measured very accurately. The experiment is expected to be operational from the beginning of the LHC run and to exploit the potential of the LHC in heavy flavour physics.

Acknowledgments

The author thanks the conference organizers for their hospitality. R. Forty is acknowledged for his careful reading of this manuscript.

REFERENCES

1. A. D. Sakharov, *JETP Lett.* **6**, 21 (1967).
2. See for example
   M. B. Gavela *et al.*, *Modern Phys. Lett.* **9A**, 795 (1994).
3. M. Kobayasi and K. Maskawa, *Prog. Theor. Phys.* **49**, 652 (1973).
4. C. Jarlskog, *Phys. Rev. Lett.* **55**, 1039 (1985) and *Z. Physik* **C29**, 491 (1985).
5. L. Wolfenstein, *Phys. Rev. Lett.* **51**, 1945 (1983).
6. N. Cabibbo, *Phys. Rev. Lett.* **10**, 531 (1963).
7. For pioneering work, see
   A. B. Carter and A. I. Sanda, *Phys. Rev. D* **23**, 1567 (1981).
   I. I. Bigi and A. I. Sanda, *Nucl. Phys.* **B193**, 85 (1981).
8. R. Aleksan, I. Dunietz and B. Kayser, *Z. Physik* **C54**, 653 (1992).
9. See for example
   P. Drell, Plenary talk given at XVIII International Symposium on Lepton Photon Interactions, Hamburg (1997).
10. See for example
    M. Neubert, Plenary talk given at International Europhysics Conference on High Energy Physics (1997).
11. M. Gronau, *Phys. Lett.* **B300**, 163 (1993).
12. G. L. Bayatian *et al.* (CMS collaboration), *Technical Proposal*, **CERN/LHCC 94-38 LHCC/P1**, 1 (1994).
    W. W. Armstrong *et al.* (ATLAS collaboration), *Technical Proposal*, **CERN/LHCC 94-43 LHCC/P2**, 1 (1994).
13. Y. Lemoigne, *Contribution to these proceedings*.
14. K. Krisebom *et al.* (LHC-B collaboration), *Letter of Intent*, **CERN/LHCC 95-5 LHCC/I8**, 1 (1995).

B Physics in Central Geometry at LHC

Yves Lemoigne
DAPNIA-SPP / CEN-Saclay
F-91191 Gif/Yvette Cedex

For the CMS Collaboration

Abstract. We discuss the potential for B physics at LHC of Central Geometry detectors like Atlas and CMS.

I INTRODUCTION

The ATLAS and CMS B-physics programmes [1] will be performed in a central geometry, as CDF does now at FNAL. They will start from the first year of LHC at a starting luminosity of 10^{32} to 10^{33} cm^{-2}s^{-1} which will be very appropriate for such studies (no event pile-up). A "young LHC" will certainly run at a modest luminosity for months (perhaps years), and at the end of each fill during all its life. This is shown in Figure 1 [2], where is shown the number of events with only one primary vertex, as a function of the luminosity, representative of about a 20-hour collider run. Between 10^{32} and 10^{33}, the number of single p-p interactions per beam crossing increases with luminosity. Beyond 10^{33}, event pile-up increases strongly, decreasing the absolute number of single-primary-vertex events. The optimum luminosity is about 10^{33}. As only one primary vertex is compulsory for tagging, there is a promising opportunity to study B physics and CP violation in good conditions.

Why is a central geometry interesting for B-physics and CP-violation studies? The first reason is that, as we have seen yesterday, central geometry is very successful at CDF [3]. Furthermore, if CDF has covered up to now a limited rapidity domain (± 0.8), Atlas and CMS will cover a much larger domain at LHC (up to ± 2.3).

The second reason is that the cost of these "B-physics detectors" will be zero as Atlas and CMS will be constructed anyway to study "hard physics." Furthermore, the need for b tagging for the main physics goals of Atlas and CMS (Higgs and SUSY searches, top physics) requires the installation of an efficient secondary vertex detector [1], which will be used for the detection of

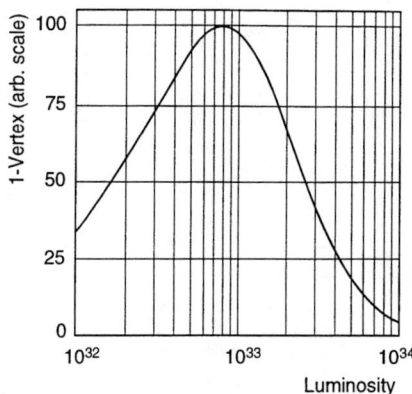

FIGURE 1. Relative number of events with only one primary vertex as a function of the initial luminosity.

B_d and B_s. The advantages of a central geometry for B physics at hadron colliders come mainly from two facts: we can trigger in the "quietest" domain in the rapidity range, and as the elements of the B pairs rarely overlap, we can use isolation criteria to recognise the B's. This talk will develop these points. We will illustrate these advantages for B physics and CP violation with three channels: $B_d^0 \to J/\Psi K_s^0$, $B_d^0 \to \pi^+\pi^-$, and $B_s^0 \to \mu^+\mu^-$.

II TRIGGERING IN A "QUIET" DOMAIN

Figure 2 shows the rapidity domain which will be covered by Atlas and CMS. The charged multiplicity will be limited to 7 or 8 per unit of rapidity. One unit of rapidity corresponds to a large angular domain here (larger than 325 mrad at rapidity 2). On this figure we can also see that it will be relatively easy to trigger on leptons as the deposited energy is really minimal in this domain. The Atlas experiment will use only a one-muon first-level trigger ($P_t \geq 6.5$ GeV). The CMS expectations for 1μ and 2μ triggers are shown in Figure 3. Trigger rates not exceeding a few kHz could be easily obtained with a P_t threshold of about 7 GeV for one muon. CMS also expects to trigger on electrons of $P_t \geq 10$ GeV, the total rate of triggers allocated to B physics being of the order of a few kHz. The electron channel will be studied in CMS because of the high granularity and quality of the PbWO$_4$ crystal calorimeter. Coupled with the central tracker (p/E), it allows identification of electrons of $P_t \geq 2.5$ GeV with an efficiency of 65% and a hadron rejection factor of ~ 400. The expected ECAL energy resolution is $\sigma/E = 2\%\sqrt{E} \oplus 0.5\% \oplus 0.2/E$.

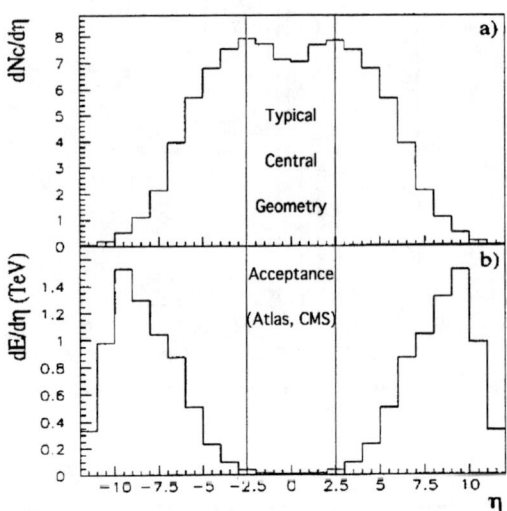

FIGURE 2. (a) Rate of charged particles per unit of rapidity at LHC. (b) Energy flow per unit of rapidity at LHC. The rapidity domain of Atlas and CMS is shown.

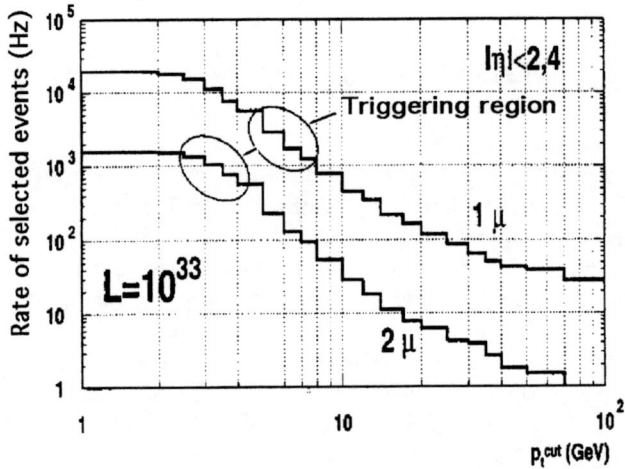

FIGURE 3. Muon trigger rate as a function of muon P_t is shown for a luminosity of 10^{33} cm^{-2}s^{-1}. The two triggering zones (1μ and 2μ) proposed for CMS are shown.

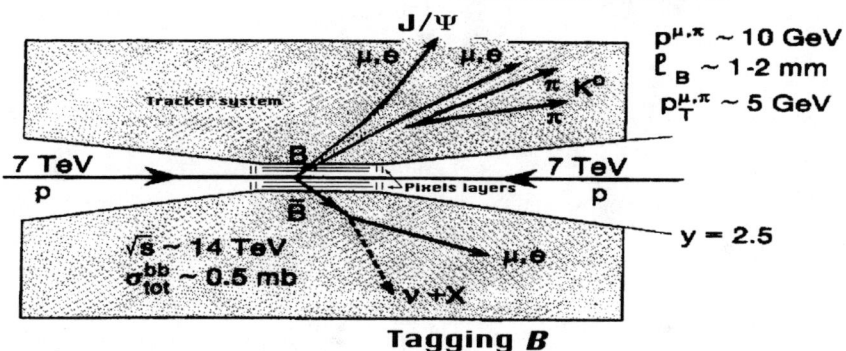

FIGURE 4. Diagram for B physics in central geometry (Atlas, CMS).

III MONTE-CARLO SIMULATIONS

Figure 4 recalls the main features of a central geometry at LHC: the interaction zone is surrounded by the tracker system which begins with a silicon vertex detector whose crucial elements are pixel layers (at 7 and 11 cm from the beam in the CMS case). Usually B and $\overline{B}$ do not overlap, one B being reconstructed from its $J/\Psi\, K_s$ decay, the $\overline{B}$ being used for tagging through its semi-leptonic decay. The typical decay lengths of B's are 1 to 2 mm, the P_t's of the decay products are 5 GeV on average, and their momenta are on the order of 10 GeV. If only a small fraction of $b\bar{b}$ is reconstructed (10^{-3} after geometry and cuts), there are still plenty of b's produced (we assume $\sigma_{b\bar{b}}$ at 14 TeV = 500 mb, giving 5×10^{12} per year at $\mathcal{L} = 10^{33}$). Note that there will be no hadron identification (π, K).

PYTHIA and JETSET are used for simulation. CMS uses PYTHIA 5.7 and JETSET 7.4 versions. Both gluon fusion and gluon splitting are simulated for beauty production. Dectector performances are estimated using GEANT-based packages. Figure 5 shows a $B_d^0 \to J/\Psi K_s^0 \to \mu^+\mu^-\pi^+\pi^- (+\mu_{tag})$ event simulated in the Atlas detector and Figure 6 the reconstruction of such a beauty meson in CMS. Note the good effective mass resolution ($\sigma = 13$ MeV) and the high signal/background ratio.

Figure 7 shows the effective mass of the beauty meson when the J/Ψ decays to e^+e^-. Two cases are envisaged according to the method used to measure the electrons: tracker and calorimeter for identification, or calorimeter only. The first method is more accurate ($\sigma = 22$ MeV) but suffers from a significant radiative tail.

Table 1 shows the sensitivity to $\sin 2\beta$ (time-integrated method) attainable by both experiments after one year at a luminosity of 10^{33}. The three-electron channel is not used because the trigger efficiency is too low (CMS) or zero (Atlas requests at least one μ). Atlas hopes to use its transition-radiation

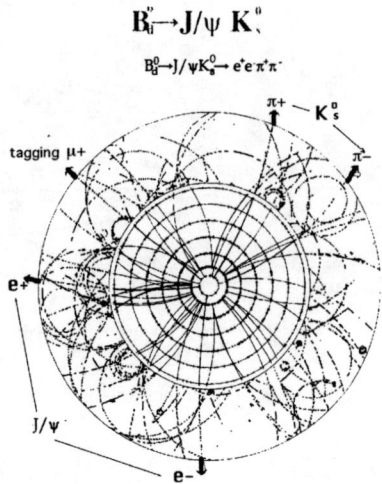

FIGURE 5. $B_d^0 \to J/\Psi K_s^0 \to \mu^+\mu^-\pi^+\pi^-(+\mu_{tag})$. Typical B_d decay event simulated in the Atlas detector (transverse view).

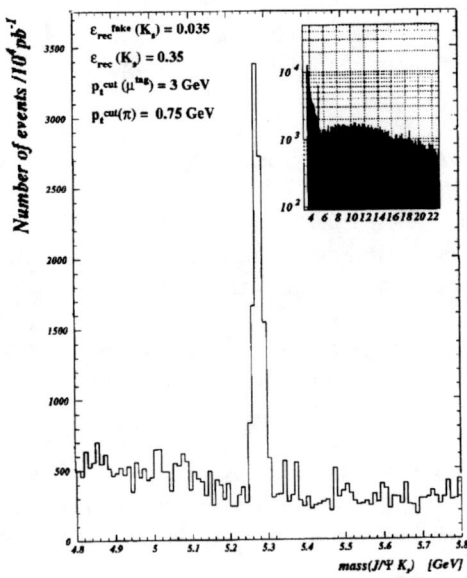

FIGURE 6. $B_d^0 \to J/\Psi K_s^0 \to \mu^+\mu^-\pi^+\pi^-$. Effective mass of B_d reconstructed in CMS.

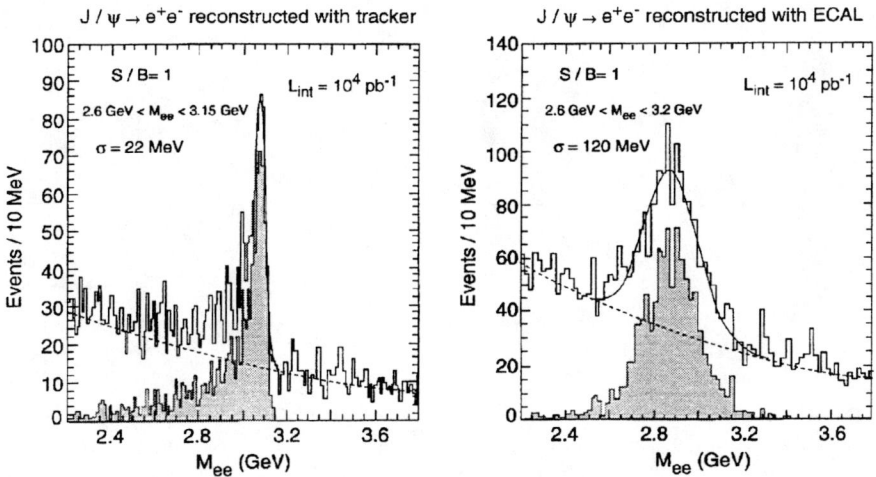

FIGURE 7. $B_d^0 \to J/\Psi K_s^0 \to e^+e^-\pi^+\pi^-$ using the Central Tracker or ECAL for reconstruction of the J/Ψ. In grey: the sample obtained in one year after background substraction.

detector to identify electrons down to 1 GeV. In this case the number of useful electrons will be large and will give good sensitivity to $\sin 2\beta$.

TABLE 1. Sensitivity to $\sin 2\beta$ in various lepton channels (Atlas and CMS).

Sensitivity to $\sin 2\beta$ in Channel	Atlas	CMS
$\overline{B}_d^0, b \Rightarrow J/\Psi K_s^0, \mu_{tag} \Rightarrow \mu^+\mu^-\pi^+\pi^-, \mu_{tag}$	0.050	0.056
$\overline{B}_d^0, b \Rightarrow J/\Psi K_s^0, \mu_{tag} \Rightarrow e^+e^-\pi^+\pi^-, \mu_{tag}$	0.040	0.098
$\overline{B}_d^0, b \Rightarrow J/\Psi K_s^0, e_{tag} \Rightarrow \mu^+\mu^-\pi^+\pi^-, e_{tag}$	0.055	0.134
$\overline{B}_d^0, b \Rightarrow J/\Psi K_s^0, e_{tag} \Rightarrow e^+e^-\pi^+\pi^-, e_{tag}$	not used	not used
Combined channels: $\delta(\sin 2\beta)$	0.022	0.046

IV ISOLATION CRITERIA

An advantage of central geometry detectors for B physics is the possibility of using isolation criteria. For instance in CMS, the decays are selected requiring:

- two opposite hadrons with $P_t > 3.5$ GeV (> 5 GeV for the electron tag sample) within $|\eta| < 2.4$; $\Delta R = \sqrt{(\Delta\eta^2 + \Delta\Phi^2)} < 1.6$ (< 1 in the electron tag sample). A pion mass is assigned to all hadrons (see Fig. 8-1).

- for both pions, a transverse impact parameter significance $\sigma_{IP} = |IP|/\sigma > 3$, where IP is the measured transverse impact parameter and σ its error

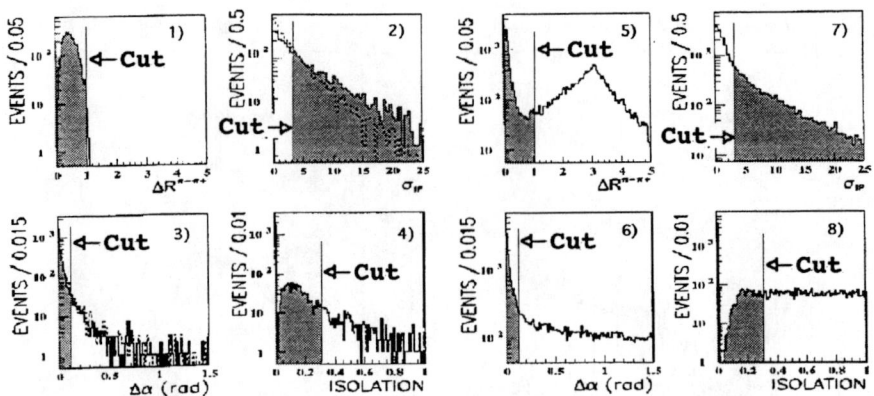

FIGURE 8. CMS cuts to select signals (see text). The selected parts of the distributions are in grey.

(see Fig. 8-2).

- in the transverse plane, the reconstructed B^0 momentum to match the primary-secondary vertex direction within 0.1 rad in Φ (see Fig. 8-3).

- the isolation parameter (see Fig. 9) $I = \sum \frac{P_t^h}{P_t^B} < 0.3$, where only hadrons with $P_t > 2$ GeV within a cone of $\Delta R < 1$ about the B^0 momentum are taken into account, and P_t^B is the transverse momentum of the B^0 candidate (see Fig. 8-4).

Figure 8 illustrates the effect of these cuts on signal distributions (sub-figures 1 to 4) and on background distributions (sub-figures 5 to 8). The grey zones are the selected parts of the distributions. ΔR and isolation cuts are particulary efficient in rejecting background without losing too much signal.

Figure 10 shows the $B_d^0 \to \pi\pi$ signal mixed with $K\pi$ and KK channels which constitute the main background, as CMS does not provide particle identification. Fortunately, except for $B_s \to K^+K^-$ decays, there is not a full overlap and, due to the good mass resolution (27 MeV), it will be possible to limit the ratio B/(S+B) to 45% by selecting $\pm 1\sigma$ in $\pi\pi$ mass. On the right part of Fig. 10, the combinatorial background is added. Figure 11 shows the ratio of total background as a function of the $\pi\pi$ mass selection.

The final sensitivity to $\sin 2\alpha$ estimated by Atlas is 0.047 and by CMS 0.041 per year at a luminosity of 10^{33}. Note that CMS adds events obtained with the electron trigger. More details can be found in Ref. [4].

ISOLATION OF B_s^0

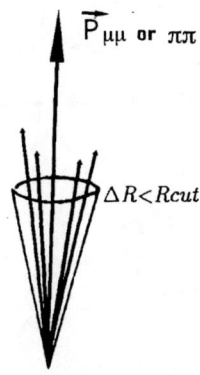

$$I = \sum_{\Delta R < Rcut} p_t^{cp} / p_t^{\mu\mu}$$

FIGURE 9. Definition of the isolation parameter I.

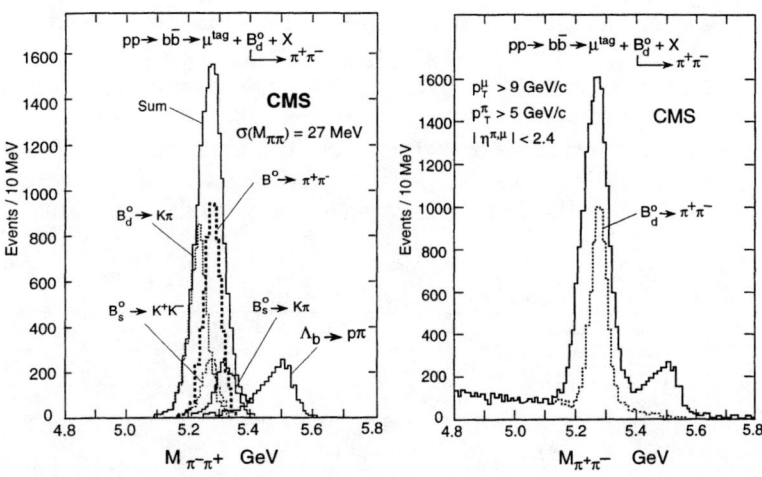

FIGURE 10. Mass distribution overlapped with 2-body decays and the signal within all the backgrounds (*i.e.* combinatorial background added).

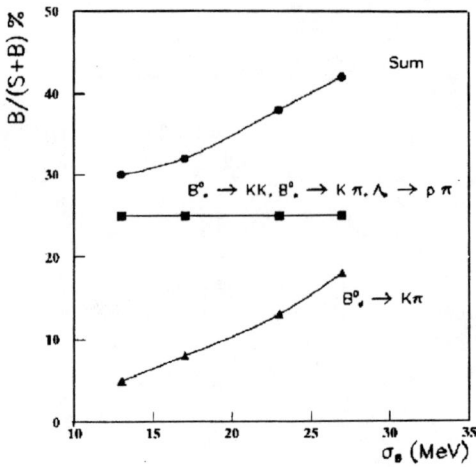

FIGURE 11. Ratio of background as a function of the selection inside the $\pi\pi$ peak.

V RARE DECAY $B_s^0 \to \mu^+\mu^-$

For this decay, which is a Flavour Changing Neutral Current (FCNC) decay, the Standard Model predicts a branching ratio of 4×10^{-9} [5]. Models beyond the Standard Model could enhance the branching ratio by a factor 10. It is clear that this could be verified only at LHC where 5×10^{12} $b\bar{b}$ will be produced each year. The background will be 7 orders of magnitude higher and the most powerful tools have to be used to fight it: secondary vertex selection, isolation cuts, and $\mu^+\mu^-$ mass resolution. CMS, for example, uses a secondary vertex cut $\sigma_{IP} > 3$ and an isolation cut which requires that $P_t^{ch} > P_{cuts}$ in a cone R_{cut} around B_s^0 not including trigger leptons. The isolation cut is $I < 0.05$ in a cone $R \leq \Delta R_{\mu\mu}/2 + 0.4$ and $P_{cuts} > 1$ GeV. They also forbid any neutrals with $E_t > 2$ GeV in the same cone. These three tools (see Table 2) give S/B = 0.14.

TABLE 2. Effect on background and signal reduction.

Tools	Main parameter cut	Background reduction	Signal reduction
Vertex cuts	$\sigma_{IP} > 3$	10^{-2}	0.60
Isolation cuts	$I < 0.05$ in cone $R < \Delta R_{\mu\mu}/2 + 0.4$	10^{-3}	0.26
Mass selection	$\mu^+\mu^-$ mass selection $\pm 1.7\sigma$	10^{-2}	0.90
Total		10^{-7}	0.14

Using similar cuts for an integrated luminosity of 10^4 pb^{-1} (one year), Atlas would obtain 35 candidate events with a background of 210 events, *i.e.* a branching ratio limit of 2.5×10^{-9} (95% CL), while CMS would obtain 11 candidate events with a background of 36 events, *i.e.* a branching ratio limit of 2.4×10^{-9} (90% CL). These numbers are close to the Standard Model predictions.

VI CONCLUSIONS

CDF has shown that a central geometry can be very powerful for B-physics study in hadron colliders. Atlas and CMS at LHC will continue this program with a larger rapidity acceptance and higher energy. Compared to other facilities, the cost will be zero as these detectors will be constructed anyway for "Hard Physics." The lepton trigger needed for B physics will be relatively easy to do and the separation of tracks allows the use of isolation criteria to select interesting channels for CP violation or for rare-decay searches. Examples have been shown. In an initial running phase the not-totally-unknown B physics will be of benefit to Atlas and CMS as a tool to calibrate the detector. After that, b tagging will be very useful, for instance in Higgs and SUSY searches.

REFERENCES

1. Atlas Collaboration, Technical Proposal, CERN/LHCC/94-43, LHCC/P2; CMS Collaboration, Technical Proposal, CERN/LHCC/94-38, LHCC/P1.
2. K. Eggert *et al.*, CERN AT/94-04 (DI), LHC Note 26.
3. B. Wicklund, *Nine Not-So-Bad Years of b physics at CDF*, in these proceedings.
4. A. P. Pralavorio, Ph.D. Thesis, IRES 97-14, Université Pasteur, Strasbourg, France.
5. A. Ali *et al.*, DESY 93-016 (1993); G. Buchalla and A. J. Buras, Nucl. Phys. **B400**, 225 (1993).

THEORY AND PROSPECTS

New Physics and the Unitarity Triangle

David London

Laboratoire de physique nucléaire, Université de Montréal,
C.P. 6128, succ. centre-ville, Montréal, QC, Canada

Abstract. After reviewing the present experimental constraints on the unitarity triangle, I discuss the various ways in which new physics can manifest itself in measurements of the parameters of the unitarity triangle. Apart from one exception, which I describe, new physics enters principally through new contributions to B^0-$\overline{B^0}$ mixing. Different models of new physics can be partially distinguished by looking at their effects on rare, flavour-changing B penguin decays.

At this conference, we have heard a number of talks discussing the prospects of various experiments for measuring CP asymmetries in B decays, *i.e.* the angles of the unitarity triangle. Ultimately, the hope is that we will find an inconsistency with the standard model (SM), which will give us some clue regarding the new physics which most of us believe must lie beyond the SM. In discussing new physics and the unitarity triangle (UT), there are basically two questions which have to be addressed:

1. What are the signals of new physics?

2. If such signals are seen, how can we identify the new physics?

The first step in answering these questions is to review our current knowledge of the UT. There are a number of measurements which constrain the UT: $|V_{cb}|$, $|V_{ub}/V_{cb}|$, B_d and B_s mixing, and $|\epsilon|$ in the kaon system. However, the problem is that there are important theoretical uncertainties in translating the experimental numbers into information about the UT. With all theoretical and experimental errors combined in quadrature, our present knowledge of the UT can be summed up in Fig. 1 [1]. As is evident from this figure, we really know rather little about the UT at present, due mainly to theoretical uncertainties.

On the other hand, the angles α, β and γ can be extracted with essentially no theoretical uncertainty from CP-violating asymmetries in B decays [2]. Due to B^0-$\overline{B^0}$ mixing, any neutral B decay to a final state f to which both B^0

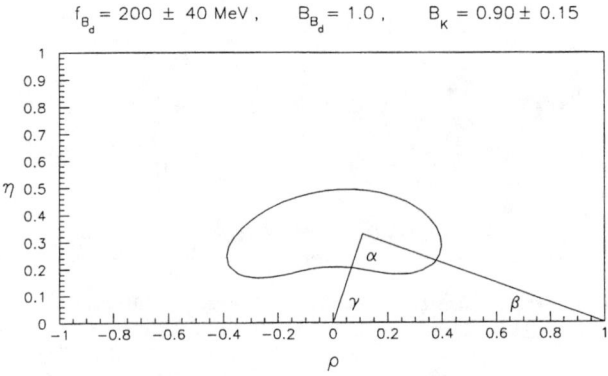

FIGURE 1. Allowed region in ρ-η space, from a simultaneous fit to both the experimental data (as given in Ref. [1]) and theoretical quantities (listed above). The theoretical errors are treated as Gaussian. The solid curve represents the 95% c.l. region. The triangle shows the best fit.

and $\overline{B^0}$ can decay can exhibit CP violation, due to the interference between the amplitudes $B^0 \to f$ and $\overline{B^0} \to f$. There are four distinct classes of CP asymmetries, involving the decays of B_d^0 or B_s^0 mesons, and the quark-level processes $b \to c$ or $b \to u$. For example, the asymmetries in $B_d^0(t) \to \pi^+\pi^-$ and $B_d^0(t) \to \psi K_S$ probe $\sin 2\alpha$ and $\sin 2\beta$, respectively, while $\sin^2 \gamma$ can be extracted from $B_s^0(t) \to D_s^\pm K^\mp$. (The decay $B^\pm \to DK^\pm$ can also be used to obtain $\sin^2 \gamma$.) The fourth CP asymmetry (e.g. $B_s^0(t) \to \psi\phi$) is expected to be zero, to a good approximation, within the SM. This is therefore a good place to look for new physics. The measurements of the three nonzero CP angles would allow us to reconstruct the UT with little theoretical error. Our present experimental knowledge constrains the angles to lie within the ranges $-1.0 \leq \sin 2\alpha \leq 1.0$, $0.26 \leq \sin 2\beta \leq 0.88$, and $0.22 \leq \sin^2 \gamma \leq 1.0$.

There are thus three distinct ways in which new physics can manifest itself in measurements of these CP asymmetries [3]:

1. $\alpha + \beta + \gamma \neq \pi$.

2. $\alpha + \beta + \gamma = \pi$, but the values of α, β and γ found disagree with the SM predictions.

3. $\alpha + \beta + \gamma = \pi$, α, β and γ are consistent with the SM, but measurements of the angles are inconsistent with measurements of the *sides* of the UT.

Now, how can new physics affect these CP asymmetries? There are basically only two ways: via new contributions to B decays ($b \to c, u$), or to B-$\overline{B}$ mixing. The first possibility can be virtually eliminated – apart from some very fine-tuned models, there are no models of physics beyond the SM in which the new contributions are competitive with the SM W-mediated decays. On the other hand, there are many models of new physics in which there are new contributions to B-$\overline{B}$ mixing, possibly with new phases [4]. Therefore *the principal way in which new physics can affect the UT is via new contributions to B mixing.* (There is an exception to this, which I will discuss below.) These new contributions will affect the experimental determinations of V_{td}, V_{ts}, α, β and γ.

In light of this, let us reconsider the three ways in which new physics can be detected. The first is to measure the three CP angles, and find $\alpha + \beta + \gamma \neq \pi$. In order for this to happen, there must be new physics, *with new phases*, in B_d or B_s mixing. However, there is an interesting twist here. Suppose there is new physics in B mixing. If β is measured in $B_d(t) \to \psi K_S$, then the phase extracted will be $\beta + \phi_{NP}^d$. And if α is obtained via $B_d(t) \to \pi^+\pi^-$, then one gets $\alpha - \phi_{NP}^d$. The key point here is that the sum $\alpha + \beta$ is *insensitive* to new physics [5]. Turning to the third angle, if γ is measured in $B^\pm \to DK^\pm$, then it is extracted with no modification, since neutral B's are not involved. However, if γ is obtained from $B_s(t) \to D_s^\pm K^\mp$, then $\gamma + \phi_{NP}^s$ will be extracted. The upshot is: since B-factories such as BaBar and Belle do not measure CP asymmetries in B_s^0 decays, they will *never* find $\alpha + \beta + \gamma \neq 0$. (Once again, there is an exception, to be discussed below.) However, hadron colliders may find $\alpha + \beta + \gamma \neq 0$ if γ is measured in B_s^0 decays. In fact, a discrepancy in the value of γ as extracted in these two ways would be a clear signal for new physics in B_s^0-$\overline{B_s^0}$ mixing.

The second way to detect new physics is if $\alpha + \beta + \gamma = \pi$, but the values of α, β and γ are in disagreement with the SM predictions. This can happen if there are new contributions, *with new phases*, to B_d or B_s mixing. Finally, the third way is if $\alpha + \beta + \gamma = \pi$ and α, β and γ are consistent with the SM, but are inconsistent with measurements of the sides of the UT. In this case, we need new contributions to B_d or B_s mixing *with the same phase* as in the SM.

Before examining which types of physics can contribute to B-$\overline{B}$ mixing, let me first discuss the exception I mentioned above. Most CP asymmetries involve tree-level B decays. However, there is another class of decays which can also be used: penguin decays [6,7]. Consider, for example, the decay $B_d^0 \to \phi K_S$, which is dominated by the quark-level $\bar{b} \to \bar{s}s\bar{s}$ penguin decay. Since the final state is a CP eigenstate, both B_d^0 and $\overline{B_d^0}$ can decay to it, thus leading to a possible CP-violating asymmetry. What does this CP asymmetry measure? The $b \to s$ penguin is dominated by internal t-quarks, so that it is proportional to the product of CKM matrix elements $V_{tb}^* V_{ts}$. Within the Wolfenstein approximation, this is real, just like $V_{cb}^* V_{cs}$, which describes the decay $B_d^0 \to \psi K_S$. In other words, the CP asymmetry in $B_d(t) \to \phi K_S$ measures β, just like $B_d(t) \to \psi K_S$. Therefore, within the SM, β as extracted from the CP asymmetry in ϕK_S equals that as found in ψK_S [6]. In fact, this is true even if there are new-physics contributions to B_d^0-$\overline{B_d^0}$ mixing.

However, since the $b \to s$ penguin is a pure loop effect, there can in principle be significant new contributions from new physics [8]. Examples of such new physics include four generations, non-minimal supersymmetry, and models with enhanced chromomagnetic dipole operators. If there is new physics, the phase of the decay amplitude may be changed. In this case, one will find β from ϕK_s is not equal to β from ψK_s. Therefore, by measuring β in $B_d(t) \to \phi K_s$, it might in fact be possible to find $\alpha+\beta+\gamma \neq \pi$, even at B-factories. Note also that, in addition to ϕK_s, the final states $\eta' K_s$, ρK_s, $\pi^0 K_s$, ηK_s, etc. may be used. In fact, recent results from CLEO, which show that the branching ratio for $B \to \eta' K$ is larger than expected, indicate that this method of measuring β may be promising [9].

The interesting thing is that what is really being probed here is new physics in the $b \to s$ flavour-changing neutral current. This same new physics will, in general, contribute to B_s^0-$\overline{B_s^0}$ mixing. Thus, this is in some sense a way of detecting new physics in B_s mixing without using B_s's at all!

Having discussed this special case, I now return to the more conventional ways of measuring the CP angles, via tree-level decays of B mesons. Suppose that the CP asymmetries are measured, and evidence for new physics is found. What could this new physics be? We know that the new physics contributes to B^0-$\overline{B^0}$ mixing. Therefore a first step is to classify models of new physics according to (i) whether they contribute to B^0-$\overline{B^0}$ mixing and (ii) if so, whether new phases are involved.

Here is a fairly extensive list of models of new physics, along with a discussion of their effects in B^0-$\overline{B^0}$ mixing [3]:

- Four generations: there are new loop-level contributions to the mixing involving internal t' quarks. Since the CKM matrix is now 4×4, new phases can be introduced.

- Z-mediated flavour-changing neutral currents (FCNC's): if the down-type quarks mix with an exotic vector singlet charge $-1/3$ quark, then the flavour-changing couplings $Zb\bar{d}$ and $Zb\bar{s}$ will be induced. In such models, there will be new contributions, with new phases, to B mixing through tree-level Z exchange.

- Multi-Higgs-doublet models:

    - with natural flavour conservation (NFC): in such models there are new contributions to B mixing involving box diagrams with internal charged Higgses. The charged Higgses couple to quarks through the CKM matrix, so no new phases are introduced.

    - without NFC: in this case there can be tree-level FCNC's involving the exchange of a neutral Higgs. Thus there are new contributions, with new phases, in B mixing.

- Left-right symmetric models: except in the most fine-tuned models, which I don't consider here, the mass of the W_R is at least 1 TeV. This renders its effects in B mixing negligible.

- Supersymmetry:
  - Minimal SUSY: there are many new contributions to B mixing involving box diagrams with internal supersymmetric particles. In the minimal model, all couplings involve the CKM matrix, so that no new phases are introduced.
  - Non-minimal SUSY: in non-minimal models, the new contributions can also have new phases.

The above list shows that there are indeed many models of physics beyond the SM which can contribute to B^0-$\overline{B^0}$ mixing, some with new phases, some without. The presence of such new physics will be detected through measurements of the CP asymmetries. However, such measurements will only tell us that new physics is present. While that would be a very exciting development, we still would want to know what the new physics is. How can we distinguish among the various possibilities listed above?

Some progress can be made through a simple observation. Any new physics which affects B^0-$\overline{B^0}$ mixing, which is a FCNC process, will also in general affect the rare, flavour-changing decays $b \to sX$ and $b \to dX$ (penguin decays). Therefore, by also looking at penguin decays, it may be possible to identify some models of new physics [3]. In fact, for certain types of new physics, if no deviation from the SM is observed in penguin decays, this would rule out there being any effects in B mixing.

To see how this works, I will examine in detail one model of new physics: Z-mediated FCNC's. As mentioned earlier, there are flavour-changing couplings $Zb\bar{d}$ and $Zb\bar{s}$, parametrized by U_{db} and U_{sb}, respectively. These couplings are constrained by $BR(B \to \mu^+\mu^- X) < 5 \times 10^{-5}$, leading to $|U_{qb}/V_{cb}| < 0.044$, or $|U_{qb}| < 0.0017$. The new couplings U_{qb} can have arbitrary phases.

In this model there are new, tree-level contributions to B^0-$\overline{B^0}$ mixing through Z exchange. Comparing to the SM, we find

$$\frac{\Delta M_d^Z}{\Delta M_d^W} = (0.9 - 26)\left[\frac{|U_{db}/V_{cb}|}{0.04}\right]^2 , \quad \frac{\Delta M_s^Z}{\Delta M_s^W} = 0.15\left[\frac{|U_{sb}/V_{cb}|}{0.04}\right]^2 . \quad (1)$$

Therefore B_d^0-$\overline{B_d^0}$ mixing can be dominated by Z-FCNC's, with new phases; B_s^0-$\overline{B_s^0}$ mixing is still due mainly to W box diagrams, but the new contribution is non-negligible, so the new phases may be important.

Let us now examine the contribution of Z-FCNC's to penguin decays. The constraint $|U_{qb}| < 0.0017$ is derived from the experimental limit $BR(B \to \mu^+\mu^- X) < 5 \times 10^{-5}$. Therefore if the new coupling takes its maximum allowed value, the model "predicts" the same branching ratio. This is, in fact, a huge effect – it is a smoking-gun signal. For the $b \to s$ decay, this is roughly 10 times bigger than in the SM, while for the $b \to d$ decay, it is an enhancement of about a factor of 100. Furthermore, if the branching ratios for the decays $B \to X_s \ell^+ \ell^-$ and $B \to X_d \ell^+ \ell^-$ are found to be consistent with the SM, this puts such stringent constraints on the $|U_{qb}|$ that it rules out the possibility of any effects in B^0-$\overline{B^0}$ mixing.

There are smoking-gun enhancements in other decays as well. For the presently-allowed values of the $|U_{qb}|$,

- $BR(B_s^0 \to \ell^+ \ell^-)$ is enhanced by about a factor of 20.
- $BR(B_d^0 \to \ell^+ \ell^-)$ is enhanced by about a factor of 300-400.
- $BR(b \to s$ EWP's$)$ is enhanced by about a factor of 25.
- $BR(b \to d$ EWP's$)$ is enhanced by about a factor of 500.

'EWP' stands for electroweak penguin decays. These are penguin decays which are dominated in the SM by a virtual Z, e.g. $B^+ \to \phi \pi^+$ and $B_s^0 \to \phi \pi^0$. There are no large effects of Z-mediated FCNC's in $b \to s\gamma$ or in other hadronic penguins.

The point of all this is to demonstrate that, if there are significant new-physics effects in B^0-$\overline{B^0}$ mixing, then this same new physics is also likely to have important effects in B penguin decays. Table 1 contains a summary of the effects of various models of new physics on both B^0-$\overline{B^0}$ mixing and penguin decays [3].

To sum up, there are many signals of new physics in CP asymmetries: $Asym(B_s^0 \to \psi\phi) \neq 0$; $\alpha + \beta + \gamma \neq \pi$; $Asym(B_d^0 \to \psi K_s) \neq Asym(B_d^0 \to \phi K_s)$ $[\beta]$; $Asym(B_s^0 \to D_s^\pm K^\mp) \neq Asym(B^\pm \to DK^\pm)$ $[\gamma]$; $\alpha + \beta + \gamma = \pi$ but α, β and γ are inconsistent with the SM (e.g. $\sin 2\beta < 0$); $\alpha + \beta + \gamma = \pi$, α, β and

TABLE 1. Contributions of models of new physics to B^0-$\overline{B^0}$ mixing and B penguin decays.

Model	Contribution to B^0-$\overline{B^0}$ Mixing?	New Phases?	Contributions to Penguins?	Modes
4 generations	Yes	Yes	Yes	EWP's
Z-FCNC's	Yes	Yes	Yes	$b \to q\ell^+\ell^-$, $B^0 \to \ell^+\ell^-$, EWP's
MHDM w/ NFC	Yes	No	Yes	$b \to s\ell^+\ell^-$, $B^0 \to \ell^+\ell^-$,
MHDM w/o NFC	Yes	Yes	No	—
Left-Right Symm.	No	—	No	—
MSSM	Yes	No	No	—
Non-min. SUSY	Yes	Yes	Yes	?

γ are consistent with the SM, but are inconsistent with measurements of the sides of the UT; etc.

The main way in which new physics can enter is via new contributions to B^0-$\overline{B^0}$ mixing. (There is an exception: for pure penguin decays, such as $B_d^0 \to \phi K_S$, there can be new decay amplitudes.) There are many models of new physics which can yield such new contributions. In this talk I have considered four generations, Z-mediated FCNC's, multi-Higgs-doublet models with and without natural flavour conservation, minimal and non-minimal supersymmetry.

Assuming that some signal for new physics is seen in the measurements of CP asymmetries, one can partially distinguish among the various models by looking at the rates for rare penguin decays. CP asymmetries and penguin decays thus give complementary information regarding the identity of the new physics.

Acknowledgements

I would like to thank Dan Kaplan for the invitation to this excellent conference. Thanks also to A. Ali, M. Gronau and A. Soni for pleasant collaborations on some of the subjects discussed here, and to S. Sharpe for correspondence regarding the appropriate value of B_K. This work was financially supported by NSERC of Canada and FCAR du Québec.

REFERENCES

1. The figure has been obtained following the procedure described in A. Ali and D. London, *Zeit. Phys.* **C65**, 431 (1995), *Nucl. Phys. (Proc. Suppl.)* **54A**, 297 (1997), except that I have taken $B_K = 0.9 \pm 0.15$ (thanks to S. Sharpe for discussions).
2. For reviews, see, for example, Y. Nir and H.R. Quinn in *B Decays*, edited by S. Stone (World Scientific, Singapore, 1994), p. 520; I. Dunietz, *ibid.*, p. 550; M. Gronau, *Proceedings of Neutrino 94, XVI International Conference on Neutrino Physics and Astrophysics*, Eilat, Israel, May 29 – June 3, 1994, eds. A. Dar, G. Eilam and M. Gronau, *Nucl. Phys. (Proc. Suppl.)* **B38**, 136 (1995).
3. For a more complete discussion, see M. Gronau and D. London, *Phys. Rev.* **D55**, 2845 (1997), and references therein.
4. C.O. Dib, D. London and Y. Nir, *Int. J. Mod. Phys.* **A6**, 1253 (1991).
5. Y. Nir and D. Silverman, *Nucl. Phys.* **B345**, 301 (1990).
6. D. London and R. Peccei, *Phys. Lett.* **B223**, 257 (1989).
7. B. Grinstein, *Phys. Lett.* **B229**, 280 (1989); M. Gronau, *Phys. Rev. Lett.* **63**, 1451 (1989), *Phys. Lett.* **B300**, 163 (1993).
8. Y. Grossman and M.P. Worah, *Phys. Lett.* **B395**, 241 (1997).
9. D. London and A. Soni, hep-ph/9704277, to appear in *Phys. Lett.* **B**.

Probing New Physics in the B System

JoAnne L. Hewett

Stanford Linear Accelerator Center, Stanford CA 94309, USA

Abstract. We determine the ability of future experiments to test the Standard Model via precision measurement of the rare decays $B \to X_s \gamma$ and $B \to X_s l^+ l^-$. A global fit to the Wilson coefficients which describe the flavor changing $b \to s$ transitions and which contribute to these decays is performed from Monte Carlo-generated data. This fit is then compared to supersymmetric predictions for the coefficients for several different patterns of the superpartner spectrum.

I OVERVIEW

The first observation of penguin-mediated processes, in both the exclusive $B \to K^* \gamma$ and inclusive $B \to X_s \gamma$ channels, by CLEO [1] has placed the study of rare B decays on a new footing. These flavor-changing neutral-current (FCNC) transitions provide an essential opportunity to test the Standard Model (SM) and offer a complementary strategy in the search for new physics by probing the indirect effects of new particles and interactions in higher-order processes. With the expected high luminosity of the B Factories presently under construction (and the associated advanced detector technology), radiative B decays will no longer be rare events, and the exploration of FCNC transitions can continue by probing decay modes with even smaller branching fractions. The cleanest rare decay which occurs at an accessible rate is $B \to X_s \ell^+ \ell^-$. In fact, experiments at e^+e^- and hadron colliders are already closing in on the observation [2] of the exclusive modes $B \to K^{(*)} \ell^+ \ell^-$ with $\ell = e$ and μ. Once this decay is observed, the utilization of the kinematic distributions of the $\ell^+\ell^-$ pair, such as the lepton-pair invariant-mass distribution and forward-backward asymmetry [3] and the tau-polarization asymmetry [4] in $B \to X_s \tau^+ \tau^-$, together with $B(B \to X_s \gamma)$, will provide a stringent test of the SM.

Several classes of new physics rapidly decouple, thus making it a challenge to search for their effects through indirect methods. However, a promising approach is to measure observables where new physics and the SM arise at the same order in perturbation theory. In this case the additional contributions do not suffer an extra $\alpha/4\pi$ reduction compared to the SM amplitudes. The relative ratio between the lowest-order SM amplitudes and those from beyond the SM could then be $\mathcal{O}(1)$ if the scale of the new interactions is not too large. $b \to s$ transitions provide just such an opportunity for discovering new indirect effects as they only occur at loop level in the SM, and they have relatively large rates for loop processes due to the massive internal top quark and the Cabbibo-Kobayashi-Maskawa (CKM) structure of the contributing penguin and box diagrams. This is in contrast to precision electroweak measurements on the Z-pole, where the virtual corrections arising from new physics must compete with large tree-level SM diagrams.

II FORMALISM FOR $B \to S$ TRANSITIONS

We begin by describing the effective field theory for $b \to s$ transitions. Incorporating the QCD corrections, these are governed by the Hamiltonian

$$\mathcal{H}_{eff} = \frac{-4G_F}{\sqrt{2}} V_{tb} V_{ts}^* \sum_{i=1}^{10} C_i(\mu) \mathcal{O}_i(\mu), \qquad (1)$$

where the $\mathcal{O}_i$ are a complete set of renormalized operators of dimension six or less which mediate these transitions and are catalogued in, e.g., Ref. [5]. The C_i represent the corresponding Wilson coefficients which are evaluated perturbatively at the electroweak scale, where the matching conditions are imposed, and then evolved down to the renormalization scale $\mu \approx m_b$. The expressions for $C_i(M_W)$ in the SM are given by the Inami-Lim functions [6].

A $B \to X_s \ell^+ \ell^-$ in the Standard Model

For $B \to X_s \ell^+ \ell^-$ this formalism leads to the physical decay amplitude (neglecting m_s and m_ℓ)

$$\mathcal{M} = \frac{\sqrt{2} G_F \alpha}{\pi} V_{tb} V_{ts}^* \left[C_9^{eff} \bar{s}_L \gamma_\mu b_L \bar{\ell} \gamma^\mu \ell + C_{10} \bar{s}_L \gamma_\mu b_L \bar{\ell} \gamma^\mu \gamma_5 \ell \right. \\ \left. - 2 C_7^{eff} m_b \bar{s}_L i \sigma_{\mu\nu} \frac{q^\nu}{q^2} b_R \bar{\ell} \gamma^\mu \ell \right], \qquad (2)$$

where q^2 represents the momentum transferred to the lepton pair. The next-to-leading order (NLO) analysis for this decay has been performed in Buras et al. [5], where it is stressed that a scheme-independent result can only be obtained by including the leading and next-to-leading logarithmic corrections to $C_9(\mu)$ while retaining only the leading logarithms in the remaining Wilson coefficients. The leading residual scale dependence in $C_9(\mu)$ is cancelled by that contained in the matrix element of $\mathcal{O}_9$, yielding an effective value C_9^{eff}. In addition, the effective value for $C_7^{eff}(\mu)$ refers to the leading-order

TABLE 1. Values of the Wilson coefficients for several choices of the renormalization scale. Here, we take $m_b = 4.87$ GeV, $m_t = 175$ GeV, and $\alpha_s(M_Z) = 0.118$.

Coefficient	$\mu = m_b/2$	$\mu = m_b$	$\mu = 2m_b$
C_7^{eff}	-0.371	-0.312	-0.278
C_9	4.52	4.21	3.81
C_{10}	-4.55	-4.55	-4.55

scheme-independent result, and we note that the operator $\mathcal{O}_{10}$ does not renormalize. The numerical estimates (in the naive dimensional regularization scheme) for these coefficients are displayed in Table 1. The reduced scale dependence of the NLO- versus the LO-corrected coefficients is reflected in the deviations $\Delta C_9(\mu) \lesssim \pm 10\%$ and $\Delta C_7^{eff}(\mu) \approx \pm 20\%$ as μ is varied. We find that the coefficients are much less sensitive to the remaining input parameters, with $\Delta C_9(\mu), \Delta C_7^{eff}(\mu) \lesssim 3\%$, varying $\alpha_s(M_Z) = 0.118 \pm 0.003$ [7], and $m_t^{phys} = 175 \pm 6$ GeV [8]. The resulting inclusive branching fractions (which are computed by scaling the width for $B \to X_s \ell^+ \ell^-$ to that for B semi-leptonic decay) are found to be $(6.25^{+1.04}_{-0.93}) \times 10^{-6}$, $(5.73^{+0.75}_{-0.78}) \times 10^{-6}$, and $(3.24^{+0.44}_{-0.54}) \times 10^{-7}$ for $\ell = e, \mu$, and τ, respectively, taking into account the above input parameter ranges, as well as $B_{sl} \equiv B(B \to X\ell\nu) = (10.23 \pm 0.39)\%$ [9], and $m_c/m_b = 0.29 \pm 0.02$.

B $B \to X_s \gamma$ in the Standard Model

The basis for the decay $B \to X_s \gamma$ contains the first eight operators in the effective Hamiltonian of Eq. (1). The next-to-leading-order logarithmic QCD corrections have been recently completed, leading to a much reduced renormalization scale dependence in the branching fraction! The higher-order calculation involves several steps, requiring corrections to both C_7 and the matrix element of $\mathcal{O}_7$. For the matrix element, this includes the QCD bremsstrahlung corrections [10] $b \to s\gamma + g$, and the NLO virtual corrections [11]. Summing these contributions to the matrix elements and expanding them around $\mu = m_b$, one arrives at the decay amplitude

$$\mathcal{M}(b \to s\gamma) = -\frac{4G_F V_{tb}V_{ts}^*}{\sqrt{2}} D \langle s\gamma | \mathcal{O}_7(m_b) | b \rangle_{tree}, \quad (3)$$

with

$$D = C_7(\mu) + \frac{\alpha_s(m_b)}{4\pi} \left(C_i^{(0)eff}(\mu) \gamma_{i7}^{(0)} \log \frac{m_b}{\mu} + C_i^{(0)eff} r_i \right). \quad (4)$$

Here, the quantities $\gamma_{i7}^{(0)}$ are the entries of the effective leading-order anomalous dimension matrix, and the r_i are computed in Greub et al. [11], for $i = 2, 7, 8$. The first term in Eq. 4, $C_7(\mu)$, must be computed at NLO precision, while it is consistent to use the leading-order values of the other coefficients. For C_7 the NLO result entails the computation of the $\mathcal{O}(\alpha_s)$ terms in the matching conditions [12], and the renormalization group evolution of $C_7(\mu)$ must be computed using the $\mathcal{O}(\alpha_s^2)$ anomalous dimension matrix [13]. The numerical value of the branching fraction is then found to be (again, scaling to semileptonic decay)

$$B(B \to X_s \gamma) = (3.25 \pm 0.30 \pm 0.40) \times 10^{-4}, \qquad (5)$$

where the first error corresponds to the combined uncertainty associated with the value of m_t and μ, and the second error represents the uncertainty from $\alpha_s(M_Z), B_{sl}$, and m_c/m_b. This is well within the range observed by CLEO [1] which is $B = (2.32 \pm 0.57 \pm 0.35) \times 10^{-4}$ with the 95% C.L. bounds of $1 \times 10^{-4} < B(B \to X_s \gamma) < 4.2 \times 10^{-4}$. We note that ALEPH has recently reported the preliminary observation of this inclusive decay, at a compatible rate [14].

III MODEL INDEPENDENT TESTS FOR NEW PHYSICS IN $B \to S$ TRANSITIONS

Measurements of $B(B \to X_s \gamma)$ constrain the magnitude, but not the sign, of $C_7(\mu)$. The coefficients at the matching scale ($\mu = M_W$) can be written in the form $C_i(M_W) = C_i^{SM}(M_W) + C_i^{new}(M_W)$, where $C_i^{new}(M_W)$ represents the contributions from new interactions. Due to operator mixing, $B \to X_s \gamma$ then limits the possible values for $C_i^{new}(M_W)$ for $i = 7, 8$. These bounds are summarized in Fig. 1. Here, the solid bands correspond to the constraints obtained from the current CLEO measurement, taking into account the variation of the renormalization scale $m_b/2 \le \mu \le 2m_b$, as well as the allowed ranges of the other input parameters. The dashed bands represent the constraints when the scale is fixed to $\mu = m_b$. We note that large values of $C_8^{new}(M_W)$ (which would yield an anomalous rate for $b \to sg$) are allowed even in the region where $C_7^{new}(M_W) \simeq 0$.

Measurement of the kinematic distributions associated with the final-state lepton pair in $B \to X_s \ell^+ \ell^-$ as well as the rate for $B \to X_s \gamma$ allows the determination of the sign and magnitude of all the Wilson coefficients for the contributing operators in a model-independent fashion. We have performed a Monte Carlo analysis in order to ascertain how much quantitative information will be obtainable at future B Factories and follow the procedure outlined in Ref. [15]. For the process $B \to X_s \ell^+ \ell^-$, we consider the lepton-pair invariant-mass distribution and forward-backward asymmetry for $\ell = e, \mu, \tau$, and the

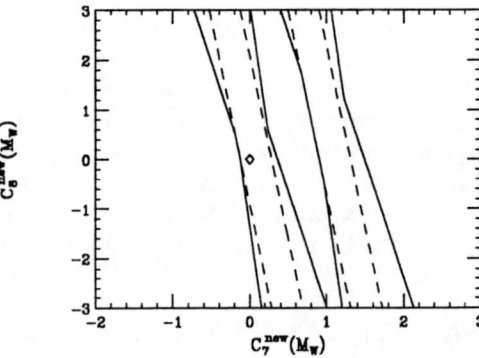

FIGURE 1. Bounds on the contributions from new physics to $C_{7,8}$. The region allowed by the CLEO data corresponds to the area inside the solid diagonal bands. The dashed bands represent the constraints when the renormalization scale is set to $\mu = m_b$. The diamond at the position (0,0) represents the standard model.

tau polarization asymmetry for $B \to X_s \tau^+ \tau^-$. A three-dimensional χ^2 fit to the coefficients $C_{7,9,10}(\mu)$ is performed for three values of integrated luminosity, 3×10^7, 10^8, and 5×10^8 $B\bar{B}$ pairs, corresponding to the expected e^+e^- B Factory luminosities of one year at design, one year at an upgraded accelerator, and the total accumulated luminosity at the end of these programs. The 95%-C.L. allowed regions (including statistical errors only for $B \to X_s \ell^+ \ell^-$ and a flat 10% error on $B \to X_s \gamma$) as projected onto the $C_9(\mu) - C_{10}(\mu)$ and $C_7(\mu) - C_{10}(\mu)$ planes are depicted in Figs. 2(a-b), where the diamond represents the central value for the expectations in the SM given in Table 1. We see that the determinations are relatively poor for 3×10^7 $B\bar{B}$ pairs and that higher statistics are required in order to focus on regions centered around the SM.

IV SUPERSYMMETRIC EFFECTS IN $B \to S$ TRANSITIONS

These model-independent bounds can be compared with model-dependent predictions for the Wilson coefficients in order to ascertain at what level specific new interactions can be probed. For the remainder of this talk, we will concentrate on Supersymmetric extensions to the SM. Other potential sources of new physics that can be readily tested in rare B decays are discussed in Ref. [16].

Supersymmetry (SUSY) contains many potential sources for flavor violation. For example, the flavor mixing angles among the squarks are *a priori* separate from the CKM angles of the SM quarks. We adopt the viewpoint here

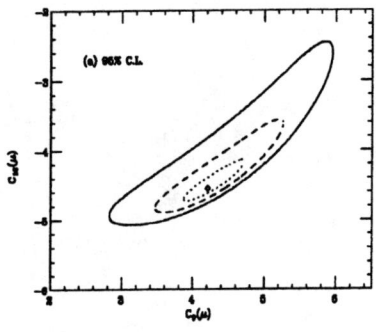 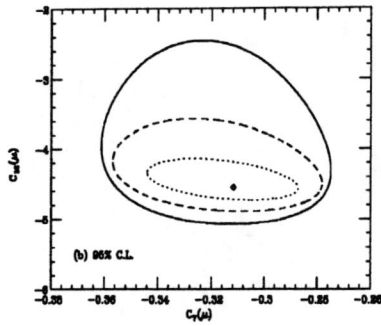

FIGURE 2. The 95% C.L. projections in the (a) $C_9 - C_{10}$ and (b) $C_7 - C_{10}$ planes, where the allowed regions lie inside of the contours. The solid, dashed, and dotted contours correspond to 3×10^7, 10^8, and 5×10^8 $B\bar{B}$ pairs. The central value of the SM prediction is labeled by the diamond.

that flavor-blind (diagonal) soft terms at the high scale are the phenomenological source for the soft scalar masses, and that the CKM angles are the only relevant flavor-violating sources. The spectroscopy of the supersymmetric states is quite model dependent and we analyze two possibilities. The first is the familiar minimal supergravity model; in this instance all the supersymmetric states follow from a common scalar mass and a common gaugino mass at the high scale. The second case is where the condition of common scalar masses is relaxed and they are allowed to take on uncorrelated values at the low scale while still preserving gauge invariance.

We analyze the supersymmetric contributions to the Wilson coefficients [15,17] in terms of the quantities

$$R_i \equiv \frac{C_i^{susy}(M_W)}{C_i^{SM}(M_W)} - 1 \equiv \frac{C_i^{new}(M_W)}{C_i^{SM}(M_W)}, \qquad (6)$$

where $C_i^{susy}(M_W)$ includes the full SM plus superpartner contributions. R_i is meant to indicate the fractional deviation from the SM value. We will search over the full parameter space of the minimal supergravity model, calculate the R_i for each generated point in the supersymmetric parameter space, and then compare with the expected ability of B Factories to measure the R_i as determined by our global fit to the Wilson coefficients. We generate [18] these supergravity models by applying common soft scalar and common gaugino masses at the boundary scale. The tri-scalar A terms are also input at the high scale and are universal. The radiative electroweak symmetry breaking conditions yield the B and μ^2 terms as output, with a sign(μ) ambiguity left over. (Here μ refers to the Higgsino mixing parameter.) We also choose $\tan\beta$

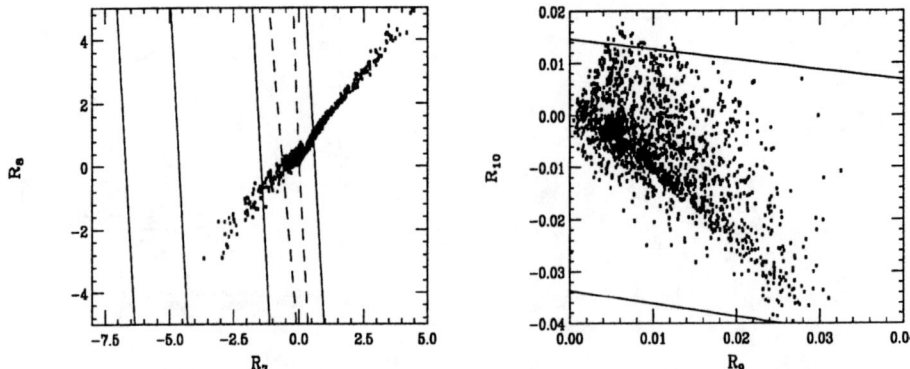

FIGURE 3. (a) Parameter-space scatter plot of R_7 vs. R_8 in the minimal supergravity model. The allowed region from CLEO data lies inside the solid diagonal bands. The dashed band represents the potential future 10% measurement of $B \to X_s\gamma$ as described in the text. (b) Parameter space scatter plot of R_9 vs. R_{10}. The global fit to the coefficients obtained with $5 \times 10^8 B\bar{B}$ pairs corresponds to the region inside the diagonal bands.

and restrict it to a range which will yield perturbative Yukawa couplings up to the GUT scale. We have generated thousands of solutions according to the above procedure. The ranges of our input parameters are $0 < m_0 < 500\,\text{GeV}$, $50 < m_{1/2} < 250\,\text{GeV}$, $-3 < A_0/m_0 < 3$, $2 < \tan\beta < 50$, and we have taken $m_t = 175\,\text{GeV}$. Each supersymmetric solution is kept only if it is not in violation of present constraints from SLC/LEP and Tevatron direct sparticle production limits, and it is out of reach of LEP II. For each of these remaining solutions we now calculate R_{7-10}. Our results are shown in the scatter plots of Fig. 3 in the (a) $R_7 - R_8$ and (b) $R_9 - R_{10}$ planes. The diagonal bands represent the bounds on the Wilson coefficients as previously determined from our global fit. We see from Fig. 3(a) that the current CLEO data on $B \to X_s\gamma$ already place signigicant restrictions on the supersymmetric parameter space, whereas the minimal-supergravity contributions to $R_{9,10}$ are predicted to be essentially unobservable.

A second, more phenomenological approach is now adopted. The maximal effects for the parameters R_i can be estimated for a superparticle spectrum, independent of the high-scale assumptions. However, we still maintain the assumption that CKM angles alone constitute the sole source of flavor violations in the full supersymmetric lagrangian. We will focus on the region $\tan\beta \lesssim 30$. The most important features which result in large contributions are a light $\tilde{t}_1$ state present in the SUSY spectrum and at least one light chargino state. For the dipole moment operators a light Higgsino state is also important. A pure higgsino and/or pure gaugino state have less of an effect than two mixed states when searching for maximal effects in R_9 and R_{10} and we have found that $M_2 \simeq 2\mu$ is optimal. Fig. 4 displays the maximum contribution to $R_{9,10}$

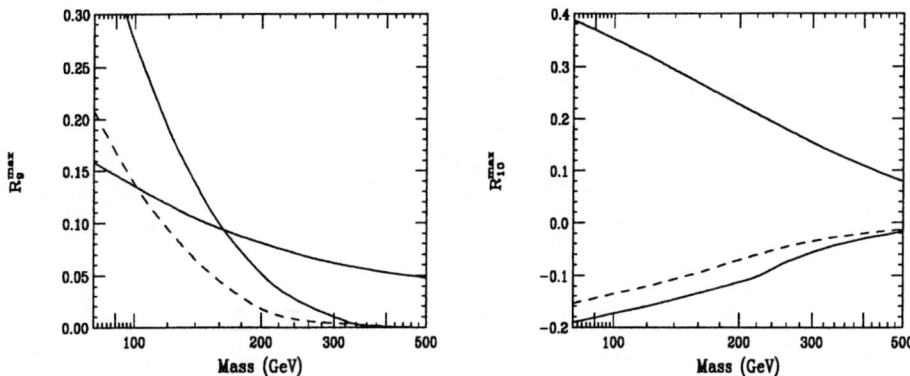

FIGURE 4. The maximum value of (a) R_9 and (b) R_{10} achievable for general supersymmetric models. The top solid line comes from the $t - H^{\pm}$ contribution and is displayed versus the $H^{\pm}$ mass. The bottom solid line is from the $\tilde{t}_i - \chi_j^{\pm}$ contribution with $\tan\beta = 1$ and is shown versus the $\chi_i^{\pm}$ mass. The dashed line is the $\tilde{t}_i - \chi_j^{\pm}$ contribution with $\tan\beta = 2$. The other mass parameters which are not plotted are chosen to be just above the reach of LEP II and the Tevatron.

versus an applicable SUSY mass scale. The other sparticle masses which are not shown ($\tilde{t}_i$, $\tilde{l}_L$, etc.) are chosen to be just above the reach of LEP II or the Tevatron, whichever yields the best bound. We see that the maximum size of $R_{9,10}$ is somewhat larger than what was allowed in the minimal supergravity model, due to the lifted restriction on mass correlations.

Given the sensitivity of the observables it is instructive to narrow our focus to the coefficient of the magnetic dipole operator. The possibility exists that one eigenvalue of the stop-squark mass matrix might be much lighter than the other squarks, and we present results for $C_7(M_W)$ in the limit of one light squark, namely the $\tilde{t}_1$, and light charginos. We allow the $\tilde{t}_1$ to have arbitrary components of $\tilde{t}_L$ and $\tilde{t}_R$ since cross terms can become very important. This is especially noteworthy in the high $\tan\beta$ limit. We note that the total supersymmetric contribution to $C_7(M_W)$ will depend on several combinations of mixing angles in both the stop and chargino mixing matrices and cancellations can occur for different signs of μ [19]. The first case we examine is that where the lightest chargino is a pure Higgsino and the lightest stop is purely right-handed: $\chi_1^{\pm} \sim \tilde{H}^{\pm}$, $\tilde{t}_1 \sim \tilde{t}_R$. The resulting contribution to R_7 is shown as a function of the $\tilde{t}_R$ mass in Fig. 5 (dashed line) for the case of $m_{\chi^{\pm}} \gtrsim M_W$. Note that the SUSY contribution to $C_7(M_W)$ in this limit always adds constructively to that of the SM. Next we examine the limit where the light chargino is a pure Wino, this contribution is shown in Fig. 5 (dotted line). The effects of a light pure Wino are small since (i) it couples with gauge strength rather than the top Yukawa, and (ii) supersymmetric models do not generally yield a light $\tilde{t}_L$ necessary to couple with the Wino. Our third lim-

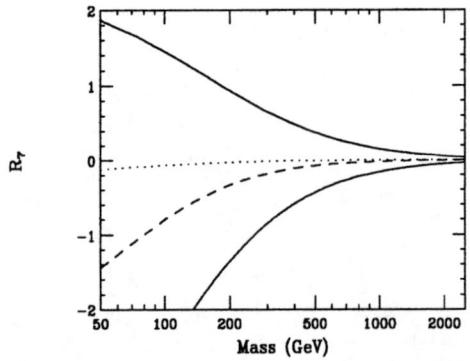

FIGURE 5. Contributions to R_7 in the different limits described in the text. The top solid line is the charged $H^\pm/t$ contribution versus $m_{H^\pm}$. The bottom solid line is the $\tilde{\chi}_1^\pm/\tilde{t}_1$ contribution versus $m_{\tilde{\chi}^\pm}$ where both the chargino and stop are maximally-mixed states with $\mu < 0$. The dashed line is the $\tilde{H}^\pm/\tilde{t}_R$ contribution, and the dotted line represents the $\tilde{W}^\pm/\tilde{t}_1$ contribution. These two lines are both shown as a function of $\tilde{\chi}_1^\pm$ mass. All curves are for $\tan\beta = 2$ and $m_t = 175\,\text{GeV}$.

iting case is that of a highly mixed $\tilde{t}_1$ state. We find that in this case large $\tan\beta$ solutions ($\tan\beta \gtrsim 40$) can yield greater than $\mathcal{O}(1)$ contributions to R_7 even for SUSY scales of 1 TeV! Low values of $\tan\beta$ can also exhibit significant enhancements; this is demonstrated for $\tan\beta = 2$ in Fig. 5 (solid line). We remark that large contributions are possible in this case in both negative and positive directions of R_7 depending on the sign of μ. We note that this is a region of SUSY parameter space which is highly motivated by $SO(10)$ grand unified theories.

Lastly, we compare the reach of rare B decays in probing SUSY parameter space with that of high energy colliders. We examine a set of five points in the minimal supergravity (SUGRA) parameter space that were chosen at Snowmass 1996 [20] for the study of supersymmetry at the NLC. Point #3 is the so-called "common" point used for a comparison of SUSY studies at the NLC, LHC, and upgraded Tevatron. Once these points are chosen the sparticle mass spectrum is obtained, as usual, via the SUGRA relations and their contributions to $B \to X_s\gamma$ can be readily computed. The results are displayed in the $R_7 - R_8$ plane in Fig. 6 (labeled $1-5$ for each SUGRA point), along with the bounds previously obtained from our fits to present and anticipated future data. We see that four of the points should be discernable from the SM in future measurements, and that one of the points is already excluded by the CLEO data!

We thus conclude that rare B decays are indeed complementary to high energy colliders in searching for supersymmetry!

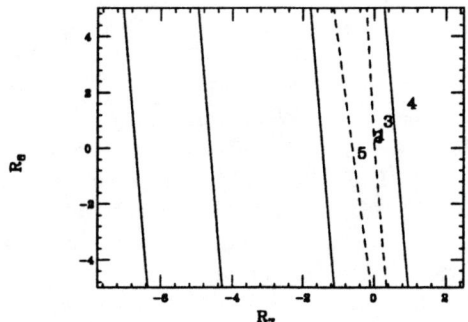

FIGURE 6. Values in the $R_7 - R_8$ plane for the five Snowmass NLC SUGRA points. The solid and dashed bands are as described in Figure 3.

REFERENCES

1. CLEO Collaboration, M. S. Alam *et al.*, *Phys. Rev. Lett.* **74**, 2885 (1995); CLEO Collaboration, R. Ammar *et al.*, *Phys. Rev. Lett.* **71**, 674 (1993).
2. D0 Collaboration, S. Abachi *et al.*, D0note-2023 (1996); CDF Collaboration, F. Abe *et al.*, Fermilab-PUB-96/1040-E; CLEO Collaboration, E. Thorndike *et al.*, in *Proceedings of the 27th Int. Conf. on HEP*, Glasgow, Scotland, 1994, edited by P. J. Bussey and I. G. Knowles (IOP, London, 1995).
3. A. Ali, T. Mannel, and T. Morozumi, *Phys. Lett.* **B273**, 505 (1991); A. Ali, G. F. Giudice, and T. Mannel, *Z. Phys.* **C67**, 417 (1995).
4. J. L. Hewett, *Phys. Rev.* **D53**, 4964 (1996).
5. A. J. Buras and M. Münz, *Phys. Rev.* **D52**, 186 (1995).
6. T. Inami and C. S. Lim, Prog. Theor. Phys. **65**, 297 (1981).
7. R. M. Barnett *et al.*, (Particle Data Group), *Phys. Rev.* **D54**, 1 (1996).
8. P. Giromini, talk presented at *XVIII International Symposium on Lepton Photon Interactions*, Hamburg, Germany, 1997.
9. J. Richman, Proceedings of the *28th International Conference on High Energy Physics*, Warsaw, Poland, 1996, ed. Z. Ajduk and A. K. Wroblewski, hep-ex/9701014.
10. A. Ali and C. Greub, *Z. Phys.* **C49**, 431 (1991); *Phys. Lett.* **B259**, 182 (1991); *Phys. Lett.* **B361**, 146 (1995); N. Pott, *Phys. Rev.* **D54**, 938 (1996).
11. C. Greub, T. Hurth, and D. Wyler, *Phys. Lett.* **B380**, 385 (1996); *Phys. Rev.* **D54**, 3350 (1996).
12. K. Adel and Y.-P. Yao, *Phys. Rev.* **D49**, 4945 (1994).
13. K. G. Chetyrkin, M. Misiak, and M. Münz, *Phys. Lett.* **B400**, 206 (1997).
14. T. Skwarnicki, talk presented at *7th International Symposium on Heavy Flavor Physics*, Santa Barbara, CA, 1997.
15. J. L. Hewett and J. D. Wells, *Phys. Rev.* **D55**, 5549 (1997).

16. J. L. Hewett, in Proceedings of the *1993 SLAC Summer Institute – Spin Structure in High Energy Processes*, Stanford, CA, 1993, ed. L. DePorcel, hep-ph/9406302.
17. S. Bertolini *et al.*, *Nucl. Phys.* **B353**, 591 (1991); F. Borzumati, *Z. Phys.* **C63**, 291 (1994); F. Gabbiani *et al.*, *Nucl. Phys.* **B477**, 321 (1996); V. Barger *et al.*, *Phys. Rev.* **D51**, 2438 (1995); D. Choudhury *et al.*, *Phys. Lett.* **B342**, 180 (1995); J. Lopez *et al.*, *Phys. Rev.* **D51**, 147 (1995); R. Barbieri, G. Giudice, *Phys. Lett.* **B309**, 86 (1993); P. Cho, M. Misiak, and D. Wyler, *Phys. Rev.* **D54**, 3329 (1996); H. Baer and M. Brhlik, *Phys. Rev.* **D55**, 3201 (1997); T. Goto *et al.*, *Phys. Rev.* **D55**, 4273 (1997).
18. For description of the procedure we follow, see G. L. Kane, C. Kolda, L. Roszkowski, J. Wells, *Phys. Rev.* **D49**, 6173 (1994).
19. R. Garisto, and J. Ng, *Phys. Lett.* **B315**, 372 (1993).
20. M. N. Danielson *et al.*, Proceedings of *New Directions for High-Energy Physics*, Snowmass, CO, 1996, ed. D.G. Cassel *et al.*.

New Physics and Enhanced Gluonic Penguin

George W.S. Hou

Department of Physics, National Taiwan University, Taipei, Taiwan 10764, R.O.C.

Abstract. We discuss the historical development of the gluonic B-penguin, its sensitivity to H^+ effects, and $b \to sg \sim 10\text{--}15\%$ as a possible solution to the $\mathcal{B}_{\text{s.l.}}$ and n_C problems. The latter and the connection of the gluonic penguin to inclusive $B \to \eta' + X_s$ production through the gluon anomaly, with the intriguing prospect of 10% inclusive CP asymmetries, bring us to topics of current interest.

I SM: HISTORICAL BACKDROP

We are here to celebrate the 20th anniversary of the Υ discovery. Because of this historic setting, I will dwell a little more on the historical aspects (from a personal perspective) of gluonic penguins, before I turn to the current.

Shortly after 1977, Bander, Silverman and Soni (BSS) [1] suggested the mechanism of (direct) CP violation in the *decay* of b quarks. The B-penguin was born, as illustrated in Fig. 1 (a). CP violation is possible because all 3 generations run in the loop, while on-shell $u\bar{u}$, $c\bar{c} \to g^* \to q\bar{q}$ rescattering provides the second i, an absorptive part. Tony Sanda told me that this work was an inspiration for his mixing dependent CP violation ideas.

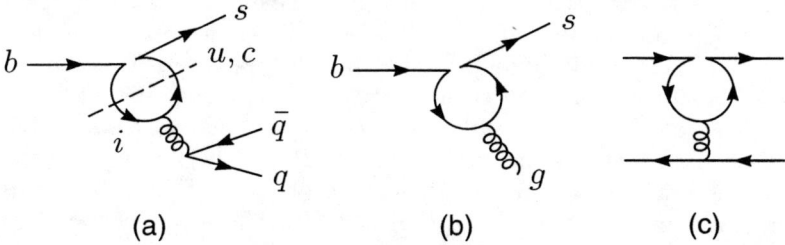

FIGURE 1. (a) Timelike, (b) lightlike and (c) spacelike gluonic penguin. The cut in (a) corresponds to on-shell $u\bar{u}$ and $c\bar{c}$ pairs, while in general $i = u,\ c,\ t$ in the loop.

Perhaps influenced by BSS, Guberina, Peccei and Rückl pointed out [2] in late 1979, using operator language, that B-penguin operators "can alter the 'natural' Cabibbo pattern of the Kobayashi–Maskawa model." Namely, $B \to K\pi > B \to \pi\pi$ is possible, since the penguin is $\propto |V_{cb}| \simeq |V_{ts}|$, while tree level $B \to \pi\pi$ could be suppressed if V_{ub} is very small. This has just turned into fact this year with CLEO's observation of a few 2-body rare decays [3].

Following the study of $b \to s\gamma$ decay in the early 80's, Eilam [4] added $b \to sg$ of Fig. 1(b) to the "parton" estimate of the inclusive penguin rate. In 1987, the so-called "large" QCD correction to $b \to s\gamma$ was discovered, leading to the "operator" approach industry. But $b \to sg$ changed only from 0.1% to 0.2% with QCD corrections, much less dramatic than the $b \to s\gamma$ case.

The inclusive gluonic penguin was clarified [5] in 1987 by noting the q^2 of g^*, i.e. $b \to sg^* \equiv b \to sq\bar{q}$, sg, and $b\bar{q} \to s\bar{q}$ (timelike, lightlike and spacelike), where the latter, given in Fig. 1(c), is the familiar looking "penguin" from kaon physics. It was found that [5] $b \to sq\bar{q} \sim 1\% \sim b \to u$ while $b\bar{q} \to s\bar{q} < b \to sg$. However, counter to one's intuition [6], $b \to sgg \ll b \to sq\bar{q}$. Interestingly, the 3-body $b \to sq\bar{q}$ at $\mathcal{O}(\alpha_s^2)$ dominates over the 2-body $b \to sg$ at $\mathcal{O}(\alpha_s)$, which comes about because of a subtlety of GIM cancellation.

There are two conserved effective bsg couplings; ignoring V_{ub} they are

$$G_F g_s v_t \bar{s} t^a \{\Delta F_1 (q^2 \gamma_\mu - q_\mu \not{q})L - F_2 i\sigma_{\mu\nu} q^\nu m_b R\} b, \tag{1}$$

where $\Delta F_1 \equiv F_1^t - F_1^c \simeq 0.25 - (-2/3 \log(m_c^2/M_b^2) - 2/3 \log(m_b^2/M_W^2)) \simeq -1.3 - 2.75$, and $F_2 \cong F_2^t \simeq 0.2$. F_1 contains large logarithms while F_2 does not, but suffers from power GIM suppression. However, F_1 cannot contribute to $b \to sg$ because of the q^2 factor. Thus, the subtle higher order dominance comes about because of having a logarithmic and a power GIM suppressed effective coupling, and only the latter leads to $b \to sg$.

II H^+ EFFECTS: F_2^γ AND F_2^g

The above subtlty leads to surprising H^+ effects: F_2^γ is very sensitive to low m_{H^+}. For the Higgs sector of minimal SUSY, the effect is always constructive and does not vanish with $\tan\beta$ (ratio of v.e.v.'s of the two Higgs doublets) [7,8], which holds similarly for F_2^g [8]. Though $b \to sg$ could not be greatly enhanced in this model, both strong enhancement/suppression of $b \to s\gamma$ are possible for a second model, and $b \to sg$ could become *very* enhanced [8].

At that time the experimental limit was $\mathcal{B}(b \to s\gamma) < 6 \times 10^{-3}$, while $b \to sg$ was practically without bound (except $b \to c$ should be dominant). The curious thing about $b \to sg$ is that it does not lead to any good, tangible signature! By 1992, however, the CLEO limit on $\mathcal{B}(b \to s\gamma)$ improved to 5×10^{-4}, entering the domain of SM predictions. This had a dramatic implication that $m_{H^+} > 250$ GeV or so [9] in SUSY type models. Unfortunately, because bsg and $bs\gamma$ couplings in Higgs models are highly correlated, the limit and eventual observation of $b \to s\gamma$ by CLEO meant that $b \to sg$ could no longer be strongly enhanced in usual charged Higgs models [10].

III $\mathcal{B}_{\text{s.l.}}$–$n_C$ PROBLEM: ENHANCED $b \to sg$

Some indirect indications for enhanced $b \to sg$ appeared, in fact, in the early 90's. As B experiments matured, the semileptonic branching ratio ($\mathcal{B}_{\text{s.l.}}$) steadily declined, from $\simeq 12\%$ in 1986, to 10.7% by early 1990. Theory predicted 12–15%, hence it appeared [11,12] that the SM had trouble with the experimental value, with a 10–15% discrepancy. The relevant QCD scale μ for B decay could be much lower than m_b [11], or one could have *new physics* $\Gamma_{\text{New}} \sim$10–15%, which drives down $\mathcal{B}_{\text{s.l.}}$ via

$$\mathcal{B}_{\text{s.l.}} = \Gamma(B \to \ell\nu + X)/(\Gamma_{\text{tot.}}^{\text{SM}} + \Gamma_{\text{New}}). \qquad (2)$$

The new process must be relatively well hidden, and low in charm content to accommodate the analogously low charm counting rate [13] (the n_C problem). Two modes were suggested [12], both from H^+ effects. The first one, $B \to \tau\nu + X \sim 10\%$, was very quickly ruled out by a superb analysis job of ALEPH, which confirmed SM predictions. The second possibility of $b \to sg \sim$10–15%, which is a charmless final state, was very difficult to rule out.

However, by 1994, the possibility of enhanced $b \to sg$ due to H^+ effects became implausible because of $b \to s\gamma$ limits/measurements. Subsequently, Kagan [14] suggested that TeV scale physics responsible for quark mass (and mixing) generation might lead to enhanced $b \to sg$. For example, gluonic insertions to $\bar{s}_L b_R$ mass terms could result in effective $s_L b_R g$ couplings. To disentangle $b \to sg$ from $b \to s\gamma$, one must employ more *color* in the loop. Note that H^+ is colorless and does not couple to gluons, hence diagrams for $b \to sg$ are only a subset of $b \to s\gamma$. But gluons could more readily couple to gluinos via SUSY $q_i \tilde{q}_j \tilde{g}$ couplings [14,15] (with flavor violation in squark mass matrix), or to techniscalars [14]. In this way $b \to sg$ could in principle be separated from $b \to s\gamma$ and be strongly enhanced.

The problems of low $\mathcal{B}_{\text{s.l.}}$ and n_C have persisted to this day, despite much theoretical and experimental effort. Two recent analyses [16,17] give $\mathcal{B}_{\text{s.l.}} = 0.105 \pm 0.005$ and $n_C = 1.10 \pm 0.06$, both low by about 10–15%. The problem could still be experimental, and in fact the latest results hint at softening of the problems, but two views are offered on enhanced charmless b decays. Kagan and Rathsman [16] think that $\mathcal{B}_{\text{s.l.}}$, n_C and kaon excess in B decays together hint at $b \to sg \sim$10–15%. Using JETSET fragmentation of the s quark, they find the K spectrum to be rather soft, hence $b \to sg$ indeed hides well. On the other hand, Dunietz et al. [17] suggest that half of $b \to sc\bar{c}$ (expected at 20–30% level) has disappeared into light hadrons. The effect has to be nonperturbative to evade the perturbative $b \to sg^* \sim 1\%$ discussed earlier. The proposed mechanism is via a $c\bar{c}g$ hybrid meson, since the $c\bar{c}$ pair is mainly in color octet configuration. The hybrid should [18] be favorably produced in $b \to sc\bar{c}$, should be relatively narrow (long lived) and should have suppressed decays into $D\bar{D} + X$ and usual charmonia. Hence [18], in a sense it is no less exotic than new physics $b \to sg \sim$ 10–15%.

IV INCLUSIVE η', $b \to sg$, GLUON ANOMALY

1997 will be remembered as the year of the strong penguin. Since the Aspen Winter Conference, CLEO has reported the first observations of a host of two-body rare B decays. $K\pi \sim 10^{-5}$ is observed, while $\pi\pi$ is not, confirming the GPR suggestion [2]. The $\omega h^{\pm}$ mode is larger than expected, while $\eta' K \sim 10^{-4}$ is huge, but ηK is not seen! All in all, we see that penguins are large.

What is even more astounding is the observation of [3]

$$\mathcal{B}(B \to \eta' + K + X) = (75 \pm 15 \pm 11) \times 10^{-5} \qquad (2.0 < p_{\eta'} < 2.7 \text{ GeV}) \qquad (3)$$

where $X = 0\text{--}4\pi\ (\leq 1\pi^0)$. While a cut on $p_{\eta'}$ is in part to suppress background, it is astonishing to see so many events in this rather unusual channel. If one extrapolates from Eq. (3), one could easily saturate $b \to sg^* \sim 1\%$.

The most prominent feature is that the η' is *fast*! Since η' is the heaviest and "stickiest" (glue rich) of the lowest lying mesons, it would have been last on the list of possible fast, leading particles in B decay searches. There is one thing unique to η', however, namely its connection to the gluon anomaly. η-η' mixing is said to be related to the axial U(1) problem, and the symmetry is broken by the $G\tilde{G}$ gluon anomaly. Indeed, in the chiral limit of $m_q \to 0$ (assuming $N_F = 3$ of light flavors), one has $\langle 0|\partial_\mu J^0_{\mu 5}|\eta'\rangle = \langle 0|(2N_F \alpha_s/4\pi)\text{tr}(G\tilde{G})|\eta'\rangle$, and it is this large, *topological* glue content of η' that makes it so heavy. So, is the η' production linked to $b \to sg$?

A η'-g-g Coupling and Need for $b \to sg \sim 10\%$

Atwood and Soni [19] (AS) have indeed made such a connection, linking $b \to sg^*$ to inclusive η' via the η'-g-g gluon anomaly. Defining the phenomenological coupling $H(q^2, k^2, m^2_{\eta'})\,\varepsilon_{\mu\nu\alpha\beta}\,q^\mu k^\nu \varepsilon^\alpha(q)\varepsilon^\beta(k)\delta^{ab}$, they extract $H(0,0,m^2_{\eta'}) \simeq 1.8$ GeV^{-1} from $J/\psi \to \eta'\gamma$ decay. Assuming that $H(q^2, 0, m^2_{\eta'}) \approx H(0,0,m^2_{\eta'})$ is constant, they find that the SM $b \to sg^* \to sg\eta'$ could account for Eq. (3). However, they seem to have mistaken $d\Gamma/dq$ for $d\Gamma/dm$, where $m = m_{X_s} \equiv m_{\text{recoil}}$. They hence have a false sensitivity to Fermi motion. Furthermore, the assumption of constant $H(q^2, k^2, m^2_{\eta'})$ is definitely too strong.

The q^2 ranges from 0 to m^2_b, way beyond the QCD scale that determines $m_{\eta'}$. We shall assume that form factor effects do not set in, which in itself is already a big assumption. But even then, for such a broad range of q^2, one does not expect couplings to stay constant. This is especially so since one finds that $d\Gamma/dq$ peaks at large $q > 3$ GeV (Fig. 2(b)).

So, let us try [18] to understand the η'-g-g coupling better. The η' problem in QCD is in itself an active field of research. The anomaly coupling comes from the Wess-Zumino term, without assuming PCAC and soft pions,

$$-i\,a_g\,c_P\,\eta'\,\varepsilon_{\mu\nu\alpha\beta}\,\varepsilon^\mu(q)\varepsilon^\nu(k)q^\alpha k^\beta, \qquad (4)$$

where $a_g(\mu^2) = \sqrt{N_F}\alpha_s(\mu^2)/\pi f_{\eta'} \equiv H(q^2, k^2, m_{\eta'}^2)$ of AS. The explicit α_s factor strongly suggests that one should use running coupling. In the case at hand, since $k^2 \to 0$ and $q^2 > m_{\eta'}^2$ is the dominant scale, we expect $\mu^2 = q^2$. As a cross check, we find that $a_g(m_{\eta'}^2) \simeq 1.9$ GeV$^{-1} \simeq H(0, 0, m_{\eta'}^2)$ of AS, but for larger q^2, this would suppress the SM $b \to sg^*$ effect, since the strong coupling changes by a factor of 2 from $m_{\eta'} \sim 1$ GeV to m_b scale. We find [18] a factor of 1/3 suppression of SM $b \to sg^* \to sg\eta'$, hence $b \to sg \approx 10\%$ is precisely what is called for. It is interesting that both $|F_1^{SM}| \sim 5$ and $|F_2^{New}| \sim 2$ effects are needed. Furthermore, because in the SM the F_1-F_2 interference effect is destructive, the sign of F_2^{New} should be opposite to that in the SM, which is precisely what is found in the SUSY example of $b \to sg \sim 10\%$ [15]!

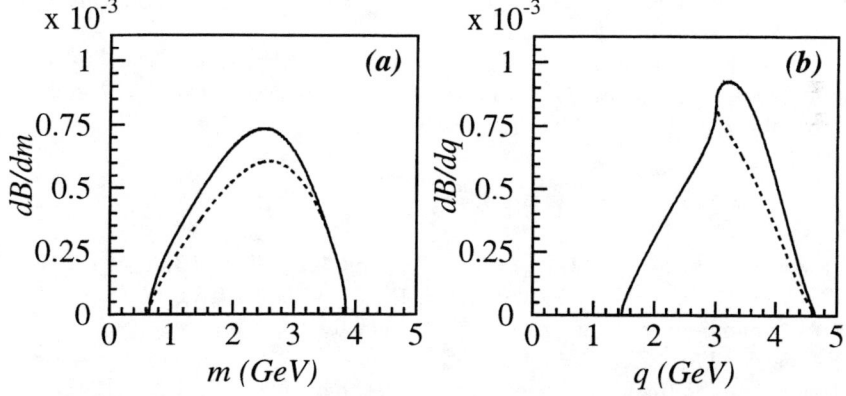

FIGURE 2. (a) $d\mathcal{B}/dm$ and (b) $d\mathcal{B}/dq$ for $b \to \eta'sg$ (solid) and $\bar{b} \to \eta'\bar{s}g$ (dashed). The $c\bar{c}$ threshold is evident in (b), while the CP asymmetry is due to new phase σ in F_2.

B BONUS: Potential for New $a_{CP} \sim 10\%$

For inclusive $b \to s$ transitions, it is known that [20] $a_{CP} \lesssim 1\%$, due to the smallness of $v_u \equiv V_{us}^* V_{ub} \sim \lambda^5$ and unitarity constraints. However, because of the appearance of the b_R field in Eq. (1), the F_2^{New} dipole coupling probes new phases that are independent of CKM phase. At the same time, the F_1^{SM} coupling, which is needed also to acocunt for the rate, provides the necessary rescattering phase, from the $c\bar{c}$ cut in Fig. 1(a). The differential BR's $d\mathcal{B}/dm$ and $d\mathcal{B}/dq$ are shown in Fig. 2, assuming phase difference $\sigma = 90°$ between F_1 and F_2. Note that, thanks to the anomaly η'-g-g coupling, although q^2 is not a physical variable, m^2 directly corresponds to the physical recoil mass against η'. Furthermore, large q^2 (hence fast η') is favored by the anomaly coupling! The upshot is that we can account for the huge branching ratios that are *already observed*, while the asymmetry $a_{CP} \sim 10\%$ in $B \to \eta' + K^{\pm} + X$. In principle, CLEO could probe this asymmetry very soon.

V CONCLUSION

The SM expectation that $b \to sq\bar{q} \sim 1\%$ and $b \to sg \sim 0.2\%$ is quite firm. However, persistent $\mathcal{B}_{\text{s.l.}}$ and n_C problems hint at the possibility of $b \to sg \sim 10\%$ from new physics. The recent observation of spectacularly large semi-inclusive $B \to \eta' + X_s \sim 0.75 \times 10^{-3}$ where $p_{\eta'} > 2$ GeV poses an additional challenge to the SM. It is proposed that large $b \to sg$ leads to large $B \to \eta' + X_s$ through the gluon anomaly. We find that, with running α_s in the g^*-g-η' coupling, both SM $b \to sg^* \sim 1\%$ and new physics $b \to sg \sim 10\%$ are needed, to feed down to $B \to \eta' + X_s$. The anomaly coupling preferentially leads to fast η' mesons. Since the new physics color dipole transition involves right-handed couplings to the b quark, one probes a new CP violating phase that is independent of CKM.

The $B \to \eta' + X_s$ mode is already observed at 0.1% level. With 10% a_{CP} possible because of the interplay of SM and new physics, perhaps CP violation could be observed before 1999, the year that B Factories turn on.

REFERENCES

1. Bander, M., Silverman, D., and Soni, A., *Phys. Rev. Lett.* **43**, 242 (1979).
2. Guberina, B., Peccei, R.D., and Rückl, R., *Phys. Lett.* **90B**, 169 (1980).
3. CLEO numbers are from the talk by Poling, R., these proceedings.
4. Eilam, G., *Phys. Rev. Lett.* **49**, 1478 (1982).
5. Hou, W.S., Soni, A., and Steger, H., *Phys. Rev. Lett.* **59**, 1521 (1987); Hou, W.S., *Nucl. Phys.* **B308**, 561 (1988).
6. Liu, J., and Yao, Y.P., *Phys. Rev.* **D41**, 2147 (1990); Simma, H., and Wyler, D., *Nucl. Phys.* **B344**, 283 (1990); and references therein.
7. Grinstein, B., and Wise, M., *Phys. Lett.* **201B**, 274 (1988).
8. Hou, W.S., and Willey, R.S., *Phys. Lett.* **202B**, 591 (1988).
9. Hewett, J.L., *Phys. Rev. Lett.* **70**, 1045 (1993).
10. Geng, C.Q., Hou, W.S., and Turcotte, P., *Phys. Lett.* **B339**, 317 (1994).
11. Altarelli, G., Petrarca, S., *Phys. Lett.* **B261**, 303 (1991).
12. Grządkowski, B., and Hou, W.S., *Phys. Lett.* **B272**, 383 (1991).
13. Cassel, D., review talk at Xth International Conference on *Physics in Collision*, Duke University, Durham, NC, June 1990.
14. Kagan, A.L., *Phys. Rev.* **D51**, 6196 (1995).
15. Ciuchini, M., Gabrielli, E. and Giudice, G.F., *Phys. Lett.* **B388**, 353 (1996).
16. Kagan, A.L. and Rathsman, J., e-print hep-ph/9701300.
17. Dunietz, I. et al., e-print hep-ph/9612421.
18. Hou, W.S., and Tseng, B., e-print hep-ph/9705304.
19. Atwood, D. and Soni, A., *Phys. Lett.* **B405**, 150 (1997).
20. Gérard, J.-M., and Hou, W.S., *Phys. Rev. Lett.* **62**, 855 (1989); *Phys. Rev.* **D43**, 2909 (1991).

Testing CPT with B Mesons

V. Alan Kostelecký

Physics Department, Indiana University, Bloomington, IN 47405

Abstract. Apparent violations of CPT and Lorentz symmetry might arise in nature as a result of spontaneous symmetry breaking in a theory beyond the standard model. This talk summarizes a few relevant theoretical and experimental issues, with some emphasis on implications for CPT tests with neutral-B mesons.

INTRODUCTION

In this talk, a brief survey is provided of theoretical and experimental issues relevant to the notion that spontaneous CPT and Lorentz breaking might occur in a theory underlying the standard model. The primary focus in the present work is the possibility that observable effects could appear at accessible energies from apparent CPT violation of this type.

For local relativistic field theories of point particles, the CPT theorem [1–7] states that the combination of the three discrete operations of charge conjugation C, parity reflection P, and time reversal T is an exact symmetry. Various experiments have tested this theorem to high precision [8]. The existence of a powerful theorem and a wide variety of accurate experimental tests suggests that apparent CPT violation is a promising signal for new physics, such as might emerge from a fundamental theory beyond the standard model [9–11]. At present, string theory is the most promising approach to the construction of a consistent and complete quantum theory of all fundamental particles and interactions. However, strings are extended objects, so the conventional axioms underlying the proof of the CPT theorem for particle models do not necessarily apply.

In the next section, a possible mechanism arising within string theory for spontaneous CPT [9,10] and Lorentz [12] breaking is briefly summarized. The mechanism can be studied explicitly in certain theories [13,14]. If spontaneous CPT breaking does arise in a realistic fundamental model, it might be apparent in nature at presently accessible energies. The subsequent section outlines the results of an effective-theory approach to incorporating possible CPT-violating interactions in a low-energy model. A general extension of the standard model [11,15] is described that includes extra terms originating from spontaneous CPT and Lorentz breaking but maintains the usual gauge invariances and power-counting renormalizability.

Experimental implications of spontaneous CPT and Lorentz breaking can be investigated within the framework of this extension of the standard model. In particular, quantitative constraints can be placed on the occurrence of CPT and Lorentz breaking in nature by bounding the coefficients of the extra terms in the theory. The remaining parts of this talk summarize a few of these possible experimental tests of CPT. Some of the most sensitive searches for apparent CPT breaking can be performed in neutral-meson systems. The associated flavor oscillations provide interferometric tests of CPT symmetry in the two B systems [11,16,17] and the D system [11,18] as well as in the more conventional setting of the K system [9–11]. Other sensitive measures of CPT breaking also exist. They include comparative measurements of anomalous magnetic moments of electrons and positrons or of protons and antiprotons [19]. It is also possible that CPT breaking is important for baryogenesis [20].

SPONTANEOUS CPT AND LORENTZ VIOLATION

The most natural mathematical setting for string theories appears to involve more than four spacetime dimensions. Assuming the fundamental theory underlying nature includes higher dimensions and is Lorentz and CPT symmetric, then it is plausible that the higher-dimensional Lorentz invariance is spontaneously broken to the four observed dimensions.

A mechanism that could trigger this effect is known in string theory [12]. The nonlocality of strings generates interactions in string field theory that do not appear in the context of usual renormalizable four-dimensional gauge theory but that are compatible with the infinite number of particle fields and the string gauge invariances. If appropriate scalar fields in the theory acquire nonzero vacuum expectation values, the static interaction potentials for Lorentz tensor fields can be destabilized by stringy interactions of this kind. Some of these Lorentz tensors may then obtain expectation values, so that Lorentz invariance is spontaneously broken in the true ground state of the theory. It can be shown that CPT is also spontaneously broken if any Lorentz tensor field with an odd number of spacetime indices acquires an expectation value [9,10].

The string field theory of the open bosonic string provides an explicit example within which the mechanism for spontaneous Lorentz and CPT violation can be studied. A level-truncation scheme permits a systematic exploration of the possible extrema of the action [13,14]. It is feasible to construct the action analytically by incorporating only particle fields with level number less than a chosen value N. The equations of motion can subsequently be found and solved for extrema of the action. The procedure can be repeated and the solutions obtained can be compared for different values of N. A given solution consists of a definite set of nonzero expectation values. It is of interest if it persists and appears to converge as N increases, since it is then plausible that the complete theory contains an extremum involving similar expectation values.

Following this procedure and using symbolic manipulation routines, it has been feasible to study aspects of the static interaction potential for the open bosonic string to a depth of over 20,000 nonzero terms. Nontrivial solutions to the equations of motion emerge, including ones violating Lorentz and CPT invariance. These exhibit properties that are to be expected from general considerations of the theoretical mechanism.

EXTENSION OF THE STANDARD MODEL

A question of immediate relevance is whether the occurrence of spontaneous CPT and Lorentz breaking in a fundamental theory could generate apparent violations at low energies. This would happen if the breaking extends to the four large spacetime dimensions, which is natural mathematically. Otherwise, a definite mechanism would be needed to explain the existence of the four-dimensional symmetry.

No CPT or Lorentz violations have been found experimentally. A high degree of suppression is therefore implied for any possible effects at low energies. The standard model is known to provide an excellent description of nongravitational physics in this regime. It can be viewed as an effective model emerging at the electroweak scale m_{ew} from a more fundamental theory, presumably governed by the Planck scale m_{Pl}. The natural suppression factor for Planck-scale effects in the standard model would then be $r \sim m_{ew}/m_{Pl} \simeq 10^{-17}$. Relatively few CPT- and Lorentz-violating effects would be accessible to experiment if this strong suppression occurs. The sections below describe a few of the possible signals.

Suppose the spontaneous CPT and Lorentz violation indeed generates minuscule effects at the level of the electroweak scale. These effects can be studied by extending the standard model to include possible terms originating in spontaneous CPT and Lorentz violation. In the fermion sector, for example, possible terms of the general form [10,11]

$$\mathcal{L} \sim \frac{\lambda}{M^k} \langle T \rangle \cdot \overline{\psi} \Gamma (i\partial)^k \chi + h.c. \tag{1}$$

could appear as a low-energy consequence of spontaneous symmetry breaking in a compactified string theory. Terms of this type are Lorentz- and possibly CPT-violating, as a result of the nonzero expectation values of Lorentz tensors T that can appear in interactions terms coupling T with fermions ψ and χ

via derivatives $i\partial$ and a gamma-matrix structure Γ. In Eq. (1), λ is assumed dimensionless, so one or more large mass scales M such as the compatification or Planck scales must also appear. The fermions ψ and χ in Eq. (1) can be identified with leptons or quarks in the standard model.

A general extension of the standard model has been obtained [15] that includes Lorentz-breaking terms both with and without CPT breaking and allows also for possible effects in the gauge and Higgs sectors. The analysis is developed around a theoretical framework for treating CPT and Lorentz breaking that appears to bypass some standard difficulties, largely because the breaking is spontaneous while the underlying theory remains Lorentz and CPT invariant. The derivation therefore includes the constraint that any new effects must be compatible with an origin in spontaneous Lorentz breaking. Invariance under the gauge group $SU(3) \times SU(2) \times U(1)$ and power-counting renormalizability are also required.

The next sections outline some possible measurable signals that are implied by this standard-model extension. Detailed treatments exist at present for effects in neutral-meson systems and for effects in certain experiments in the context of quantum electrodynamics.

TESTS WITH NEUTRAL MESONS

In the four neutral-meson systems, K, D, B_d, B_s, the small mass differences between weak-interaction eigenstates offer an interferometric sensitivity to highly suppressed effects such as might arise from Planck-scale physics [11]. This section has three subsections. One summarizes theoretical issues, one outlines general experimental issues and some established results, and the third discusses the case of the B_d system.

Theoretical Issues

The time evolution of a neutral-P meson, where the symbol P denotes any of the four neutral mesons, is governed by a 2×2 effective hamiltonian Λ_P. Assuming conventional quantum mechanics, there are two types of indirect CP violation that might be described by phenomenological parameters appearing in Λ_P. The first is the usual CP-violating parameter ϵ_P that breaks T but preserves CPT. The second is a complex CP-violating parameter δ_P that preserves T but breaks CPT. Note that this parametrization of CP violation is independent of any underlying model and is at a purely phenomenological level.

In the context of the usual standard model, parameters in the CKM matrix can be regarded as controlling the parameters ϵ_P for each P. No such understanding exists for possible nonzero values of δ_P. However, the CPT-violating extension of the standard model outlined in the previous section does provide a theoretical basis for nonzero δ_P. For instance, terms of the type shown in Eq. (1), which appear in the quark sector of the general standard-model extension when the fermions ψ and χ are identified with quark fields, act to change the time evolution of a neutral-P meson in a δ_P-dependent way. The point is that the propagators of the (valence) quarks are affected and generate CPT-violating effects.

A general expression for the the quantity δ_P for a given P system can be obtained in the context of spontaneous CPT and Lorentz breaking. It is [10,11]:

$$\delta_P = i\frac{h_{q_1} - h_{q_2}}{\sqrt{\Delta m^2 + \Delta\gamma^2/4}}e^{i\hat{\phi}} \quad . \tag{2}$$

In this equation, Δm and $\Delta\gamma$ are the P-meson mass and rate differences, which are experimentally observable. They are used to define the angle $\hat{\phi}$ by $\hat{\phi} \equiv \tan^{-1}(2\Delta m/\Delta\gamma)$. The quantities h_{q_j} are determined by parameters in the standard-model extension arising from the spontaneous CPT violation in the fundamental theory and by the effects r_{q_j} of the quark-gluon sea: $h_{q_j} = r_{q_j}\lambda_{q_j}\langle T\rangle$.

The hermiticity of the underlying theory and the extension of the standard model implies that the quantities h_{q_j} are real. This in turn can be used to show that the real and imaginary parts of δ_P are proportional, with proportionality constant determined by experimentally measurable rate and mass differences. Explicitly, the result is

$$\text{Im}\,\delta_P = \pm\frac{\Delta\gamma}{2\Delta m}\,\text{Re}\,\delta_P \quad . \tag{3}$$

This can be regarded as a determining signature for CPT breaking arising within the present framework. Assuming a suppression factor of $r \simeq 10^{-17}$ along the lines of the discussion above, it also follows that direct CPT violation arising in the P-meson decay amplitudes is too small to observe.

Within the present framework, it is plausible that the values of the CPT-violating quantities δ_P could be significantly different for distinct P-meson systems. The point is that the dimensionless coupling constants λ_{q_j}, appearing in terms of the form given in Eq. (1), might depend on the quark flavor q_j. The corresponding CPT-violating quark couplings within the extension of the standard model would then also be flavor dependent. A related effect occurs for the Yukawa couplings, which take values for different quark flavors that range over about six orders of magnitude. The possibility of flavor-dependent CPT violation means that the values of δ_P might vary with P, so it might be crucial to perform experimental tests of CPT symmetry in more than one neutral-meson system. Moreover, under some circumstances the experimental signals could be startling. For instance, only relatively weak limits have been obtained as yet on B_d-meson CP violation, which means it is possible that CP violation parametrized by the CPT-violating quantity δ_{B_d} could be *larger* than conventional CP breaking parametrized by ϵ_{B_d} and therefore could produce unexpected results in proposed experiments at B factories.

Experimental Issues and Results

This subsection provides a short outline of some experimental issues and the present status of established tests of indirect CPT violation with neutral-P mesons. Tests with neutral-B mesons are discussed in the following subsection.

Searches for indirect CP violation in a neutral-P system can be performed with experimental data taken from decays either of uncorrelated tagged P mesons or of correlated P-$\overline{P}$ pairs produced through prior quarkonia decay. Both indirect T and CPT violation are in principle accessible. Time-dependent and fully integrated decay-probability asymmetries, sensitive to the various CP parameters, have now been established in each P-meson system for both correlated and uncorrelated situations. These can be applied to analyses of real data, or can be used in theoretical estimates of CP reach performed either analytically or through detailed Monte-Carlo simulations with acceptances and background effects.

To date, the sharpest bound on CPT violation in a neutral-meson system comes from analyses of K oscillations. There already exist published bounds on $|\delta_K|$ of order 10^{-3} [8,21,22]. Analyses currently underway or experiments being performed or planned are anticipated to provide even tighter limits.

In the D system, mass mixing has yet to be detected. Furthermore, the expected strong dispersive effects and the complication of dominant contributions arising from physics beyond the standard model makes theoretical predictions difficult and subject to uncertainties potentially of orders of magnitude. Nonetheless, in circumstances that are favorable theoretically, certain tests of CPT invariance in the D system might produce signals when performed with current techniques and perhaps even with data that already exist [18]. The expected increase in reconstructed events to be obtained in various future machines provides an interesting arena for establishing CPT bounds from the D system.

The Neutral-B System

Tests of CPT with neutral-B_d mesons are of interest both theoretically and experimentally. Theoretically, if indeed the dimensionless couplings corresponding to λ in Eq. (1) are flavor dependent and follow the same general pattern as the Yukawa couplings, then the strength of the CPT violation is related to the mass of the valence quarks that are bound in the neutral meson. In this case, since the b quark is involved, it is possible that any CPT violation would be larger in the B system than in other neutral-meson systems [11,17].

On the experimental front, large numbers of B_d events have already been obtained, and future machines and detectors under construction are expected to produce high-statistics event samples.

Several studies have been performed to estimate the likely constraints on CPT violation that could be obtained from experiments with B_d mesons [11,16,17]. The most detailed treatment uses Monte-Carlo simulations to model experiments performed with uncorrelated, correlated unboosted, or correlated boosted mesons [17]. Backgrounds, resolutions, and acceptances are incorporated in simulating realistic experimental data that might be obtained at typical detectors at LEP, CESR, and the future B factories. One result from this analysis is that data already taken suffice to place meaningful bounds on δ_{B_d}.

Until recently, no bound existed on δ_{B_d}. Early in 1997, the OPAL collaboration at LEP obtained the first experimental constraint on CPT violation in the neutral-B_d system [23]. The relevant experimental observable is an asymmetry derived in Ref. [17] that is sensitive to $\text{Im}\,\delta_{B_d}$ and $\text{Re}\,\epsilon_{B_d}$. The time evolution of this asymmetry can be extracted from the experiment and used to bound both these quantities. The result is a constraint on the value of $\text{Im}\,\delta_{B_d}$ of less than 3×10^{-2} at the 95% confidence limit.

An interesting feature of the B_d system is that Eq. (3) predicts that the real part of δ_{B_d} is greater than the imaginary part because the value of $2\Delta m/\Delta\gamma$ is believed to be large. In contrast, the real and imaginary parts of δ_P for the K system and perhaps also the D system would be comparable. The analysis of Ref. [17] suggests that data already taken with the CLEO detector at CESR could be used to bound $\text{Re}\,\delta_{B_d}$. The expected relatively large size of this quantity compared to $\text{Im}\,\delta_{B_d}$ implies that even a limit of order 20% would be of interest.

The B factories and other B-dedicated experiments now under construction should be capable of improving on bounds obtained from current data. Moreover, the corresponding detectors are also expected to be sensitive to both $\text{Re}\,\delta_{B_d}$ and $\text{Im}\,\delta_{B_d}$ [17]. This opens the possibility in principle of testing Eq. (3) in a single experiment, should CPT violation indeed be discovered.

EFFECTS IN OTHER SYSTEMS

Effects from spontaneous CPT and Lorentz violation could also be manifest in contexts other than neutral-meson oscillations. In the standard-model extension, distinct quantities govern CPT and Lorentz breaking in, for example, the quark and lepton sectors. A wide variety of experiments outside the neutral-meson systems is therefore potentially crucial for uncovering effects. This section briefly describes some possibilities of this type.

The standard description of baryogenesis requires CP- and C-violating interactions and nonequilibrium processes [24] as well as baryon-number violating effects. In grand-unified theories, for example, the CP breaking is selected in a range suitable for reproducing the known baryon asymmetry and is unrelated to the observed CP violation in the standard model. The presence of CPT-breaking processes of the type given in Eq. (1) suggests an alternative possibility for baryogenesis that could occur in thermal equilibrium without the need for additional CP violation. An analysis [20] shows that under suitable circumstances a mechanism of this kind could result in a large baryon asymmetry at grand-unification scales that diminishes to the observed value by a process such as sphaleron dilution.

Another implication of the CPT- and Lorentz-violating extension of the standard model is a modification of some conventional results in quantum electrodynamics [15]. One example concerns comparative measurements of the anomalous magnetic moments of the electron and positron. It has recently been shown [19] that the standard figure of merit used in these experiments is misleading. However, a more appropriate measure can be defined that is directly sensitive to some of the additional terms appearing in the modified version of quantum electrodynamics. With current experimental techniques, constraints on CPT violation could be attained with a precision similar to those from neutral-meson systems. Related experiments with protons and antiprotons may provide interesting limits on CPT violation. Bounds on CPT and Lorentz breaking are also possible from precision experiments using cyclotron frequencies [19] and photon properties [15].

ACKNOWLEDGMENTS

I thank Orfeu Bertolami, Robert Bluhm, Don Colladay, Rob Potting, Neil Russell, Stuart Samuel, and Rick Van Kooten for collaborations. This work was supported in part by the United States Department of Energy under grant number DE-FG02-91ER40661.

REFERENCES

1. J. Schwinger, *Phys. Rev.* **82** (1951) 914.
2. G. Lüders, *Det. Kong. Danske Videnskabernes Selskab Mat.-fysiske Meddelelser* **28**, no. 5 (1954).
3. J.S. Bell, Ph.D. thesis (Birmingham University, England, 1954); *Proc. Roy. Soc. (London)* A **231** (1955) 479.
4. W. Pauli, in *Niels Bohr and the Development of Physics,* ed. W. Pauli, (McGraw-Hill, New York, 1955), p. 30.
5. G. Lüders and B. Zumino, *Phys. Rev.* **106** (1957) 385.
6. R.F. Streater and A.S. Wightman, *PCT, Spin and Statistics, and All That* (Benjamin Cummings, Reading, 1964).
7. R. Jost, *The General Theory of Quantized Fields* (AMS, Providence, 1965).

8. See, for example, R.M. Barnett *et al.*, *Review of Particle Properties, Phys. Rev. D* **54** (1996) 1.
9. V.A. Kostelecký and R. Potting, *Nucl. Phys.* B **359** (1991) 545.
10. V.A. Kostelecký, R. Potting, and S. Samuel, in *Proceedings of the 1991 Joint International Lepton-Photon Symposium and Europhysics Conference on High Energy Physics*, eds. S. Hegarty *et al.* (World Scientific, Singapore, 1992); V.A. Kostelecký and R. Potting, *Gamma Ray–Neutrino Cosmology and Planck Scale Physics*, ed. D.B. Cline (World Scientific, Singapore, 1993) (hep-th/9211116).
11. V.A. Kostelecký and R. Potting, *Phys. Rev.* D **51** (1995) 3923.
12. V.A. Kostelecký and S. Samuel, *Phys. Rev.* D **39** (1989) 683; *ibid.*, **40** (1989) 1886; *Phys. Rev. Lett.* **63** (1989) 224; *ibid.*, **66** (1991) 1811.
13. V.A. Kostelecký and S. Samuel, *Nucl. Phys.* B **336** (1990) 263; *Phys. Rev. Lett.* **64** (1990) 2238; *Phys. Rev.* D **42** (1990) 1289.
14. V.A. Kostelecký and R. Potting, *Phys. Lett.* B **381** (1996) 389.
15. D. Colladay and V.A. Kostelecký, *Phys. Rev.* D **55** (1997) 6760; Indiana University preprint IUHET 359 (1997).
16. D. Colladay and V.A. Kostelecký, *Phys. Lett.* B **344** (1995) 259.
17. V.A. Kostelecký and R. Van Kooten, *Phys. Rev.* D **54** (1996) 5585.
18. D. Colladay and V.A. Kostelecký, *Phys. Rev.* D **52** (1995) 6224.
19. R. Bluhm, V.A. Kostelecký, and N. Russell, *Phys. Rev. Lett.* **79** (1997) 1432; Indiana University preprint IUHET 368 (1997).
20. O. Bertolami *et al.*, *Phys. Lett.* B **395** (1997) 178.
21. L.K. Gibbons *et al.*, *Phys. Rev.* D **55** (1997) 6625.
22. R. Carosi *et al.*, *Phys. Lett.* B **237** (1990) 303.
23. OPAL Collaboration, R. Ackerstaff *et al.*, CERN preprint CERN-PPE/97-036 (April 1997).
24. A.D. Sakharov, *JETP Lett.* **5** (1967) 24.

The Mystery of Flavor

R. D. Peccei

Department of Physics and Astronomy, UCLA, Los Angeles, CA 90095-1547

Abstract. After outlining some of the issues surrounding the flavor problem, I present three speculative ideas on the origin of families. In turn, families are conjectured to arise from an underlying preon dynamics; from random dynamics at very short distances; or as a result of compactification in higher dimensional theories. Examples and limitations of each of these speculative scenarios are discussed.

I THE QUESTION OF FLAVOR

Flavor is an old problem. I. I. Rabi's famous question about the muon: "who ordered that?" has now been replaced by an equally difficult question to answer: "why do we have three families of quarks and leptons?" Although qualitatively we understand the issues connected to flavor a lot better now, quantitatively we are as puzzled as when the muon was discovered.

When thinking of flavor, it is useful to consider the standard model Lagrangian in a sequence of steps. At the roughest level, neglecting both gauge and Yukawa interactions, the Standard Model Lagrangian $\mathcal{L}_o = \mathcal{L}_{SM}(g_i = 0; \Gamma_{ij} = 0)$ has a $U(48)$ global symmetry corresponding to the freedom of being able to interchange any of the 16 fermions of the 3 families of quarks and leptons with one another. If we turn on the gauge interactions, the Lagrangian $\mathcal{L}_1 = \mathcal{L}_{SM}(g_i \neq 0; \Gamma_{ij} = 0)$ has a much more restricted symmetry $[U(3)]^6$ corresponding to interchanging fermions of a given type (*e.g.* the $(u,d)_L$ doublet) from one family to the other. When also the Yukawa interactions are turned on, $\mathcal{L}_2 = \mathcal{L}_{SM}(g_i \neq 0; \Gamma_{ij} \neq 0)$, then the only remaining symmetry of the Lagrangian is $U(1)_B \times U(1)_L$. In fact, because of the chiral anomaly [1], at the quantum level the symmetry of $\mathcal{L}_2$ is just $U(1)_{B-L}$.

The above classification scheme serves to emphasize that there are really three distinct flavor problems. There is a **matter problem**, a **family problem** and a **mass problem**. The first of these problems is simply that of understanding the origin of the different species of quarks and leptons (*i.e.* why does one have a ν_L^c and a u_L^c state?). The second problem is related to the triplication of the quarks and leptons. What physics forces such a triplication? Finally, the last problem is related to understanding the origin of the observed peculiar mass pattern of the known fermions.

The usual approach when thinking about flavor is to try to decouple the above three problems from one another. Thus, for example, one assumes the existence of the quarks and leptons in the Standard Model and asks for the physics behind the replication of families. Although it is difficult to argue cogently on this point, it is certainly true in the examples which we will discuss that the matter problem seems to be unrelated to the question of family replication. Indeed, quite often one also assumes the reverse, namely, that the family replication question is independent of the types of quarks and leptons one has. In fact, it is possible that there is other matter besides the known quarks and leptons and that this matter is also replicated. Certainly, even in the minimal Standard Model there is **other matter** besides the quarks and leptons, connected to the symmetry breaking sector. This raises a host of questions including that of possible family replication of the ordinary Higgs doublet. One knows, empirically, that this cannot happen if one is to avoid flavor changing neutral currents (FCNC) [2]. However, some replication is needed if there is supersymmetry, but the two different Higgs doublets needed in supersymmetry are connected with different quark charges and need not replicate as families.

The above remarks suggest that there are some perils associated with trying to seek the origin for family replication independently from that of the quarks and leptons themselves. Nevertheless, that is the approach usually taken and the one I will follow here. Similarly, one also usually tries to disconnect the problem of mass from that of matter and family. That is, one generally assumes the existence of the three observed families of quarks and leptons, and then tries to postulate (approximate) symmetries of the mass matrices for quarks and leptons which will give interrelations among the masses and mixing parameters for some of these states.

This approach usually involves some kind of **family symmetry** and is sensible provided that:

i) There is some misalignment between the mass matrix basis and the gauge interaction basis for the quarks and leptons. Only through such a misalignment will there result a nontrivial mixing matrix: $V_{\text{CKM}} \neq 1$.

ii) The family symmetries of the mass matrices are broken (otherwise $V_{CKM} \equiv 1$) either explicitly or spontaneously. Furthermore, if the breaking is spontaneous, it must occur at a sufficiently high scale to have escaped detection so far.

Although the origin of flavor remains a mystery, I want to discuss here three speculative ideas for the origin of families. These ideas are realized up to now only in incomplete ways, in what amount essentially to toy models. Thus, for instance, the issue of family generation is in general disconnected from the question of $SU(2) \times U(1)$ breaking and, often, also from trying to explicitly calculate the Yukawa couplings. As a result, in all of these attempts at trying to understand flavor, the question of mass is approached from a much more phenomenological viewpoint. One guesses certain family or GUT symmetries, and their possible patterns of breaking, and then one checks out these guesses by testing their predictions experimentally. In all of these considerations, the top mass, because it is the dominant mass in the spectrum, plays a fundamental role.

In what follows, I will not discuss in any detail the issue of mass generation [3]. Rather, here I will concentrate on describing three speculative ideas for the origin of families. Specifically, I will consider in turn the generation of families dynamically; through short distance chaotic dynamics; and as a result of geometry. After this speculative tour, I will conclude with some general remarks.

II GENERATING FAMILIES DYNAMICALLY

The underlying idea behind this approach to the flavor problem is that families of quarks and leptons result because they are themselves composites of yet more fundamental ingredients–**preons**. There is a nice isotope analogy [4] which serves to illustrate this point. Think of the three isotopes of hydrogen as three distinct families. Just like the families of quarks and leptons, all three isotopes have the same interactions–their chemistry being determined by the electromagnetic interactions of the proton. Deuterium and tritium, however, have different masses than the proton because they have, respectively, 1 and 2 neutrons. Of course, the analogy is not perfect since ^1H and ^3H are fermions and ^2H is a boson! Nevertheless, it is tempting to suppose that the 3 families of quarks and leptons, just like the hydrogen isotopes, result from the presence of different "neutral" constituents.

I will illustrate how to generate families dynamically by using as an example some recent work of Kaplan, Lepeintre, and Schmaltz [4]. By using essentially the isotope analogy, these authors constructed an interesting toy model of flavor. Their simplest toy model is based on an underlying supersymmetric gauge theory based on the symplectic group $Sp(6)$. The fundamental constituents in this model are 6 preons Q_α transforming according to the fundamental representation of $Sp(6)$ and one preon $A_{\alpha\beta}$ transforming according to the 2-rank antisymmetric representation. Such a theory has three families of bound states distinguished by their $A_{\alpha\beta}$ content, plus a pair of (neutral) exotic states. To wit, the bound states of the model are the 15 flavor states

$$F_3^{[i,j]} \sim Q_\alpha^i Q_\alpha^j; \quad F_2^{[i,j]} \sim Q_\alpha^i A_{\alpha\beta} Q_\beta^j; \quad F_1^{[i,j]} \sim Q_\alpha^i A_{\alpha\beta} A_{\beta\gamma} Q_\gamma^j \qquad (1)$$

plus the two neutral exotic states

$$T_2 \sim \text{Tr} A^2 \; ; \quad T_3 \sim \text{Tr} A^3 \; . \tag{2}$$

The six Q_α preons act as the protons in the isotope analogy. In principle, one could imagine having the $SU(3) \times SU(2) \times U(1)$ interactions act on the Q_α states, while the $A_{\alpha\beta}$ preons act as the neutrons. Furthermore, there is clearly a family $U(1)_F$ in the spectrum which counts the number of $A_{\alpha\beta}$ fields. Finally, one should note that, because of the supersymmetry, each of the states in Eqs. (1) and (2) contain both fermions and bosons.

Although the number of bound states per family (15) is encouraging, these states cannot really be the ordinary quarks and leptons (minus the right-handed neutrinos). It turns out that one cannot properly incorporate the $SU(3) \times SU(2) \times U(1)$ gauge interactions with only 6 Q_α preons. To do that, in fact, one has to at least **triplicate** the underlying gauge theory [4] from $Sp(6)$ to $Sp(6)_L \times Sp(6)_R \times Sp(6)_H$. Each of these $Sp(6)$ groups has again six Q_α and one $A_{\alpha\beta}$ preon. To obtain the desired quarks and leptons the Q preons are assumed to have the following $SU(3) \times SU(2) \times U(1)$ assignments:

$$Q_L : (3,1)_0 \oplus (1,2)_{1/6} \oplus (1,1)_{-1/3}$$
$$Q_R : (\bar{3},1)_0 \oplus (1,1)_{-2/3} \oplus 2(1,1)_{1/3}$$
$$Q_H : (1,2)_{-1/6} \oplus (1,1)_{1/3} \oplus (1,1)_{2/3} \oplus 2(1,1)_{-1/3} \tag{3}$$

Because of the preon group triplication, instead of having 15 $F^{[i,j]}$ bound states per family, one now has 45 such states. Per family, these states now include 16 states with the quantum numbers of the observed quarks and leptons, plus 29 exotic states which, however, sit in vector-like representations of the Standard Model group. Specifically, the quark doublet $(u,d)_L$ is a bound state of $Sp(6)_L$; u_L^c and d_L^c are bound states of $Sp(6)_R$; while the lepton states $(\nu, e)_L$, ν_L^c and e_L^c are bound states of $Sp(6)_H$. Among the exotic states one finds as bound states of $Sp(6)_H$ two states with the quantum numbers of the Higgs doublets of a supersymmetric theory: $H_1 \sim (h_1^o, h_1^-)$ and $H_2 \sim (h_2^+, h_2^o)$. So, in this model, there is a natural family repetition of the Higgs states. Naively, this could cause problems with FCNC. It turns out, however, that when one calculates the dynamical superpotential of the theory [5] one can show [4] that there is a ground state where only one of the three families of Higgs states is left light. So, in fact, there are no FCNC problems.

This nice result is tempered by other troublesome features of the model which render it unrealistic–but not uninteresting. For example, to break the $[U_F(1)]^3$ family symmetry of the model, it is necessary to introduce by hand some heavy fields (with masses $\mu > \Lambda$–the dynamical scale of the preon theories) which serve to couple the preon groups together. The simplest possibility is afforded by having 3 such fields, $v^1_{\alpha_H \beta_R}$, $v^2_{\alpha_R \beta_L}$, $v^3_{\alpha_L \beta_H}$, with indices spanning 2 of the preon groups, interacting through a superpotential

$$W = a v^1 v^2 v^3 + b_H^1 v^1 v^1 A_H + b_R^1 v^1 v^1 A_R + b_R^2 v^2 v^2 A_R + b_L^2 v^2 v^2 A_L$$
$$+ b_L^3 v^3 v^3 A_L + b_H^3 v^3 v^3 A_H \; . \tag{4}$$

The a-term above ties the preon theories together, while the various b-terms serve to break the family symmetries. Although Eq. (4) is introduced by hand, integrating out the effects of the heavy v^i fields gives effective Yukawa couplings of different strengths, much in the way originally suggested by Froggatt and Nielsen [6]. This is illustrated schematically in Fig. 1 for the Yukawa coupling of u_L^c with $(c,s)_L$ via the Higgs state $H_2^{(3)}$ of the third family–which is the only one which is assumed to get a VEV.[1] One finds [4]

$$\Gamma_{12}^{(3)} \sim a\, b_R^1 b_R^2 b_L^3 (\Lambda/\mu)^6 \sim \epsilon^6 \ . \tag{5}$$

Although the various elements in the up-and-down quark mass matrices are hierarchial, unfortunately there is no resulting quark mixing since $M_u \sim M_d$. This follows because the model has an unbroken global $SU(2)$ symmetry at the preon level corresponding to the interchange of the $(1,1)_{-2/3}$ and $(1,1)_{1/3}$ assignments in Eq. (3). Furthermore, for the lepton sectors there is a dynamically generated set of Yukawa couplings [5] which are typically unsuppressed. As a result, naively, one expects $m_\tau \gg m_t$. Both of these results make the $[Sp(6)]^3$ model as presented above unrealistic. By further complicating the model, Kaplan, Lepeintre, and Schmaltz [4] are able to obtain both a nontrivial CKM matrix and re-establish the top as the heaviest bound state. However, these "improved" models are not particularly attractive and represent, more than anything else, a "proof of principle." In addition, even after these problems are resolved, the models still lack mechanisms for breaking $SU(2) \times U(1)$ and supersymmetry, features which must be understood to make contact with reality.

These negative remarks should not obscure the considerable achievement of these dynamical models for understanding the origin of flavor. Families in these models arise as a result of hidden degrees of freedom in some underlying confining dynamics. Furthermore, the presence of heavy excitations in this same dynamics can result in hierarchial patterns of Yukawa couplings, once all family symmetries are explicitly broken. Unfortunately, it is difficult to see how one can obtain real evidence for these kinds of schemes, barring the discovery of some of the exotic bound states they predict–in the example discussed, the T_2 and T_3 states or the vector-like partners of the quarks and leptons.

III FAMILIES FROM SHORT-DISTANCE RANDOM DYNAMICS

A radically different scheme for the origin of families has been proposed and elaborated by Holger Nielsen and his collaborators [7]. The basic idea

---

[1] In the model [4] the lightest family has the most $A_{\alpha\beta}$ fields–c.f. Eq. (1).

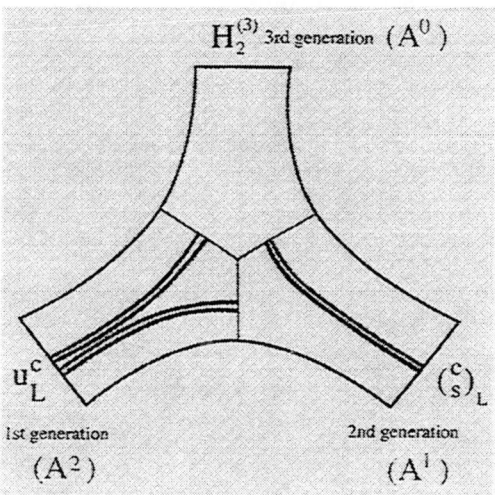

FIGURE 1. Effective Yukawa coupling generated in the $[Sp(6)]^3$ preon model.

that Nielsen has put forth is that there exist both order and chaos at very short distances. He imagines that at scales much smaller than the inverse of the Planck mass there is actually a lattice structure of scale length $a \ll 1/M_{\text{Planck}}$. However, both the dynamics on the lattice as well as the structure of the lattice are random. In particular, the lattice is amorphous with sites at random positions. Furthermore, characteristic of the random dynamics, the interactions on each of the links are governed by different groups, with the groups varying from link to link.

Remarkably, even starting from these very general assumptions, one can arrive at some conclusions. Generally, one naively would imagine that no group could survive the random dynamics. That is, that the gauge group will end up by breaking down spontaneously, producing supermassive fields of mass $M \sim a^{-1} \gg M_{\text{Planck}}$. In fact, as Brene and Nielsen [8] showed, there are special groups $G_{\text{surv.}}$ on the links which survive the random dynamics–$i.e.$, the associated vector bosons are massless. What Brene and Nielsen [8] showed is that the groups which survive must have a center which is non-trivial and connected. By taking values in the center the links are effectively gauge-invariant. However, the center cannot be simply the unit matrix because the random nature of the dynamics would then end up by averaging out the effects of all links. The connectedness of the center, finally, is necessary to insure that the Bianchi identities are satisfied. Specifically, it turns out that $G_{\text{surv.}}$ is a product of "prime" groups with a certain discrete group D_{prime}, generated from the center, removed:

$$G_{\text{surv.}} = U(1) \times SU(2) \times SU(3) \times SU(5) \times \ldots SU(\text{prime})/D_{\text{prime}}. \quad (6)$$

From the above, it appears that Nielsen's random dynamics allows the Standard Model group to survive, with a restriction:

$$G^*_{SM} = SU(3) \times SU(2) \times U(1)/D_3 . \tag{7}$$

Here the discrete group D_3 is given by powers of the center element $h = \left\{ e^{\frac{2\pi i}{3}} I_3, -I_2, e^{\frac{2\pi i}{6}} \right\}$:

$$D_3 = \{h^n | n \subset Z_6\} . \tag{8}$$

In practice, this imposes a restriction on the matter states which are placed on the random lattice sites

$$h\psi = \psi , \tag{9}$$

which fixes the hypercharge of the quarks relative to the leptons. Eq. (9) effectively imposes the familiar charge quantization, giving the quarks third-integral charges. This is a very nice result!

In this scheme the origin of family replication occurs through what Bennett, Nielsen, and Picek [9] call "confusion" in the random dynamic processes. This can be understood as follows. At some step in the random dynamics what survives is not simply the group G^*_{SM} but a number N_F of copies of G^*_{SM}, each with one family of quarks and leptons. Subsequently, this product group collapses to its diagonal subgroup $[G^*_{SM}]_{\text{diag}}$. This collapse, through "confusion," results in N_F replicas of a Standard Model family of quarks and leptons. Thus, schematically, family generation occurs in random dynamics when N_F Standard Model surviving groups collapse:

$$G^*_{SM} \times G^*_{SM} \times \ldots \times G^*_{SM} \to [G^*_{SM}]_{\text{diag}} . \tag{10}$$

Bennett, Nielsen, and Picek [9] try to estimate N_F–the number of families which arise from random dynamics confusion–by making a number of assumptions. Although some of these assumptions are questionable, they are not unreasonable. First, Bennett *et al.* suppose that the lattice scale associated with the random dynamics is of order the Planck scale: $a = M_P^{-1}$. This allows the calculation of the coupling constants of the Standard Model group $[G^*_{SM}]_{\text{diag}}$ from their low energy values via the renormalization group:

$$[g_i]_{\text{diag}} \simeq g_i[M_P] . \tag{11}$$

Second, by identification of the gauge fields in $[G^*_{SM}]_{\text{diag}}$ with the individual fields in each of the SM groups in Eq. (10), it follows that the individual couplings of each of the groups in the "confused" configuration $G^*_{SM} \times G^*_{SM} \times \ldots \times G^*_{SM}$ is given by

$$g_i^{\text{conf}} = \sqrt{N_F} [g_i]_{\text{diag}} . \tag{12}$$

A knowledge of g_i^{conf} then provides an estimate for N_F. What Bennett, Nielsen, and Picek [9] assume is that

$$g_i^{\text{conf}} = g_i^* \, , \qquad (13)$$

with the g_i^* being the mean field theory critical coupling for each of the groups in the Standard Model. This assumption guarantees that in the confusion stage there is no confinement of quark and lepton states at Planck length scales–a reasonable boundary condition.

The result for N_F which follows from the three assumptions (11)-(13) are rather remarkable, given the spare theoretical framework! One finds [7]

$$N_F = \begin{cases} 3.4 & U(1) \\ 3.5 & SU(2) \\ 3.1 & SU(3) \end{cases} \qquad (14)$$

This result notwithstanding, however, it is not clear how one proceeds further in developing a consistent theoretical framework from random dynamics. For instance, it is totally unclear how through this scheme one induces the breakdown of the $SU(2) \times U(1)$ electroweak group at scales of $\mathcal{O}(100 \text{ GeV})$, or how one even generates the Yukawa couplings which can provide the quarks and leptons eventually with some mass.

IV A GEOMETRICAL ORIGIN FOR FAMILIES

Perhaps the most interesting way to get family replications is through the compactification of extra dimensions. One starts with a theory in $d > 4$ dimensions but then assumes that the extra dimensions somehow compactify, leaving a 4-dimensional theory. The earliest example of such a theory was the 5-dimensional Kaluza-Klein theory of gravity [10], which when compactified to 4 dimensions gave rise, in addition to gravity, also to electromagnetic interactions. More modern examples are superstring theories [11] which are known to be consistent only in $d > 4$ dimensions, but where again the extra dimensions can compactify leaving an effective 4-dimensional theory.

It is quite easy to understand how one can generate families in these types of theories. The general idea was first sketched out by Wetterich [12] and Witten [13] in the early 1980's. Consider chiral fermions in a d-dimensional space-time.[2] Such fermions, by definition, obey a massless Dirac equation

$$\Gamma^\alpha D_\alpha \psi = 0 \, . \qquad (15)$$

Here $\alpha = 1, 2, \ldots d$, and D_α is the Dirac operator in the background of whatever other fields (gravity, Yang-Mills) are present in the theory. Suppose now $(d - 4)$ dimensions compactify. Then Eq. (15) can be written as

---

[2] Chiral fermions occur naturally in $d = 2$ mod 4 dimensions.

$$(\Gamma^\mu D_\mu + \Gamma^a D_a)\psi = 0, \tag{16}$$

with $\mu = 0, 1, 2, 3$ and $a = 4, \ldots, d-1$. Clearly the $(d-4)$ operator $\Gamma^a D_a$ in Eq. (16) acts as a 4-dimensional mass $[\Gamma^a D_a \equiv M]$ **unless** it vanishes when applied on ψ:

$$\Gamma_a D^a \psi = 0. \tag{17}$$

If Eq. (17) holds, corresponding to a chirality constraint on the $(d-4)$-dimensional compact space K, then also the 4-dimensional fermions will be chiral. If (17) does not hold, then the resulting 4-dimensional fermions have a mass.

Since the quarks and leptons are chiral, if they are produced from $d > 4$ chiral fermions via compactification of the extra dimension, a constraint equation like (17) on the compact space K must hold. Now, in general, such constraint equations have a number of solutions,[3] which depend on the intrinsic properties of the compact space K. So, in these kinds of theories, families and family number are intrinsically related to the topological properties of compact spaces associated with the original $d > 4$ theory.

Perhaps the best known example of family replication using these ideas is the one considered by Candelas, Horowitz, Strominger, and Witten [14] involving the Calabi-Yau compactification of the $d = 10$ heterotic superstring. This string theory [15] has an associated $E_8 \times E_8$ gauge symmetry and is supersymmetric. The chiral fermions in the $d = 10$ theory are gauginos of one of the E_8 groups (the other E_8 acts as a hidden sector), sitting in the 248-dimensional adjoint representation.[4] Candelas et al. [14] assumes that the 10-dimensional space of the theory compactifies down to $d = 4$ Minkowski space times a 6-dimensional Calabi-Yau space, whose principal property for our purposes is that it possesses an $SU(3)$ holonomy. This means that the chiral zero modes in K–those that obey the constraint equations (17)–have non-trivial $SU(3)$ properties, even though this $SU(3)$ is broken in the compactification. By decomposing E_8 into its $E_6 \times SU(3)$ subgroup one identifies the chiral zero modes in the gauginos which are candidates for the surviving chiral matter in 4 dimensions. Since, under this decomposition,

$$248 = (78, 1) \oplus (27, 3) \oplus (\overline{27}, \bar{3}) + (1, 8), \tag{18}$$

one sees that, after Calabi-Yau compactification, the 4-dimensional chiral matter involves fermions in either the 27 or $\overline{27}$ reprentations of E_6. So, in general, one expects to have $n_F = 27$ plus $\delta(27 + \overline{27})$ states in the spectrum. The numbers n_F and δ are related to topological indices characteristic of the Calabi-Yau space K that compactified. In particular, Candelas et al. [14] showed that n_F–the number of families–is connected to the Euler number of K:

---

[3] Think of solving a differential equation in a periodic box.
[4] Majorana fermions exist in $d = 2$ mod 8 dimensions.

$$n_{\rm F} = \frac{1}{2} n_{\rm Euler} . \tag{19}$$

Note that, in this example, the families one obtains have the right stuff. The 27-dimensional representation of E_6 when decomposed in terms of its $SO(10)$ subgroup contains the 16-dimensional representation, appropriate for a family of quarks and leptons, plus a 10 and a singlet. The 10 itself, since the theory is supersymmetric, contains the two needed Higgs doublets, which in this case also come in family repetitions. In principle, the $\delta(27 + \overline{27})$ states (as well as the 10 and 1) are vectorlike, and one can imagine these states getting masses of the order of the compactification scale–presumably of $\mathcal{O}(M_{\rm P})$. So in this example, the light states are just $n_{\rm F}$ replications of the chiral quarks and leptons!

Connecting family replication to the geometry of a compact space is a beautiful idea. Furthermore, there is another advantage. Through compactification, Yukawa couplings are naturally produced, arising from the fermion-gauge field interactions in $d > 4$ dimension along the gauge field components in the $(d-4)$ compact dimensions. Unfortunately, however, one cannot in general compute these couplings explicitly. Nevertheless, often one can infer some useful symmetry restrictions among the Yukawa couplings in these schemes [16].

In my view, obtaining families from compactification is the most appealing solution to the origin of the mysterious repetitions we see in nature. It is not, however, easy to arrive at the correct theory. Basically, even believing that superstrings are the right theory, we still do not understand how to choose among the many possible compactifications available for these theories, since we have no idea what is the underlying physics principle that drives the compactification. At the same time, we are also ignorant of how these schemes can give rise to terms which break supersymmetry and eventually $SU(2) \times U(1)$. Until such problems are solved, these ideas will remain just speculations that are appealing but untested.

V CONCLUDING REMARKS

This brief tour of some of the extant speculations on the origin of families should give one reason to pause. Although the problem of flavor is clearly one of the most fundamental problems in elementary particle physics, its resolution is far from obvious. Furthermore, unless we are very lucky, it is unlikely that new experimental data will be of much help sorting out the possible alternatives. Clearly, the discovery of some exotic matter would be very helpful. So would be the discovery of supersymmetry, as it would provide the death-knell for Nielsen's random-dynamics conjecture.[5]

---

[5] Supersymmetry alters substantially the evolution of the gauge couplings and ends up by disfavoring a world of only three families based on random dynamics.

VI ACKNOWLEDGEMENTS

I would like to thank Dan Kaplan for giving me the opportunity to help celebrate Leon Lederman's birthday. This work is supported in part by the Department of Energy under Grant # DE-FOO3-91ER40662, Task C.

REFERENCES

1. S. L. Adler, Phys. Rev. **177**, 2426 (1969); J. Bell and R. Jackiw, Nuovo Cimento **51A**, 47 (1969); W. A. Bardeen, Phys. Rev. **184**, 1848 (1969).
2. S. Glashow and S. Weinberg, Phys. Rev. **D15**, 1958 (1977).
3. This matter is discussed in some detail in an extended version of this talk, presented at the Enrico Fermi Summer School, Varenna, Italy, July 1997, and will be published in the School's Proceedings.
4. D. B. Kaplan, F. Lepeintre, and M. Schmaltz, hep-ph 9705411.
5. P. Cho and P. Kraus, Phys. Rev. **D54**, 7640 (1996); C. Czaki, W. Skiba, and M. Schmaltz, Nucl. Phys. **B487**, 128 (1997).
6. C. D. Froggatt and H. B. Nielsen, Nucl. Phys. **B147**, 727 (1979).
7. These ideas are discussed in considerable detail in the monograph **Origin of Symmetries** by C. D. Froggatt and H. B. Nielsen (World Scientific, Singapore, 1991).
8. H. B. Nielsen and N. Brene, Nucl. Phys. **B236**, 167 (1984); see also, Proceedings of the XVIII International Ahrenshoop Symposium (Akademie der Wissenschaften der DDR, Berlin-Zeuthen, 1985).
9. D. L. Bennett, H. B. Nielsen, and I. Picek, Phys. Lett. **208B**, 275 (1988).
10. Th. Kaluza, Sitzungber Preuss. Akad. Wiss Berlin Math. Phys. **K1**, 966 (1921); O. Klein, Nature **118**, 516 (1926).
11. M. Green, J. Schwarz, and E. Witten **Superstring Theory**, Vols. 1 & 2 (Cambridge University Press, Cambridge, U.K. 1987).
12. C. Wetterich, Nucl. Phys. **B223**, 109 (1983).
13. E. Witten, in Proceedings of the 1983 Shelter Island Conference II (MIT Press, Cambridge, Mass., 1984).
14. A. Candelas, G. Horowitz, A. Strominger, and E. Witten, Nucl. Phys. **B258**, 46 (1985).
15. D. Gross, J. Harvey, E. Martinec, and R. Rohm, Nucl. Phys. **B255**, 257 (1985); **B267**, 75 (1986).
16. See for example B. R. Greene, K. H. Kirklin, P. J. Miron, and G. G. Ross, Nucl. Phys. **B278**, 667 (1986); **B292**, 606 (1987).

CONFERENCE SUMMARY

20 Years of Beauty Physics and 50 Years of Search for Discoveries

A. I. Sanda

Department of Physics, Nagoya University
Nagoya 464-01, Japan

Abstract. I was fortunate enough to live through – as a physicist – the period during which beauty physics was born. In this talk, I review the material presented at this conference with a historical prospective. As this is also the 75th birthday celebration of Professor Leon Lederman, I will summarize what I have learned about his lifelong chase after discoveries, and interject some of my own encounters.

I INTRODUCTION

I learned a great deal from this conference. Looking back I see that we have made tremendous progress during the past 20 years. To me, the purpose of studying the history of physics is to learn from the past and use it as a springboard to the future. I will present my talk with this spirit. Since I limit myself to the material presented at this conference, I can not claim that my discussion is historically complete nor do I claim it to be totally accurate, as what I present is based on the words of principal participants.

In 1980, $\Upsilon(4S)$ was discovered. Looking at what was presented, I found it convenient to split my discussion into two parts: (1) the pre-$\Upsilon(4S)$ era, which was the period of anticipation; and (2) the post-$\Upsilon(4S)$ era, the fun time.

II PRE-$\Upsilon(4S)$ ERA

Discoveries relevant to our discussion during this era are given in Table 1.

TABLE 1. Discoveries before 1980.

58	Upper limit of ϵ
60	Lepton quark symmetry
62	two neutrinos
64	CP violation
70	Missing the J/ψ
73	KM model
74	J/ψ
75	Charm particles
76	Υ
80	Prediction of Large CP violation in B decay

A Non-discovery of J/ψ

Back in 1970, I was a post-doc in the physics department at Columbia University where Lederman was a professor. Three post-docs shared an office on the 8th floor of Pupin. Lederman came in one day to our office and showed me the dimuon mass spectrum from the proton beam scattered off a Uranium target, shown in Fig. 1. The data were taken at Brookhaven National Laboratory. A curious thing about this spectrum is the shoulder at around 3 GeV. Lederman used to tell us, jokingly, "You see, there was a huge resonance at 6 GeV! It appears like a shoulder because there is not enough center of mass energy to actually produce a peak." He suggested that we work on computing the mass spectrum.

After this experiment, Lederman went on to ISR to investigate the same process at higher energies, and also high-$P_\perp$ hadron final states. Sam Ting, as we all know, stayed at low energy but with much better electron energy resolution – and he made history.

B The discovery of the charm particle

Soon after arriving at Nagoya University in 1992, I found that people there talked about the discovery of charm in a cosmic ray exposure of emulsion. The Nagoya group had invented emulsion electronic hybrid detectors. I decided to look into this. Here is what I found. Back in 1971, Niu [1] and his collaborators had one clean event with a kink in the tracks left in the emulsion detector. They called it the "X" particle. They obtained $m_X = 1.78$ GeV by assuming the two-body decay to be $X \to \pi\pi$. Ogawa and his collaborators [2] reached the conclusion that X carried a new quantum number beyond isospin and strangeness – now known as charm. X decayed into two particles. If we assume that the decay is $X \to K\pi$, this event leads to the correct D-meson mass.

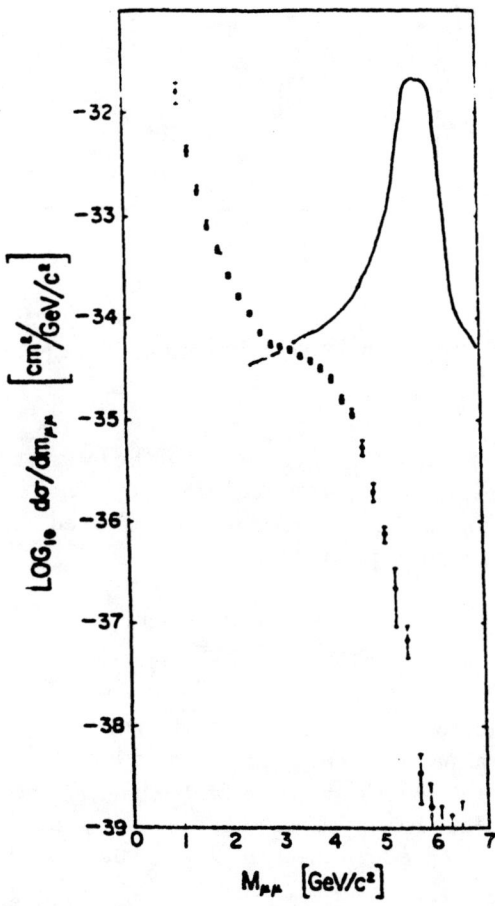

FIGURE 1. Invariant mass spectrum of $p+$ Uranium $\to \mu^+\mu^-$ + anything. Also shown is Lederman's imaginative resonance.

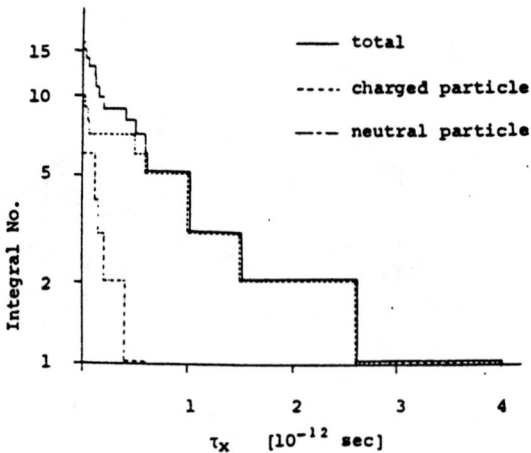

FIGURE 2. The lifetime distribution of charged and neutral D mesons found by Niu and his collaborators in 1975 [3].

While the event is interesting, this should not be taken as the charm discovery. However, the search continued. In 1975, they announced that 20 events of this type were found [3]. Furthermore, they showed that the lifetimes of the charged and neutral X particles differed:

$$\tau_\pm = (1 \sim 2) \times 10^{-12} \text{ sec}$$
$$\tau_0 = (3 \sim 4) \times 10^{-13} \text{ sec.} \qquad (1)$$

The lifetime distribution of these 20 events showed a clear difference between charged and neutral particle decays, as shown in Fig. 2.

This discovery was published in the proceedings of the *14th Int. Cosmic Ray Conference* (Munich), but never published in a journal as it was rejected repeatedly. The evidence for a neutral D with mass of 1865 ± 15 MeV was also found at SLAC [4].

Niu and his group should be recognized as co-discoverers, if not *the* discoverers, of the charm particle.

C The discovery of Υ

Lederman pushed on. A new state had to be there somewhere! It didn't come easily. But, he finally found one! I shall not go into this and instead refer to many discussions at this conference – for example, that of Yoh [5].

D Theoretical developments

1 KM model

In 1960 quark-lepton symmetry was proposed by Sakata and his collaborators [6]. They were the first to propose that neutrinos of different flavors will mix, just as do quarks. This was 35 years ago [7].

It is interesting to compare the atmosphere to which graduate students were exposed at Nagoya with that of a typical US graduate school around 1970. In Nagoya, Sakata insisted that quarks were real objects. His younger collaborator Ohnuki was a strong advocate of the field theoretic approach. Kobayashi and Maskawa were both graduate students at Nagoya. They had known about Niu's strange event and, following Ogawa, interpreted it as the charm particle. They had no resistance in going from a model with four quarks to that with six quarks. Also, taking the viewpoint that field theory had something to do with nature, Weinberg's model of electroweak interaction, written in terms of quarks, was taken very seriously. The fact that CP violation must be introduced within this framework came naturally to them.

In a typical US graduate school, there was an attitude that field theory has very little to with nature. Nuclear democracy was fashionable. Many of us investigated such topics as "non-sense-wrong-signature fixed-pole on the complex angular momentum plane." In fact Gell-Mann himself advocated that quarks might be mere mathematical objects [8].

It illustrates the point that the physics atmosphere we create has a profound influence over younger generations.

2 J/ψ

Just before the discovery of J/ψ, we had

1. The GIM mechanism.

2. Asymptotic freedom to explain the scaling in deep inelastic scattering data.

3. Lederman's shoulder, fig 1.

4. CEA data [9] in which we see a definite jump in

$$R = \frac{\sigma(e^+e^- \to \text{hadrons})}{\sigma(e^+e^- \to \mu^+\mu^-)}. \qquad (2)$$

So, if we had a firm belief in the validity of field theory and the quark model, it is obvious that something major is about to show its face. In fact there were some who worked along this direction. Around this time the Drell-Yan process was born. The existence of a narrow resonance [10] at 3 GeV was predicted. This led to an extensive investigation of quarkonium by the Cornell group, as discussed by Gottfried [11] and Quigg [12].

3 CP violation in the B system

Around 1978, Pais gave a seminar at Rockefeller University. The seminar was entitled "CP violation in charmed-particle decays" [13]. My recollection of how Pais started his seminar is as follows:

> There is good news and bad news. The good news is that CP violation in a heavy-meson system is quite similar to that of the K-meson system. The bad news is that there is little distinction like that of the K_L and K_S mass eigenstates. For a heavy-meson system, lifetimes are both short.

This was the paper which stimulated the search for large CP violation in B decays. Soon afterwards came the paper by Bander, Silverman, and Soni [14]. They discussed CP violation in b quark decay generated by penguin amplitudes.

Around that time, if one were familiar with K-meson physics in the context of the Standard Model, it did not take much imagination to extend it to the B-meson system. We were all convinced that $B^0 - \overline{B}^0$ mixing existed at some level, though it might not be observable in experiments. So, it made sense to search for CP-violating effects which depended on the mixing. Indeed, it could be shown that in spite of a lack of experimental knowledge of $B - \overline{B}$ mixing, CP violation in neutral as well as charged B's could be large, depending on the relative size of the KM matrix elements [15]. Soon afterwards, we realized [16] that $B \to J/\psi K_S$ decay leads to almost an ideal experimental signature.

III POST-$\Upsilon(4S)$ ERA

$\Upsilon(4S)$ was found by the CLEO [17] and CUSB [18] collaborations at CESR. In this section, I deviate from a chronological discussion and discuss discoveries (Table 2) according to their importance in influencing physics which followed. Of course deciding what is most important requires some judgement and this is my personal view.

A Longevity of B Mesons

The key discovery, which is at the foundation of many interesting phenomena in B physics, is that B mesons live for long time. This discovery was made possible by the development of vertex detectors to study charm mesons

TABLE 2. Discoveries after 1980

80	$\Upsilon(4S)$
81	$\Upsilon(4S) \not\to B\overline{B}^*$
82	Longevity of B mesons
86	$B - \overline{B}$ mixing
91	$b \to ul\nu$
95	$b \to s\gamma$
97	$b \to K^{\pm}\pi^{\mp}; K^+K^-; \pi^+\pi^-$

in hadronic colliders. In a hadronic environment, where there are large numbers of background tracks, neutral particles which travel some distance before they decay to charged particles leave a signature – a secondary vertex. A secondary vertex which is separated by few hundred microns from the interaction point can be identified.

The vertex detector turned out to be ideal for study of B mesons, which have about the same lifetime as D mesons. Experiments at SLAC, the MAC and MARK II collaborations, have established the long lifetime of the B mesons [19]. Their result was $1.8 \pm 0.6 \pm 0.4$ psec.

To appreciate the fact that this is a long time for a particle like the B meson, let us give a naive estimate of the lifetime. The total width of the b quark is given by just scaling the expression for the width of the μ meson:

$$\Gamma_\mu = \frac{G_F^2}{192\pi^3}m_\mu^5$$
$$\Rightarrow \frac{G_F^2}{192\pi^3}m_b^5|V_{bc}|^2 \times (2 \times 3 + 3) \qquad (3)$$

There are altogether 9 channels (Fig. 3). For this rough estimate, the phase space difference due to finite quark masses is ignored. So, if $|V_{bc}| \sim 1$,

$$\tau \sim \tau_\mu \left(\frac{m_\mu}{m_b}\right)^5 \frac{1}{9} \sim 10^{-15} \text{sec.} \qquad (4)$$

B's live 1000 times longer than our expectations! ¿From this suppression, we deduce that

$$|V_{cb}| \sim \frac{1}{30} \sim (\sin\theta_c)^2. \qquad (5)$$

B $B^0 - \overline{B}^0$ Oscillations

Now that we know a B meson lives for a long time, we might hope that it would do something interesting while it is alive. Indeed, $B - \overline{B}$ mixing was discovered by the ARGUS collaboration by establishing the existence of same-sign dilepton events:

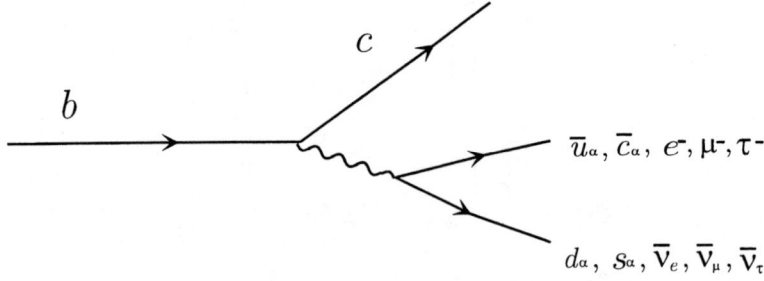

FIGURE 3. Feynman diagram which is responsible for b-quark decay.

$$\begin{aligned} e^+e^- &\to \Upsilon(4S) \\ &\to B^0 \overline{B}^0 \\ &\to \mu^\pm \mu^\pm + \text{anything}. \end{aligned} \quad (6)$$

Of course if there is no mixing, we expect the leptons from $B - \overline{B}$ decay to have opposite signs. The rate for dimuon events corresponds to

$$\frac{\Delta m}{\Gamma} = \frac{\text{lifetime}}{\text{mixing time}} \sim 0.7. \quad (7)$$

When we were making predictions for this quantity, we had a notion that the top quark mass can not be more than 50 GeV. Since the rate for dimuon events goes like $\left(\frac{m_t}{M_W}\right)^4$, we failed to predict that the mixing could be seen.

C Observation of $b \to ul\nu$

A careful analysis of the endpoint region in the inclusive lepton spectrum in $B \to l\nu X$, performed by the CLEO and ARGUS collaborations in 1990, had revealed the existence of events where the lepton energy is beyond the kinematic limit for final states containing charm hadrons [20]. This discovery led to

$$\left|\frac{V_{ub}}{V_{cb}}\right|_{incl} \sim 0.1. \quad (8)$$

If we go back to Ref. [15], where the CP asymmetry is plotted against $\frac{V_{ub}}{V_{cb}}$, with $\frac{V_{ub}}{V_{cb}} \sim 0.1$ and $m_t \sim 20$ GeV, our calculation shows that there is as much as 10% CP asymmetry in B decay.

D Do we have $\Upsilon(4S) \to B\overline{B}^*$?

The time dependence of CP asymmetry is proportional to

$$\sin \Delta m(t_1 \pm t_2). \tag{9}$$

Here t_1 is the time at which one of the B's decayed to ψK_S and t_2 is the time at which the other B decayed leptonically. The "+(−)" sign applies to the CP asymmetry for the even (odd) angular momentum $B\overline{B}$ state before they decayed. In particular, for the $B\overline{B}$ pair from $\Upsilon(4S)$ decay, the asymmetry vanishes when we integrate over t_1 and t_2, i.e. if we don't make any time measurement.

The "−" sign is disappointing. It means that we can't detect the asymmetry at, for example, CLEO, since the B's travel only about $20\,\mu m$ before they decay and this is too short to be measured with available detectors. One way to get out of this is if some $B\overline{B}$ pair is produced in an even angular momentum state. If the decay

$$\Upsilon(4S) \to B\overline{B}^* \tag{10}$$

occurs, we are all set. The mass of B^*, the $J = 1$ state of a $\bar{b}$ with a u or d quark, was determined to be only 50 MeV heavier than the B meson. This was an important fact – it implies that

$$B^* \to B + \gamma \tag{11}$$

is the only decay mode. In the $B\overline{B}$ production process,

$$\Upsilon(4S) \to B^0 + \overline{B}^{0*} \to B^0 \overline{B}^0 + \gamma \tag{12}$$

the resulting $B^0 \overline{B}^0$ state is in a $C = +$ state, i.e. an even angular momentum state. We have a "+" sign for the time dependence of the asymmetry, and thus the asymmetry would not be washed out even if we have no information on decay times. This wishful thinking did not come through. As the decay products of $\Upsilon(4S)$ are nearly at rest, it was easy to search for B^* production. The CUSB detector was set up to detect photons. They looked for monoenergetic 50 MeV photons and in 1981, they concluded that

$$\Upsilon(4S) \to B\overline{B} \tag{13}$$

is the unique decay mode [21]. So, we are stuck with the "−" sign. This means that we must measure times at which B mesons decay. This is impossible at the existing e^+e^- colliders as mentioned above. We must do something to enhance the decay path.

The wild idea of building an asymmetric e^+e^- collider was first suggested by P. Oddone during our conversation on the difficulty mentioned above. If the

$\Upsilon(4S)$ were moving so that the B mesons have large enough β, they would leave long enough tracks to be measured by existing vertex detectors. The fact that we need not have much energy asymmetry, for example 8 GeV on 3.5 GeV or 7 GeV on 4 GeV will do the job, was first pointed out by Hitlin et al. [22].

In the beginning accelerator physicists both at SLAC and at KEK told us that an asymmetric collider with a luminosity 10^{34} cm^{-2} sec^{-1} cannot be built.

At this very moment, both KEK and SLAC are building e^+e^- asymmetric colliders.

E Discovery of Penguins in B decay

- $b \to s\gamma$

In 1990, this decay was discovered by CLEO. This is the first discovery of $\Delta B \neq 0$ in a neutral current decay amplitude [20]. It involves a loop effect in the standard electroweak theory.

- $B \to \pi\pi$ and $K^\pm \pi^\mp$

These two-body decays have been seen for a number of years. The particle identification of the CLEO detector in the high end of the momentum spectrum is poor and they have not been able to separate out the decay channels. CLEO seems to have solved this problem.

- How big are penguins?

Recent CLEO data are consistent with observing both $B \to \pi^+\pi^-$ and $B \to K^+\pi^-$. The quark diagrams which generate these decays are shown in Fig. 4. The amplitudes for these decays can be written as

$$A(B \to \pi\pi) = V_{ub}^* V_{ud} T + V_{tb}^* V_{td} P$$
$$\sim \lambda^3 T + \lambda^3 P$$
$$A(B \to K^+\pi^-) = V_{ub}^* V_{us} T + V_{tb}^* V_{ts} P$$
$$\sim \lambda^4 T + \lambda^2 P \tag{14}$$

where we have indicated the suppression factor in powers of $\lambda \equiv \sin\theta_c$. So, $B \to \pi\pi$ is dominated by the tree graph and $B \to K^+\pi^-$ is dominated by the penguin. If the two decay rates are same order of magnitude,

$$\lambda^3 T \sim \lambda^2 P \quad \text{or} \quad \frac{P}{T} \sim \lambda. \tag{15}$$

This estimate of the penguin contribution is much larger than a naive estimate of a loop graph:

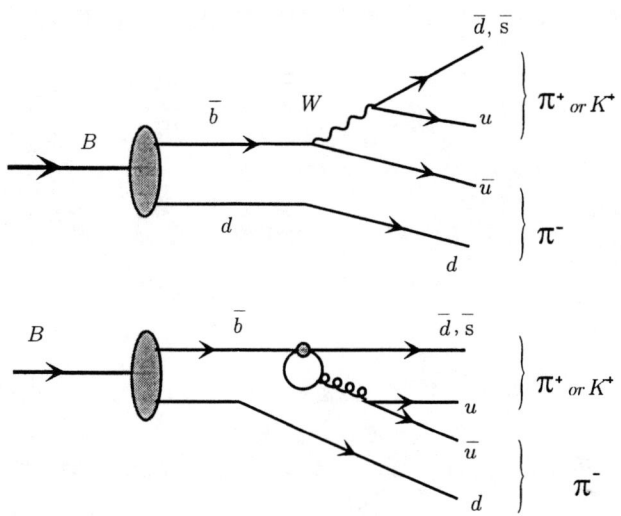

FIGURE 4. Quark diagrams which contribute to $B \to \pi\pi$ and $B \to K^+\pi^-$.

$$\frac{P}{T} \sim \frac{\alpha_s}{12\pi^3} \ln \frac{m_t}{m_c} \sim \lambda^2 \tag{16}$$

Such a large penguin contribution is a problem for measuring the angles of the unitarity triangle. Various methods for getting around this problem have been devised. In particular, such decays as

$$B \to \pi^0 \pi^0 \tag{17}$$

must be measured. A challenge for the future is to design an experiment to get at this decay and determine its decay rate.

IV FUTURE PROSPECTS

We expect considerable progress in the experimental study of B physics during the next 20 years. Even on a much shorter time scale, some new results are expected during this century.

1. SLD has advertised that they can measure $B_S - \overline{B}_S$ mixing [23].

2. KEKB and PEPB will turn on in the second half of 1998. $10^7 \sim 10^8$ $B\overline{B}$ pairs will be produced. The Belle and BaBar collaborations will look for CP violation in B decays. There is a race between these collaborations as to which will succeed in measuring $B \to \psi K_S$ asymmetry first.

3. HERA-B is also in the race. Results from this experiment should be watched carefully.

4. At this moment CDF has the most identified $B \to \psi K_S$ events. The problem is to get the tagged events. With the main ring injector, CDF may have a chance to detect the asymmetry by 2003.

5. LHC-B and BTeV are second-generation B physics experiments at LHC [24] and Fermilab [25], respectively. These experiments will keep us informed about nature for decades to come.

V 50 YEARS OF CHASING AFTER DISCOVERIES

Here is a list of Lederman's major accomplishments:

1. Observation of long-lived neutral V particle (1956).

2. Observation of failure of conservation of parity and charge conjugation in meson decays: the magnetic moment of the free muon (1957).

3. Observation of the high energy neutrino reaction and the existence of two kinds of neutrinos (1962).

4. Observation of massive muon pairs in hadron collisions (1970).

5. Observation of π mesons with large transverse momentum in high energy proton-proton collisions (1973).

6. Observation of a dimuon resonance at 9.5 GeV in 400 GeV proton-nucleon collisions (1977).

7. Observation of the $\Upsilon(4S)$ at CESR (1980).

Lederman claims to have been chasing after fame for all of his life [26]. Some people bump into fame by luck. Lucky, he was not. In fact sometimes he was unlucky! Here is a list of Lederman's missed opportunities:

1. Missed discovering CP violation in 1958.

2. Missed discovering J/ψ.

It reminds me of a famous phrase:

> Lederman got his fame in the old fashioned way – he earned it.

REFERENCES

1. K. Niu et al., *Prog. in Theor. Phys.* **46**, 1644 (1971).
2. T. Hayashi et al., *Prog. in Theor. Phys.* **47**, 280 (1972).
3. K. Hoshino et al., Conf. Papers, 14th Int. Cosmic Ray Conf. (Munich), **7**, 2442 (1975).
4. G. Goldhaber et al., *Phys. Rev. Lett.* **37**, 255 (1976).
5. Lecture by J. Yoh, this conference.
6. Z. Maki et al., *Prog. in Theor. Phys.* **23**, 1174 (1960).
7. Z. Maki et al., *Prog. in Theor. Phys.* **28**, 870 (1962).
8. M. Gell-Mann and Y. Ne'eman, *The Eightfold Way*, W. A. Benjamin, Inc., New York, 1964.
9. A. Litke et al., *Phys. Rev. Lett.* **30**, 1189 (1973); *ibid.*, 1394 (1973).
10. T. Appelquist and D. Politzer, *Phys. Rev. Lett.* **34**, 43 (1974).
11. Lecture by K. Gottfried, this conference.
12. Lecture by C. Quigg, this conference.
13. A. Pais and S. B. Treiman, *Phys. Rev. D* **12**, 2744 (1975).
14. M. Bander, D. Silverman, and A. Soni, *Phys. Rev. Lett.* **43**, 242 (1979).
15. A. B. Carter and A. I. Sanda, *Phys. Rev. D* **23**, 1567 (1981).
16. I. I. Bigi and A. I. Sanda, *Nucl. Phys.* **B193**, 85 (1981).
17. D. Andrews, et al. *Phys. Rev. Lett.* **45**, 291 (1981).
18. L. Spencer et al. *Phys. Rev. Lett.* **47**, 771 (1981).
19. Lecture by J. Jaros, this conference.
20. Lecture by R. Poling, this conference.
21. Lecture by J. Lee-Franzini, this conference.
22. D. Hitlin, T. Nakada, and A. I. Sanda, in *High Energy Physics in the 1990's*, Proc. of the 1988 Snowmass Summer Study on High Energy Physics, S. Jensen, ed., Singapore: World Scientific, 1989.
23. Lecture by D. Su, this conference.
24. Lecture by T. Nakada, this conference.
25. Lecture by P. McBride, this conference.
26. Lecture by L. Lederman, this conference.

Symposium Participants

Adams, Mark	University of Illinois, Chicago
Berger, Edmond L.	Argonne National Laboratory
Bernard, Claude	Washington University
Bigi, Ikaros	University of Notre Dame
Blankman, Alan J.	LeCroy Corporation
Brown, Bruce	Fermilab
Brown, Chuck	Fermilab
Bruner, Nichelle L.	University of New Mexico
Burnstein, Ray A.	Illinois Institute of Technology
Cacciari, Matteo	DESY
Chakravorty, Alak	Illinois Institute of Technology
Christian, David C.	Fermilab
Cooper, William E.	Fermilab
Crittenden, James	Universität Bonn
Dighe, Amol S.	University of Chicago
Dunietz, Isard	Fermilab
Eichten, Estia	Fermilab
El-Khadra, Aida X.	University of Illinois, Urbana-Champaign
Erber, Thomas	Illinois Institute of Technology
Franzini, Paolo	Università di Roma
Frisch, Henry	University of Chicago
Garelick, David	Northeastern University
Gottfried, Kurt	Cornell University
Hans, Randal	University of Illinois, Urbana-Champaign
Herb, Steve W.	DESY
Hewett, JoAnne	Stanford Linear Accelerator Center
Honscheid, Klaus	The Ohio State University
Hou, W. S.	National Taiwan University
Innes, Walter	Stanford Linear Accelerator Center
Introzzi, Gianluca	Fermilab
Ito, Albert	Fermilab
Jaros, John A.	Stanford Linear Accelerator Center
Jöstlein, Hans	Fermilab
Johnson, Porter W.	Illinois Institute of Technology
Kaplan, Daniel M.	Illinois Institute of Technology
Kaplan, David E.	University of Washington
Kephart, Bob	Fermilab
Khalfin, Leonid	Russian Academy of Sciences
Khelashvili, Anzor	Tbilisi State University
Kobayashi, Makoto	KEK
Kolb, Adrienne	Fermilab
Kostelecký, Alan	Indiana University

Lederman, Leon	Illinois Institute of Technology
Lee-Franzini, Juliet	Laboratori Nazionali di Frascati
Lemoigne, Yves	CEA Saclay
Lipkin, Harry	Weizmann Institute
Lipton, Ronald	Fermilab
London, David	Université de Montréal
Luebke, William	Illinois Institute of Technology
Marlow, Daniel R.	Princeton University
McBride, Patricia L.	Fermilab
McCarthy, Bob	State University of New York, Stony Brook
Mishra, Shekhar	Fermilab
Nakada, Tatsuya	Paul Scherrer Institute
Onuchin, Alexei	Budker Institute of Nuclear Physics
Pakvasa, Sandip	University of Hawaii
Paulini, Manfred	Lawrence Berkeley National Laboratory
Peccei, Roberto	University of California, Los Angeles
Peoples, John	Fermilab
Piskounova, Olga	JINR, Dubna
Poling, Ronald A.	University of Minnesota
Pope, Bernard G.	Michigan State University
Quigg, Chris	Fermilab
Raja, Rajendran	Fermilab
Rosen, Jerry	Northwestern University
Rubin, Howard A.	Illinois Institute of Technology
Rutherfoord, John	University of Arizona
Sanda, Anthony I.	Nagoya University
Schael, Stefan	Max Planck Institut, Munich
Schmidt-Parzefall, Walter	DESY
Sens, Johannes	CNRS
Sharma, Vivek	University of California, San Diego
Silverman, Albert	Cornell University
Slaughter, Jean	Yale University
Smizanska, Maria	CEA Saclay
Spiegel, Leonard G.	Fermilab
Stech, Berthold	Universität Heidelberg
Stone, Sheldon	Syracuse University
Su, Dong	Stanford Linear Accelerator Center
Tafoya, Jessie M.	University of California, Santa Cruz
Ueno, Koji	National Taiwan University
Wegener, Dietrich	Universität Dortmund
White, Christopher G.	Illinois Institute of Technology
Wicklund, Barry	Argonne National Laboratory
Wise, Mark	California Institute of Technology
Yoh, John	Fermilab

Author Index

B

Baru, S. E., 97
Bernard, C., 227
Blinov, A. E., 97
Blinov, V. E., 97
Blum, T., 227
Bondar, A. E., 97
Bukin, A. D., 97

C

Cacciari, M., 189

D

DeGrand, T., 227
DeTar, C., 227
Dunietz, I., 235

E

Eidelman, Yu. I., 97
El-Khadra, A., 198

G

Gottfried, K., 3
Gottlieb, S., 227
Groshev, V. R., 97

H

Heller, U. M., 227
Hetrick, J., 227
Hewett, J. L., 328
Honscheid, K., 125
Hou, G. W. S., 339

J

Jaros, J. A., 106

K

Kiselev, V. A., 97
Klimenko, S. G., 97
Kobayashi, M., 15
Kolachev, G. M., 97
Kostelecký, V. A., 345

L

Lederman, L. M., 26
Lee-Franzini, J., 85
Lemoigne, Y., 308
Lipkin, H. J., 246
Lipton, R., 269
London, D., 321

M

Marlow, D. R., 161
McBride, P., 287
McNeile, C., 227
Mishnev, S. I., 97

N

Nakada, T., 298

O

Onuchin, A. P., 97

P

Panin, V. S., 97
Peccei, R. D., 354
Petrov, V. V., 97

Poling, R. A., 113
Protopopov, I. Ya., 97

Q

Quigg, C., 173

R

Rummukainen, K., 227
Rutherfoord, J. P., 43

S

Sanda, A. I., 367
Schmidt-Parzefall, W., 279
Shamov, A. G., 97
Sidorov, V. A., 97
Silverman, A., 65
Skovpen, Yu. I., 97
Skrinsky, A. N., 97
Stech, B., 207
Stone, S., 75
Su, D., 147
Sugar, R., 227

T

Tayursky, V. A., 97
Telnov, V. I., 97
Tikhonov, Yu. A., 97

Toussaint, D., 227
Tumaikin, G. M., 97

U

Undrus, A. E., 97

V

Vorobiov, A. I., 97

W

Wegener, D., 50
Wicklund, A. B., 251
Wingate, M., 227
Wise, M. B., 217

Y

Yoh, J. K., 29

Z

Zhilich, V. N., 97